T0229765

PRACTICAL ASPECTS
OF
ION TRAP MASS
SPECTROMETRY

Volume I

*Fundamentals of
Ion Trap
Mass Spectrometry*

Modern Mass Spectrometry

*A Series of Monographs on
Mass Spectrometry and Its Applications*

EDITOR-IN-CHIEF
Thomas Cairns, Ph.D., D.Sc.

Forensic Applications of Mass Spectrometry
Edited by Jehuda Yinon

*Practical Aspects of Ion Trap Mass Spectrometry, Volume I
Fundamentals of Ion Trap Mass Spectrometry*
Edited by Raymond E. March and John F. J. Todd

*Practical Aspects of Ion Trap Mass Spectrometry, Volume II
Ion Trap Instrumentation*
Edited by Raymond E. March and John F. J. Todd

*Practical Aspects of Ion Trap Mass Spectrometry, Volume III
Chemical, Environmental, and Biomedical Applications*
Edited by Raymond E. March and John F. J. Todd

EDITORIAL ADVISORY BOARD

PRACTICAL ASPECTS OF ION TRAP MASS SPECTROMETRY

Volume I

Fundamentals of Ion Trap Mass Spectrometry

CRC PRESS

Boca Raton   London   New York   Washington, D.C.

Library of Congress Cataloging-in-Publication Data

Practical aspects of ion trap mass spectrometry / edited by Raymond E. March, John F.J. Todd.
    p. cm.—(Modern mass spectrometry)
    Includes bibliographical references and index.
    ISBN 0-8493-4452-2 (vol. 1)
    1. Mass spectrometry. I. March, Raymond E. II. Todd, John F.J. III. Series.
QD96.M3P715   1995
539.7'.028'7—dc20
                                            95-14146

Visit the CRC Press Web site at www.crcpress.com

© 1995 by CRC Press LLC

No claim to original U.S. Government works
International Standard Book Number 0-8493-4452-2
Library of Congress Card Number 95-14146
Printed in the United States of America   3 4 5 6 7 8 9 0
Printed on acid-free paper

DEDICATION

To ion trappers, young and old, everywhere.

FOREWORD

Publication of the volumes in the new series entitled *Modern Mass Spectrometry* represents a milestone for scientists intimately involved with the practice of mass spectrometry in all its various forms. Forthcoming monographs in this series will focus on various selected topics within the rapidly expanding realm of mass spectrometry. Individual volumes will provide in-depth reports on mainstream developments where there is an urgent need for a specific mass spectrometry treatise in an active and popular area.

While mass spectrometry as a field is quite well served by several publications and a number of societies, the application of mass spectrometric techniques across the basic scientific disciplines has not yet been recognized by existing journals. The present distribution of research and application papers in the scientific literature is widespread. There is a multi-disciplinary audience requiring access to concise reports illustrating the latest successful approaches to difficult analytical problems.

The distinguished members of the Editorial Advisory Board all agreed that a platform exists for a premier book series with high standards to cover comprehensively general aspects of developing mass spectrometry. Contributing authors to the series will provide concise reports together with a bibliography of publications of importance, selecting worthy examples for inclusion. Due to the rapid and extensive growth of the literature in mass spectrometry, there is a need for such reports by authorities who critique the entire subject area. There is an increasing urgency to provide readers with timely, informative, and cogent reviews stripped of outdated material.

I believe that the decision to publish the series *Modern Mass Spectrometry* reflects the realization that increasing numbers of mass spectrometrists are applying nascent state of the art approaches to some fascinating problems. Our challenge is to develop the best forum in which to present these emerging issues to encourage and stimulate other scientists to comprehend and adopt similar strategies for other projects.

Lofty ideals aside, our immediate goal is to build a reputable series with scientific authority and credibility. To this end we have assembled an excellent Editorial Advisory Board that mirrors the prerequisite multi-disciplinary exposure required for success.

I am confident that the mass spectrometry community will be pleased with this "new publication" approach and will welcome the opportunities it presents to foster the development of interactions between the various scientific disciplines. No doubt there is a long and difficult road ahead of us to ensure that the series grows into a position of leadership, but I am convinced that the hard work of our outstanding Editorial Advisory Board, and the enthusiasm of our authors and readers, will achieve the degree of success for which we all seek.

Thomas Cairns
Editor-in-Chief

PREFACE

It was a great privilege for us to be invited, in 1991, to undertake the preparation of a monograph on quadrupole ion trap mass spectrometry for CRC Press. It was even more gratifying to discover that so many of our colleagues were willing to contribute to such a monograph. The ion trapping field has been, and continues to be, a very active one. Clearly, the principal players in the field, the very ones whom we invited to contribute, were extremely busy not only with their researches and teaching but also in writing applications for funds for further research; as competition for such funds has become so keen, one must now spend an inordinate amount of time in preparing and writing such applications. It is to the enormous credit of our contributors that they responded so positively to our invitations to contribute to this monograph.

In mid-1992, as arrangements were well advanced for the preparation of the monograph, we learned that CRC Press had decided to launch a new series, entitled *Modern Mass Spectrometry*. An outcome of this decision was that we were invited to undertake the preparation of not one, but two, monographs on quadrupole ion trap mass spectrometry, which would form Volumes I and II of this new series. Following discussions among ourselves, with CRC Press, and with our contributors, it was agreed that we would arrange for the preparation of two volumes on practical aspects of ion trap mass spectrometry: to Volume I (the original monograph) entitled *Fundamentals and Instrumentation* was to be added a second volume entitled *Chemical, Environmental, and Biomedical Applications*.

Meanwhile, during these protracted negotiations, work on the original monograph was continuing apace. All of the research groups and individuals invited initially to contribute to the original monograph had accepted their invitations with enthusiasm that matched, if not exceeded, that which they displayed for ion trap research. This enthusiasm for research had not gone unrewarded in that significant advances were made during the early 1990s to our understanding of ion trap behavior, new traps, trapped ion trajectory control, and trapped ion behavior. The net result was a burgeoning of new material beyond that anticipated in our planning for the original monograph, and in excess of that accommodated normally in a single monograph. The decision was made to proceed with dispatch and to publish *Fundamentals and Instrumentation* as a set of two volumes (I and II) entitled *Fundamentals* and *Instrumentation*, respectively, with each volume being complete within itself. This monograph is Volume I; it is accompanied by Volume II and will be followed by yet a third monograph entitled *Chemical, Environmental, and Biomedical Applications*.

The history of the quadrupole ion trap has been presented, in Chapter 2, in tabular form as "The Ages of the Ion Trap". There are two landmarks in this history, of which we should like to make special mention. The first is the invention of the ion trap by Wolfgang Paul and Hans Steinwedel, which was recognized by the award of the 1989 Novel Prize in Physics, in part, to Wolfgang Paul and Hans Dehmelt; the second is the discovery, announced in 1983, of the mass-selective instability scan by George C. Stafford, Jr. On these two landmarks rests the entire field of ion trap mass spectrometry. The award of the Nobel Prize has focused attention more sharply on the ion trapping field and has possibly been responsible, though indirectly, for the commissioning of both this monograph and the one to follow.

On July 15, 1992, Professor Paul was awarded an Honorary Doctor of Science degree by the University of Kent. The following day, an "Ion Trap Seminar Day" was held, at which Professor Paul spoke on the development of quadrupole mass spectrometry and the ion trap; other presentations were made by Richard Thompson, Hugh Klein, Colin Creaser, and ourselves. It was our great pleasure and privilege to become better acquainted with Professor Paul during this memorable period with its accompanying festivities very much enlivened by his demonstrations, reminiscences, humor, and conversation; thus, we note with sadness the death of Professor Paul in November 1993. Since the award of the Nobel Prize, chemists have tended to follow the lead of the physicists in referring to the quadrupole ion trap as the Paul trap. The discovery by George Stafford has made possible the commercial development of a powerful mass spectrometer based on the extraordinarily versatile ion trap.

The principal objective of this monograph is to present an account of the development and theory of the quadrupole ion trap and its utilization as an ion storage device, a reactor for ion/molecule reactions, and as a mass spectrometer. A secondary objective is to expand the reader's appreciation of ion traps from that of a unique arrangement of electrodes of hyperbolic form (and having a pure quadrupole field) to a series of ion traps having fields with hexapole and octopole components; furthermore, the reader is introduced to the practical ion trapping device in which electrode spacing has been increased. This book is composed of eight chapters arranged in two parts, namely Fundamentals and Ion Activation and Ion/Molecule Reactions.

Part 1: Fundamentals. The fundamentals of the ion trap are covered in four chapters, commencing with Chapter 1, which provides an introduction to the origin of the ion trap, its development and operating principles, and improvements in performance. This chapter touches lightly on many of the aspects of ion trap use described in later chapters and in Volume II. In Chapter 2 is presented the theory of quadrupole mass spectrometry, which includes the origin of the stability region and the derivation of the trapping parameters, a_z and q_z, in terms of the practical device. Commercial versions of the ion trap all have a stretched geometry which introduces a nonlinear component to the normal quadrupole field, and so we come to Chapter 3, which is a detailed treatment of nonlinear ion traps, including a full theoretical exposition of nonlinear traps, the superposition of multipole fields, and nonlinear resonances. Accounts of simulation studies of nonlinear effects and experimental studies in nonlinear traps are also given. The character of Chapter 4 differs from that of the other chapters in that it is more in the narrative tradition; here, we have the story of the development and launching of the first commercial version of the ion trap. The road to commercialization was far from smooth, and there were occasions when the entire project almost floundered; however, we will not spoil the story for the reader except to say that, as we all know, the story has a happy ending.

Part 2: Ion Activation and Ion/Molecule Reactions. This part, which also includes four chapters, focuses on the environment within the ion trap; that is, the movement of ions within the trap and how this movement is modified by repeated collisions of the ions with buffer gas atoms of helium, and on the collisions of ions with molecules that lead to chemical change. A particular challenge in ion trap mass spectrometry is the ability to bring about endoergic processes, particularly ion dissociation. Chapter 5 deals with the effects of collisions of ions with buffer gas inside the trap on the detection of ions external to the trap and emphasizes the critical role of collisions in focusing the ion cloud for subsequent operations. With the advent of powerful desktop computers, simulations of ion movements within the ion trap may now be carried out readily, as explained in Chapter 6. Here, an account is given of two different, yet complementary, approaches to ion trap simulation studies. First, multiparticle simulations are described that can be compared directly with ion signals detected in ion trap experiments. Second, single particle simulations of trap operations are described from which ion behavior can be deduced. In Chapter 7, the new technique of boundary excitation is explored; this technique combines ion isolation (which is a necessity for tandem mass spectrometry) with nonresonant ion activation. Early indications are quite promising that the boundary excitation technique may be applicable to tandem mass spectrometry on the gas chromatographic time scale. Finally in Part 2, Chapter 8 is a combined contribution that consists of a detailed examination of ion/molecule reactions in the ion trap; such reactions are of both fundamental interest and of applied interest, particularly with respect to chemical ionization within the ion trap.

It is our hope that the contributions in this monograph, in presenting a coherent picture of the present status of research in the ion trapping field, will both expedite the entrée of potential ion trappers to this field and provide a backdrop for ion trap research and development in the next decade. We leave it to the reader to judge.

Finally, our thanks are due to many people who have assisted us in one way or another with the many tasks that make up completion of a manuscript. First of all, to our contributors whose efforts have made possible this book; we thank them for the fruits of their labors and for their patient toleration of anxious faxes and telephone calls. We thank the staff of CRC Press for their ready cooperation and encouragement. We note, with appreciation, the guidance and assistance of the Series Editors, Dr. Thomas Cairns and Dr. M. Allen Northrup. In the Trent laboratory, REM thanks Jenny Bazdikian for cheerfully and quickly handling disks from contributors and translating a variety of word processing programs into WordPerfect 5.1; in this way, the editorial work could begin on the same day that each contribution was received. Thanks, also, to Oscar Vega, Jeff Plomley, Peter Popp, Dr. Richard Hughes, and Jenny Bazdikian for their enormous contribution in proofreading and checking manuscripts and in preparing the final copies of each of the chapters and the appropriate figures. To Dr. Richard Hughes, for redrawing some of the figures for Chapter 2 and for his ready cooperation throughout the preparation of this volume. To Bonnie MacKinnon, for her ever-ready assistance with the preparation of tables and equations and with a myriad of word processing minutiae.

Raymond E. March
John F. J. Todd

THE EDITORS

Raymond E. March, Ph.D., is presently Professor of Chemistry at Trent University in Peterborough, Ontario, Canada. He obtained his B.Sc. degree from the University of Leeds in 1957; his Ph.D. degree, which he received in 1961 from the University of Toronto, was supervised by Professor John C. Polanyi.

Dr. March has conducted independent research for over 28 years and has directed research in gas phase kinetics, optical spectroscopy, gaseous ion kinetics, analytical chemistry, and mass spectrometry.

He has published and/or coauthored over 130 scientific papers in the above areas of research with emphasis on mass spectrometry, both with sector instruments and quadrupole ion traps. Dr. March is a coauthor with Dr. Richard J. Hughes of *Quadrupole Storage Mass Spectrometry*, published in 1989.

Professor March is actively engaged in the supervision of graduate student research and is an Adjunct Professor of Chemistry at Queen's and York Universities in Ontario; he is the Associate Director of the Trent/ Queen's Cooperative Graduate Program. Professor March is a Fellow of the Chemical Institute of Canada and a member of the American, British, and Canadian Societies for Mass Spectrometry. Research in Dr. March's laboratory is supported by the Natural Sciences and Engineering Research Council of Canada, the Ontario Ministry of the Environment, and Varian Associates. Dr. March has enjoyed long-term collaborations with the co-editor, John Todd, and with colleagues at the University of Provence and Pierre and Marie Curie University in France, and with colleagues in Italy.

John F. J. Todd, Ph.D., is currently Professor of Mass Spectroscopy and was, until recently, Director of the Chemical Laboratory at the University of Kent, Canterbury, England. He obtained his B.Sc. degree in 1959 from the University of Leeds, from whence he also gained his Ph.D. degree and was awarded the J. B. Cohen Prize in 1963, working in the radiation chemistry group led by Professor F. S. (now Lord) Dainton, FRS. He was a postdoctoral research fellow in the laboratory of the late Professor Richard Wolfgang at Yale University, Connecticut, from 1963 through

1965 and was one of the first appointees to the academic staff of the, then new, University of Kent in 1965.

Since arriving in Canterbury Professor Todd's research interests have encompassed mass spectral fragmentation studies, gas discharge chemistry, ion mobility spectroscopy, analytical chemistry, and ion trap mass spectrometry. His work on ion traps commenced in 1968, and he first coined the name QUISTOR for **qu**adrupole ion **stor**e, to describe a trap coupled to a quadrupole mass filter for external mass analysis. Acting as a consultant to Finnegan MAT, he was a member of the original team that developed the ion trap commercially. Research in Professor Todd's laboratory has been supported by the Science and Engineering Research Council, the Engineering and Physical Sciences Research Council, the Defense Research Agency, the Chemical and Biological Defence Establishment, and Finnegan MAT, Ltd.

Dr. Todd has published and/or co-authored over 100 scientific papers, concentrating on various aspects of mass spectrometry. With Dr. Dennis Price, he co-edited several volumes of *Dynamic Mass Spectrometry,*; he edited *Advances in Mass Spectrometry 1985*; and he is currently editor of the *International Journal of Mass Spectrometry and Ion Processes*.

Professor Todd is actively engaged in the supervision of full-time and industrial-based part-time graduate students, all working in the field of mass spectrometry. He is a Chartered Chemist and a Chartered Engineer and is currently an elected member of the Council of the Royal Society of Chemistry. He recently completed a four-year term as Treasurer of the British Mass Spectrometry Society. Outside the immediate confines of academic work, he was for ten years Master of Rutherford College at the University of Kent and also for four years was appointed as the Chairman of the Canterbury and Thanet Health Authority. He is presently a Governor of the Clergy Orphan Corporation and of Canterbury Christ Church College. He has enjoyed long-term collaborations with co-editor Professor Raymond March, with colleagues at Finnigan MAT in the United Kingdom and the United States and with groups in Nice (France) and Padova and Torino (Italy).

CONTRIBUTORS

Jennifer Brodbelt, Ph.D.
Associate Professor of Chemistry
Department of Chemistry
University of Texas
Austin, Texas

Thomas Cairns, Ph.D.
Vice President
Psychemedics
Culver City, California

R. Graham Cooks, Ph.D.
Distinguished Professor
Department of Chemistry
Brown Laboratory
Purdue University
West Lafayette, Indiana

Colin S. Creaser, Ph.D.
Professor and Head,
Department of Chemistry and
    Physics
Nottingham Trent University
Nottingham, United Kingdom

Jochen Franzen, Ph.D.
Managing Director
Bruker-Franzen Analytik Gmb.
Fahrenheitstrasse 4
Bremen, Germany

R.-Holger Gabling, Ph.D.
Senior Developer
Ion Trap Development
Bruker-Franzen Analytik Gmb.
Fahrenheitstrasse 4
Bremen, Germany

Randall K. Julian, Jr.
Senior Scientist
Eli Lilly and Co.
Indianapolis, Indiana

Raymond E. March, Ph.D.
Professor of Chemistry
Department of Chemistry
Trent University
Peterborough, Ontario
Canada

Michael Schubert, Ph.D.
Senior Scientist
Ion Trap Development
Bruker-Franzen Analytik Gmb.
Fahrenheitstrasse 4
Bremen, Germany

John E. P. Syka
Product Engineer
Finnigan MAT
San Jose, California

John F. J. Todd, Ph.D.
Professor of Mass Spectrometry
Chemical Laboratory
University of Kent
Canterbury, Kent
England

Pietro Traldi, Ph.D.
Consiglio Nazionale Ricerche
Area di Ricerca di Padovs
Corso Stati Uniti 4
Padova, Italy

Fernande Vedel, Ph.D.
Professor
Physique des Interactions
Ioniques et Moléculaires
Université de Provence
Centre de St. Jerome Case A61
Marseille, France

Michel Vedel, Ph.D.
Professor
Physique des Interactions
Ioniques et Moléculaires
Université de Provence
Centre de St. Jerome Case A61
Marseille, France

Yang Wang, Ph.D.
Ion Trap Development
Bruker-Franzen Analytik Gmb.
Fahrenheitstrasse 4
Bremen, Germany

TABLE OF CONTENTS

PART 1

Fundamental Aspects

Chapter 1

Introduction to Practical Aspects of Ion Trap Mass Spectrometry

John F. J. Todd

CONTENTS

0-8493-4452-2/95/$0.00+$.50

3

I. ORIGIN OF THE QUADRUPOLE ION TRAP

This introductory survey does not purport to be a comprehensive review of the history and development of the ion trap, nor does it provide a full treatment of the theory underlying its operation; the reader is referred to Chapter 2 for a detailed exposition of the mathematical basis of the operation of the ion trap. Rather, the aim here is to set the scene and present a brief account of how the ion trap operates in its current mode of use and to indicate the key stages of development that have been reported over the past 10 years.

The history of the ion trap dates back to the pioneering work of Paul and Steinwedel,[1] which was recognized by presentation of the shared award of the 1989 Nobel Prize for Physics to Wolfgang Paul.[2] Detailed accounts of the early development of the quadrupole-type devices as mass spectrometers were published by Dawson and Whetten[3] and by Dawson.[4] A full treatment of the theory of the ion trap is found in the now-standard text *Quadrupole Storage Mass Spectrometry*, by March and Hughes.[5] This text includes a historical account by the present author, which was expanded recently into a full-scale review.[6] Other reviews on specific topics have been contributed by Cooks and co-workers,[7,8] and a special collection of papers reporting upon recent developments has also appeared;[9] to this should be added the comprehensive Keynote Lecture given at the 12th International Mass Spectrometry Conference in Amsterdam.[10]

II. THEORY AND OPERATING PRINCIPLES OF ION TRAP MASS SPECTROMETRY

As implied by its name, this instrument operates on the basis of first storing ions and then facilitating their detection according to their mass/charge (m/z) ratio. In all, there have been three essentially different means by which this has been achieved. Initially, *mass-selective detection* was employed, in which the motion of the ions was sensed by means of tuned circuits,[11] such that a response was obtained for each *m/z* value in turn; the approach had certain similarities to ion cyclotron resonance, with the merit that the ions were detected nondestructively. This method, which was not generally appreciated by mass spectroscopists at the time, then gave way to mass-selective storage, in which the ions were trapped according to their *m/z* ratios and then detected by pulse ejecting them from the trap into an external detector.[12,13] While this arrangement gave satisfactory mass spectra, albeit over a limited mass range, the instrumentation was somewhat complex and did not appear to offer particular advantages over, for example, the quadrupole mass spectrometer. One development which did prove to be of some interest was the combination of the ion trap, termed the QUISTOR (for *quadrupole ion store*), with the quadrupole mass analyzer.[14,15] In this system, ions could be trapped in a broad-band or in a mass-selective mode[16] for a predetermined period of time before ejection and external analysis. In this way, various physical and chemical studies could be performed on the trapped ions.

Current analytical use of the ion trap, however, relies upon the more recently developed technique of the *mass-selective ejection* of ions,[17,18] pioneered commercially by Finnigan MAT (San Jose, CA)[19] and marketed under the name "Ion Trap Detector" (ITD™); other instruments available are the Finnigan ITS40™ and the Varian Saturn™ gas chromatograph/mass spectrometer (GC/MS) systems. The operational details of these instruments are best considered in terms of the schematic diagram of the instrument shown in Figure 1 and the description that follows. The ion trap consists of three cylindrically symmetric electrodes, two end-caps, and a ring. Each of these electrodes has accurately machined hyperbolic internal surfaces and, in the normal mode of use, the end-cap electrodes are connected to earth potential while a radiofrequency (RF) oscillating "drive" potential, typically around 1 MHz, is applied to the ring electrode. Ions are created within the ion trap by injection of electrons or may be injected from an external source, and a range of *m/z* values may be held in bound or so-called stable orbits by virtue of the RF potential; alternatively, a single or a range of *m/z* values may be stored by the superposition of an appropriate DC potential on top of the RF drive potential (see below for further details). As the amplitude of the RF potential is increased, the motion of the ions becomes progressively more energetic, such that eventually they develop unbound (unstable) trajectories along the axis of symmetry (the z-axis). Then, in order of increasing *m/z* values, the ions exit the device through holes in one of the end-cap electrodes and impinge on a detector. In this way, a mass spectrum is generated; usually, several such mass spectra, termed microscans, are obtained in succession and are summed prior to display and recording as

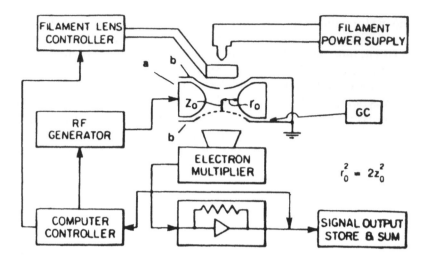

FIGURE 1
Schematic diagram of the ITD™. The ring electrode and the end-cap electrodes are labeled a and b, respectively. (From Glish, G. L. and McLuckey, S. A., *Int. J. Mass Spectrom. Ion Processes*, 106, 1–20, 1991. With permission.)

a macroscan. The sequence of operations may be seen from the timing diagram, or scan function, shown in Figure 2.

The theory behind the operation of the ion trap is best considered by examining the equations for the electric field within the ion trap of so-called perfect quadrupolar geometry and for the resulting motion of the ions. The shape of the potential developed within the trap when the electrodes are coupled to the RF and DC potentials, as indicated above, is described by

$$\phi = \frac{1}{2}(U - V \cos\Omega t)\frac{(x^2 + y^2 - 2z^2)}{r_0^2} + \frac{U - V \cos\Omega t}{2} \tag{1}$$

where U represents the maximum DC potential and V the maximum RF potential applied between the ring and the end-cap electrodes, Ω is the angular frequency of the RF drive potential, and r_0 is the internal radius of the ring electrode. For the perfect quadrupole field the arrangement of the electrodes corresponds to the case in which $r_0^2 = 2z_0^2$, where $2z_0$ is the closest distance between the two end-cap electrodes, although it has been found that a different relationship may perform better in practice (see below and Chapter 4). The oscillation of the RF potential causes the field to reverse direction periodically so that the ions are focused and defocused alternately along the z-axis and vice versa in the radial plane.

The force, F, acting upon an ion of mass m and charge e is given by

$$\vec{F} = -e \cdot \nabla\phi = m\vec{A} \tag{2}$$

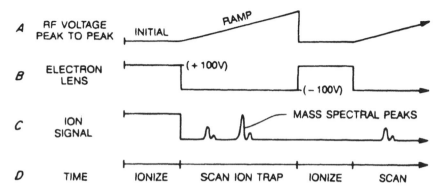

A RF VOLTAGE PEAK TO PEAK

INITIAL · RAMP

B ELECTRON LENS

(+100V) (−100V)

C ION SIGNAL

MASS SPECTRAL PEAKS

D TIME

IONIZE · SCAN ION TRAP · IONIZE · SCAN

FIGURE 2

Timing sequence (scan function) for the operation of the ion trap in the mass-selective ejection mode. (From Glish, G. L. and McLuckey, S. A., *Int. J. Mass Spectrom. Ion Processes*, 106, 1–20, 1991. With permission.)

from which the forces acting upon the ion in each of the perpendicular directions are given by

$$\left(\frac{m}{e}\right)\ddot{x} + (U - V \cos\Omega t)\frac{x}{r_0^2} = 0 \tag{3}$$

$$\left(\frac{m}{e}\right)\ddot{y} + (U - V \cos\Omega t)\frac{y}{r_0^2} = 0 \tag{4}$$

$$\left(\frac{m}{e}\right)\ddot{z} - 2(U - V \cos\Omega t)\frac{z}{r_0^2} = 0 \tag{5}$$

Note that none of these expressions contains cross-terms between x, y, and z, with the result that the motion may be resolved into each of the perpendicular coordinates, respectively. The x- and y-components are identical and may be treated independently, provided that any angular momentum which the ions may have around the x-axis is ignored. Because of the cylindrical symmetry, the x- and y-components are often combined to give a single radial r-component using $x^2 + y^2 = r^2$.

The z-component of motion is out-of-phase by half a cycle with respect to the x- and y-motion (hence the minus sign), and the factor of two arises because of the asymmetry of the device brought about by the need to observe the Laplace condition $V^2\phi = 0$ when applied to Equation 1. These equations are all examples of the Mathieu equation, which has the generalized form

$$\frac{d^2 u}{d\xi^2} + (a_u + 2q_u \cos 2\xi) u = 0 \tag{6}$$

where

$$u = x, y, z \tag{7}$$

$$\xi = \frac{\Omega t}{2} \tag{8}$$

$$a_z = -2a_x = -2a_y = -\frac{16eU}{m\left(r_0^2 + 2z_0^2\right)\Omega^2} \tag{9}$$

and

$$q_z = -2q_x = -2q_y = \frac{8eV}{m\left(r_0^2 + 2z_0^2\right)\Omega^2} \tag{10}$$

Thus, the transformations, Equations 7 to 10, relate the Mathieu parameters a_z and q_z to the experimental variables and also to the "time" variables ω and t. The a_u and q_u parameters are quite fundamental to the operation of the ion trap because they determine whether the ion motion is stable (i.e., the ions remained trapped) or unstable. The diagram shown in Figure 3 (actually only a small portion of a much larger family of curves) defines the areas within which the axial (z) and radial (r) components of motion are stable; the region of overlap indicates the (a_z, q_z) coordinates corresponding to those ions that are held in the ion trap.

The scan function (Figure 2) for the operation of the ion trap may be seen, therefore, as comprising a time period during which, in Figure 3, the (a_z, q_z) coordinates for the ions remain constant at points lying on the q_z-axis close to the origin, followed by movement of the coordinates along the axis until they reach the right-hand boundary. Here, the ions develop unstable trajectories along the z-axis of the trap.

The lines drawn across the stability region in Figure 3 are called iso-β lines and describe the detailed trajectories of the ions at that point; the boundaries of the diagram correspond to $\beta_r, \beta_z = 0$ and $\beta_r, \beta_z = 1$, with the boundary $\beta_z = 1$ being that at which mass-selective ejection is normally achieved during a mass spectral scan. Further details concerning the calculation of β-values are given in Chapter 2. The resulting ion trajectories have the general appearance of Lissajous' figures, in which a high frequency ripple (micromotion) is superimposed upon a lower frequency secular motion. When the values of β_r and β_z are increased, for example, by increasing the amplitude of the RF drive potential, then the nature of the motion becomes much more violent. This characteristic motion of the trapped ions plays an important part in the operation of the ion trap, especially in experiments in which it is desired to pump kinetic energy into the ions by resonance excitation through application of auxiliary oscillating fields applied between the end-cap electrodes (see below).

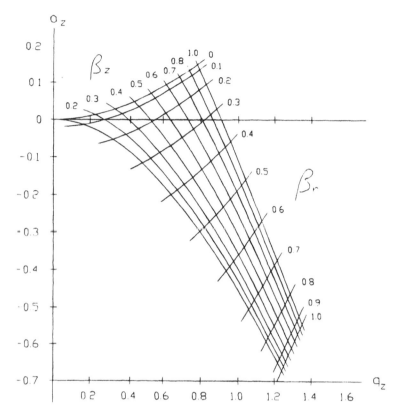

FIGURE 3
Stability diagram for the ion trap. The value of q_z = 0.91 corresponds to q_{ej} in the mass-selective ejection mode; the lines labeled β_r and β_z describe the oscillatory characteristics of ion motion.

III. IMPROVEMENTS IN PERFORMANCE

We learned in Section II how the ion trap may be operated in the mass-selective ejection mode. The chief advantage over the two previous methods of generating mass spectra is that the equipment is considerably simpler. Thus detection relies on the use of relatively cheap *channeltron* electron multipliers rather than complex circuitry, and the ejection of ions from the trap occurs "naturally", through the effects of unstable trajectories as opposed to the application of DC pulses to the end-cap electrodes required by the mass-selective storage mode. The instrument is also readily amenable to computer control using a personal computer. However, in its simplest implementation, mass-selective ejection does not yield particularly high quality mass spectra, and much effort was devoted in the early days of the commercialization of the ion trap (see Chapter 4) to improving the performance of the trap so that it was at least as good as other, more conventional instruments aimed at the same market, for example, benchtop GC/MS. This section describes five especially important improvements.

A. Effect of a Light Buffer Gas

Very early on in the development of the ion trap, long before it was introduced to the public, it was discovered that the presence of the significant background pressure of a light buffer gas (for example, 10^{-3} torr of helium) produces a dramatic improvement both in mass resolution and in sensitivity.[17] This apparent contradiction of effects may be explained in terms of a momentum-moderating effect due to collisions between the ions and the helium atoms, resulting in a "cooling" of the kinetic energy of the former and the migration of the ions toward the center of the trap in both axial and radial directions. Thus, simple modeling calculations[20] have shown that on reaching the threshold for trajectory instability, the ions tend to start from essentially the same position in the trap. As a result, not only are they all well bunched as they leave, but they are also focused tightly along the z-axis, so that they are transmitted to the detector efficiently.

B. Stretched Geometry

Louris et al.[21] announced that rather than conforming to the pure quadrupole geometry in which $r_0^2 = 2z_0^2$, the commercial traps have a *stretched geometry*, in which the distance between the end-caps has been extended by some 11% beyond the ideal value. During the early stages in the commercial development of the ion trap it was discovered that the assignment of m/z values to the ejection ion could be compound specific, and an intensive search for a solution led to the empirical optimization of the dimensions as above. This modification to the design remained a trade secret until the announcement; further detail of the drama associated with this aspect of the evolution of the trap is found in Chapter 4.

C. Automatic Gain Control

Despite the improvements brought about by the use of the buffer gas, the early mass spectra reported with the ion trap showed significant distortions as the sample concentration was changed. Clearly, these distortions were unsatisfactory for analytical applications such as GC/MS, in which quantitative integrity of the output as the analyte peaks are being eluted is essential. In particular, there could be a significant mismatch between the electron ionization (EI) mass spectra recorded with the ion trap compared to those obtained with so-called standard instruments and held in mass spectral library collections.

The problems are essentially twofold. On the one hand, high sample concentrations combined with significant trapping times of, e.g., several milliseconds, led to the occurrence of ion/molecule reactions,[14] thus changing the identities of the ions being analyzed and also causing a loss of quantitative response. Second, the buildup of ion density within the ion trap can lead to space-charge effects, substantially modifying the electric fields to which the ions are being subjected (thereby causing shifts in the positions of the boundaries of the stability diagram,[22] for example) resulting in changes in the m/z ratio assignments of the ions. To overcome these problems, the method of automatic gain control (AGC) was introduced.[23] The idea of AGC is to incorporate two ionization stages into the scan function. The first ionization time is of a fixed duration, for example, 0.2 ms, after which ions formed from the background gases (typically up to m/z 44) are removed, and the remaining analyte ions are detected without further mass analysis. This *total ion signal* is used then to calculate the optimum ionization time for the second stage in order to avoid the effects noted above. This procedure occurs each time the scan function is repeated, and the resulting ionization times are recorded along with spectral intensities in order to normalize the data before retrieval. This extremely elegant method, in which a degree of machine intelligence is employed, has established ion trap mass spectrometry as a standard quantitative analytical method, as indicated by the work of Yost et al.[24]

D. Axial Modulation

A further substantial improvement in performance has been obtained through the technique of axial modulation. One of the inherent features of the ion trap in this mode of operation is that while the ions of lower m/z ratio are being "scanned out" of the ion trap into the detector, the higher m/z ions are still in the trap, and the space charge potential that they contribute causes a broadening of the peaks arising from the ions being ejected. This effect can be demonstrated by observing the improvement in the spectral quality when the immediately higher mass ions are first removed from the trap, e.g., by employing a superimposed DC field before the lower mass ions are analyzed.[25] This deleterious effect on peak shape can be reduced dramatically by applying a supplementary oscillating field of approximately 6 $V_{(p-p)}$ at a frequency of about half of the RF drive potential between the end-cap electrodes during the analytical portion of the scan function.[26] At this point, just as the ions are being ejected, their secular motion enters into resonance with the supplementary field so that the ions are energized as they suddenly "come into step," and are therefore much more tightly bunched as they are ejected. This technique of axial modulation has been employed with spectacular success as a means of extending the m/z range of the trap (see below and Chapter 9); the application of a supplementary oscillating field is also the basis of studying the collision-induced dissociation of ions in the trap (see below and Chapter 5).

E. High Resolution

The final major improvement in performance, and one which may ultimately lead to a successful challenge of the sector instruments by the ion trap as the standard analytical mass spectrometer, is the substantial improvement in resolution that has been obtained.[27] By slowing down the speed with which the ions are scanned out of the trap, e.g., by a factor of up to 2000, and applying the axial modulation technique, resolutions in excess of 10^6 have been obtained. A detailed account of this work, together with a discussion of the problems of accurate mass measurement, are given in Volume II, Chapter 1.

IV. THE USE OF ALTERNATIVE SCAN FUNCTIONS

With the exception of the scan function for AGC, the techniques described thus far have relied on a time variation of the RF potential substantially the same as that shown in Figure 2. However, one of the merits of having a system that works under computer control is that it is relatively easy to re-program the time profile of the RF potential and to regulate the application of DC potentials superimposed upon the RF potential in order to provide alternative modes of operation.

A. Chemical Ionization

It has long been realized[28] that because of the long storage times within the ion trap, ion/molecule reactions of the type employed in chemical ionization (CI) may be facilitated, but at much lower reagent gas pressures (typically 10^{-5} torr) than those used in conventional high pressure sources on magnetic sector and quadrupole instruments. The idea is to incorporate an ionization period at constant low amplitude RF potential into the scan function, during which the reagent ion concentration is established, and to follow this by a second reaction period of a few milliseconds' duration, at a slightly higher RF potential, in which the analyte ions can then be formed and stored. The RF drive potential is then ramped in order to mass selectively eject the ions, as for the EI scan function described earlier. Several papers describing detailed ion trap CI studies have appeared, including a very thorough account by Brodbelt et al.[29] and a comparison between CI and the QUISTOR/quadrupole system, the ITD™, and a high-pressure source by Boswell et al.[30] Dorey[31] commented on the increased fragmentation observed in ion trap CI with methane as compared to that obtained with a conventional source, and has attributed this to the increased kinetic energies of the reagent ions in the former.

The merits of performing CI with the ion trap include the fact that there are no additional pumping requirements, in contrast to conventional CI, and that under computer control it is relatively easy to program alternate EI/CI spectral scans. In addition, ionization conditions can be optimized using automatic reaction control (ARC), analogous to AGC. Finally, using the ion isolation technique described below, it is possible to mass select the reagent ion and thus make CI much more specific.[32]

B. Ion Isolation

It is evident from the above description of the operation of the ion trap in the mass selective mode that it is possible to apply DC and RF potentials in such a manner that a single value, or narrow range of values, of m/z may be stored. The application of ion isolation to the study of ion chemistry in the QUISTOR was first proposed by Bonner[33] and applied by Fulford and March[34] under the title "selective ion reactor". More recently, this approach was incorporated into the scan functions employed for the mass-selective ejection mode as a means of isolating a single ion prior to performing a subsequent experiment.[35] A common means of isolating a specified value of m/z is for the amplitude of the RF drive potential to be adjusted so that the value of q_z for the ion is 0.78; i.e., the "working" (a_z, q_z) coordinate lies on the q_z-axis below the upper apex of the stability diagram in Figure 3. A negative DC potential is then applied to the ring electrode such that the working point moves to a value of a_z just below the apex. In this way, ions of lower m/z are lost through trajectory instability at the $\beta_z = 1$ boundary (in the axial direction), while ions of higher m/z are unstable in the radial direction at the $\beta_r = 0$ boundary. After about 2 ms, the DC potential is returned to zero and the RF potential reduced to a lower value for the next stage of the experiment. Alternative means of ion isolation include the use of a more complex scan function to render low m/z ions unstable at $\beta_z = 1$ and high m/z ions unstable at the $\beta_z = 0$ boundary[36] (see also Chapter 6). McLuckey et al.[37] described a combination of DC and RF potential scans designed to isolate a specified range of m/z values. Kelley[38] described an axial modulation method involving the application of a filtered noise field between the end-caps. The idea is to excite resonantly and to eject the complete range of trapped ions, with the exception of the m/z value(s) of interest, by notching out the appropriate secular frequency components from the applied noise field. This approach has the advantage that by avoiding moving the working (a_z, q_z) point of the isolated ion to a boundary, fewer of these ions are lost, and hence inherent sensitivity of the instrument is increased.

Via this simple means of ion isolation it is easy to increase the versatility of the ion trap, and it is indeed essential for the tandem mass spectrometry (MS/MS) studies described later. Analytical applications include the selection of specific reagent ions for CI, a technique that has been exploited by Strife and Keller[39] and Berberich et al.[40] Creaser et al.[41] showed how multiple ionization–isolation steps and simultaneous ionization–isolation may be used to enhance the population of selected ions.

V. ALTERNATIVE SCANNING MODES FOR THE ION TRAP

Using mass-selective ejection we have seen that mass analysis is achieved by rendering unstable the trajectories of ions with successively greater m/z values at the $\beta_z = 1$ boundary. Experiments utilizing the $\beta_z = 0$ boundary as a means of extending the m/z range are described below. Two alternative means of generating mass spectra with the ion trap have been reported recently. One[42,43] involves the application of a swept supplementary frequency to the end-cap electrodes, followed by Fourier transformation of the image currents of the motion of the kinetically excited ions induced in the electrodes. This approach, which is analogous to the Fourier transform-ion cyclotron resonance (FT-ICR) experiment, gave recognizable mass spectra, although the quality of the data recorded in these early experiments did not suggest that there was much good to be gained from this mode of operation at this stage.

A second method of scanning was described by Griffiths and Heesterman.[44] This method may be imagined as combining the mass-selective storage and the mass-selective ejection methods. Ions are created by a gated electron beam while the RF and DC potentials are stepped in the form of a staircase. The amplitudes of the potentials are maintained in a constant ratio and adjusted so that the (a_z, q_z) coordinates of each m/z value in turn are held just under the apex of the stability diagram (see Figure 3) in a manner analogous to that employed in the ion isolation technique described above. As the potentials are then incremented to the next step, any trapped ions are expelled toward an external detector because the coordinates now lie beyond the $\beta_z = 1$ stability boundary. This mode of operation possesses some limitations when compared to analysis by mass-selective detection, although the system does have the merit of simplicity and ease of operation. The authors suggest that the device should be well suited to use as a low-cost gas analyzer.

VI. TANDEM MASS SPECTROMETRY

One of the most exciting developments in the field of ion trap mass spectrometry has been the adaptation of the device for tandem (MS/MS) mass spectrometry, especially in conjunction with collision-induced dissociation (CID).[45] Here, the idea is to create or inject the ions, isolate the parent ion (see above) which is to be dissociated, and then excite resonantly the axial component of the ion motion by applying across the end-cap electrodes a supplementary sinusoidal "tickle" potential, which is tuned to the fundamental secular frequency of the ion. This process of resonance excitation is analogous to the axial modulation experiment described above, but here the amplitude of the tickle potential is adjusted carefully so as not to cause the ions to be ejected. Under these conditions, the ions are effectively pulled away from the center of the trap so that they acquire energy from the RF drive potential and undergo energetic collisions with the helium buffer gas. The fragment ions resulting from the dissociation are then analyzed by increasing the amplitude of the RF drive potential in the normal manner. This experiment is tandem in time, again similar to MS/MS using FT-ICR, and is clearly less expensive to implement than is tandem-in-space analysis with a quadrupole or magnetic sector instrument. Indeed, because the system operates under software control, once the initial hardware modifications have been made, it is extremely easy to modify the scan function in order to implement higher-order MS/MS experiments, especially as the fragment efficiencies are so high with the ion trap.[46] One possible limitation is the extent of energy deposition that may be achieved during resonant excitation with the ion trap as compared to the triple quadrupole instrument. Thus, in an elegant early series of experiments, Brodbelt et al.[47] showed that the upper limit of internal excitation is approximately 5.8 eV. Alternative techniques of collisional excitation have been described recently, including the application of a rapid DC pulse in order to effect surface-induced dissociation (SID)[48] of the ions through collision with the ring electrode, pulsed axial excitation,[49] and the use of helium mixed with heavier-mass target gases.[50]

Another area of difficulty is that efficient resonance excitation of the selected ion is only possible when the frequency of the tickle potential is accurately tuned. As a result, MS/MS analysis with the ion trap can involve a fairly lengthy tuning procedure in order to optimize the performance for each m/z value being studied. Software modifications described by Pannell et al.[51] and by Todd et al.[52] have shown ways in which this problem may be overcome. The latter account describes the technique of dynamically programmed scans, which enables the ion trap to be used for the tandem mass analysis of unknown samples, in which one is not aware of which species to select as parent ions until the first conventional mass spectrum has been run. Parent ions are then selected according to preset threshold criteria and their collisionally induced MS/MS spectra are recorded in turn. Another advantage of this approach is that the data can be presented in a form that readily allows parent ion (fixed product) and neutral loss spectra to be deduced; the "intelligent" recording of MS/MS/MS spectra is also possible. An alternative technique was developed by Johnson et al.[53] for performing MS/MS parent ions scans with the ion trap by the simultaneous resonant excitation of multiple ions.

The laser photodissociation of trapped mass-selected ions[54] is considered in some detail in Section V. An evident advantage over collisional dissociation by resonant excitation is that, provided the ions absorb at the photon frequency employed, there is no requirement to fine tune the operating conditions of the ion trap (for example, the tickle frequency) in order for the ions to be energized, a potentially advantageous situation for analysis by GC/MS/MS with the ion trap.[55] The problem of the need to tune the tickle frequency prior to resonance excitation has been addressed by several groups of workers. Penman et al.[56] showed how the use of dynamically programmed scans to generate conditions under which a ramped DC potential is applied to the ring electrode during the application of the tickle potential broadens the absorption peak; Yates et al.[57] described an optimization routine analogous to the AGC scan function discussed above. Broad-band excitation through the application of noise signals has also been reported by Vedel et al.[58] and by McLuckey et al.[59]

The modeling of the application of supplementary electric fields has been examined by March et al.,[60,61] and three distinct modes of resonant excitation have been defined. In the conventional mode, equal but out-of-phase RF potentials are applied to each end-cap electrode to give what has been termed *dipolar excitation*. Alternatively, the end-caps can be connected in-phase so as to produce a *quadrupolar excitation* field superimposed upon the quadrupolar field of the RF drive potential. Finally, there is *monopolar excitation*, in which the supplementary potential is connected to one end-cap while the other is grounded; this method was employed in the initial QUISTOR resonant ejection (QRE) experiments, described by Fulford et al.,[62] in which ions could be removed selectively from the ion trap in a manner akin to axial modulation. One outcome of these simulation studies was the realization that other series of resonant frequencies exist, especially in the quadrupolar mode (including some that lead to radial excitation of the ions), and these effects have been observed experimentally.[63,64] Attempts to quantify the dissociation of ions through resonance excitation of ions in the trap using a kinetic approach have been made by Todd and co-workers,[65] and further details of the treatment are described in Volume III.

VII. EXTENSION OF THE MASS/CHARGE RANGE

As marketed commercially by Finnigan MAT[19] the Ion Trap Detector has an upper mass/charge limit—$(m/z)_{max}$—of 650 Da e^{-1}, which is determined by rearrangement of Equation 10 to give

$$(m/z)_{max} = \frac{8eV_{max}}{q_{ej}\left(r_0^2 + 2z_0^2\right)\Omega^2} \tag{11}$$

where q_{ej} (= 0.908) is taken as being the value of q_z at which ion ejection occurs and V_{max} is the maximum value (zero to peak) of the RF drive potential (approximately 16 kV). Thus, in order to increase the value of $(m/z)_{max}$ for a given maximum RF amplitude one may reduce the values of r_0 and/or Ω, or reduce the value of q_{ej}. In the initial attempts to increase the m/z range, Todd et al.[66] tried to avoid changing the basic physical and electronic configuration of the trap and developed the technique of *reverse scans* in which a positive DC potential is superimposed upon the RF drive potential fed to the ring electrode. Reducing these potentials together, keeping the ratio between them constant, makes the (a_z, q_z) coordinates move along a scan line which cuts the $\beta_z = 0$ boundary (Figure 3), thus causing mass-selective instability to occur in the reverse order (i.e., ions of high m/z value before those of low m/z value) compared to the conventional mode of scanning. Various other alternative scan functions making use of the $\beta_z = 0$ boundary are possible, and values of m/z up to about 2000 have been analyzed using this approach.[67]

In their attack on the problem of extending the m/z range, Cooks and co-workers[68] showed how lowering the values of both r_0 and Ω could be employed to yield quite acceptable mass spectra up to m/z 2600 (i.e., from Equation 11, reducing the value of r_0 or Ω by 2 increases $(m/z)_{max}$ by 4). However, the major breakthrough has come from their use of axial modulation (see above), but applying supplementary frequencies corresponding to ion ejection at very low values of β_z, for example, 0.01, rather than β_z = 1.[69] This form of resonance ejection has the effect of increasing $(m/z)_{max}$ 100 times, and the method has been verified by recording mass spectra of clusters of cesium iodide, generated using an external Cs$^+$ ion bombardment source, up to values in excess of m/z 45,000 Da e^{-1}. Comparative details of the three methods they employed have been published by Kaiser et al.,[70] and the topic is considered further in Chapter 1 of Volume II.

VIII. USE OF EXTERNAL ION SOURCES

With the exception of the references to cesium iodide cluster ions mentioned in the preceding section, all the work described thus far has involved the use of ions created within the trap, either by EI or CI; two reports of the use of laser desorption have also appeared in the literature.[71,72] However, there are clear advantages to be gained from creating ions externally (for example, with a fast atom bombardment ionization source) and injecting them into the trap for subsequent analysis. The difficulty of trapping ions formed in this way is that, unless the ions enter the trap at the correct phase angle of the RF drive potential, they will not have the correct combination of velocity and displacement to remain in stable orbits, as has been noted in several theoretical discussions on the subject[73-75] and experimentally by March and co-workers.[76]

This problem is overcome, however, when helium buffer gas is present, and Louris et al.[77] described a successful system in which an external EI source (mounted in place of the conventional filament assembly of the ITD) was combined with an Einzel lens to gate the ion beam. The efficiency of trapping was clearly mass dependent, and the authors rationalized their results in terms of the model of pseudo-potential wells,[78] which approximates the trajectories of the ions to that of simple harmonic motion about the center of the trap. Pedder et al.[79] described a system in which an off-axis CI source was employed to create negative ions which were then injected into the trap. At this point it should be noted that the in-trap production of negative ions by direct electron attachment is very inefficient because thermal electrons and heavy negative ions cannot be held in the trap simultaneously, on account of their difference in mass. The use of ion/molecule reactions for the production of negative ions overcomes this difficulty, and the in-trap CI of nitroaromatics using OH^- was first demonstrated by McLuckey et al.[80] Schlunegger and co-workers[81] described a hybrid MS system consisting of magnetic and electric sectors, followed by a retardation system in which the 3 keV ions are decelerated down to 5 eV before injection into a QUISTOR (supplied with helium buffer gas); after trapping, during which ion chemistry can be allowed to occur, the ions are pulse extracted for analysis with a quadrupole mass filter in a manner analogous to the original QUISTOR/quadrupole.[14,15] A sector (BE)/ion trap mass spectrometer also was reported by Schwartz et al.[82]

Another system that involves the trapping of injected ions makes use of an atmospheric sampling glow discharge ionization (ASGDI) source, and has been described by McLuckey et al.[83] Using a differentially pumped arrangement, atmospheric gases enter a region through an orifice at a pressure of 0.2 to 0.8 torr, and in which a DC discharge is maintained between a pair of electrodes. The ions are then drawn through a second orifice and a gating lens to the ion trap where they enter through a hole in an end-cap electrode. This source has been shown to be capable of detecting organic explosives at the parts per 10^{12} level; therefore, combination of this mode of ionization with the capabilities of the ion trap has great potential in the field of trace contaminant monitoring. A detailed report on the negative ion fragmentation processes occurring in 2,4,6-trinitrotoluene using an ASGDI/ion trap system has been published by Langford and Todd.[84]

As with the external EI source, the trapping efficiency with the ASGDI source has been characterized in terms of the operating conditions of the trap, as well as the pressure and nature of the buffer gas. One important result is that in arresting the motion of the injected ions, CID may occur, possibly on the electrode surfaces themselves. While the ions still may be mass selected for subsequent MS/MS analysis, the intensities of the higher mass species are inevitably decreased. An alternative approach, which is especially useful in the production of negative ions in the trap, is to inject externally created reagent anions and then form the analyte ions via chemical ionization.[85] A recent study demonstrated that NO_3^- ions formed in the glow discharge have at least two different stable forms.[86] A detailed description of the glow discharge source and its coupling to an ion trap is given in Volume III.

A further, very exciting development is the coupling of electrospray ionization[87] to the ion trap.[27,88] Thus, multiply protonated biomolecules may be examined at low m/z values, as with triple quadrupole instruments,[89] but with the added feature that through the addition of a reagent species, such as dimethylamine, the relative rates of proton transfer between the charge states of the analyte ions and the reagent may be determined.[90] As a result, the possibility of using chemical means for additional characterization of, e.g., peptide ions and their fragments, is opened up. A more detailed account of electrospray ionization is included in Chapter 3 of Volume II and in Volume III of this series. An atmospheric pressure ionization interface for a benchtop quadrupole ion trap has been described recently.[91]

IX. NEW DESIGNS OF ION TRAP

The work described thus far has involved the use of ion traps having the ideal geometry so that the ions are subjected to pure quadrupolar electric fields. Because of field imperfections, due to nonideal spacing of the electrodes or contamination of the electrode surfaces, it is sometimes possible to observe nonideal effects, such as the unexpected ejection of ions through the influence of nonlinear resonances observed during reverse scanning experiments.[67] An interesting development is the deliberate incorporation of higher-order hexapole and octopole field contributions[92,93] into the field geometry of the QUISTOR as a means of enhancing its performance. A detailed account of the simulation and operation of the ion trap under these conditions is given in Chapter 3.

X. ION ENERGETICS AND ION CHEMISTRY

It was noted above that CI may be performed easily with the ion trap by simply modifying the scan function. Indeed, unintentional self-chemical ionization may occur[94,95] if care is not taken to control the sample pressure. Rate constant data for ion/molecule reactions may be acquired by varying the storage time during which the reactions are occurring, and from these data an idea of the energy distribution of the ions (i.e., is, the temperature) is obtained.[96,97] A detailed review of ion chemistry by Nourse and Cooks has already been cited,[7] and the same authors have presented an account of proton affinity determinations with the ion trap.[98]

XI. CHROMATOGRAPHY COMBINED WITH THE ION TRAP

The original purpose for developing the ITD was for use as a benchtop mass spectrometer in combination with capillary GC and, indeed, the bulk of the sales of the instrument have been for this application. However, preliminary experiments were reported with other types of systems, using both internal and external ionization. Thus, Todd and co-workers were able to combine successfully an ion trap with a supercritical fluid chromatography column (SFC) operating with carbon dioxide as the mobile phase[99,100] and obtained EI mass spectra of polycyclic aromatic compounds which matched the National Bureau of Standards' library spectra. The limiting factor on performance appeared to be the pressure of carbon dioxide in the ion trap, which increased as the pressure-programmed elution of the sample took place. Reference experiments, in which the mass spectral intensities were monitored as a function of CO_2 pressure, indicated an improvement in signal level at increased pressure, presumably because of charge-transfer effects between the ionized carbon dioxide and the analyte molecules.

A thermospray liquid chromatography/ion trap system also has been described[101] in which very fast pumping was employed to keep the operating pressure at 2×10^{-4} torr at an effluent rate of 0.5 ml min^{-1}. Phenylalanine and adenosine, both dissolved in 80:20 methanol:water, gave recognizable spectra, but the performance was evidently degraded by space charge due to ions arising from the high background pressure of the solvent. It was suggested that the application of a supplementary oscillating field between the end-cap electrodes to resonantly eject these ions should lead to an improvement in spectral quality, and that such problems could presumably be overcome by using the filtered noise field method of ion isolation[38] described earlier.

XII. CONCLUSIONS

The aim of this introductory background chapter was to present a brief outline of the recent developments in ion trap technology and applications. Inevitably, many important aspects were omitted, and nothing was mentioned about sensitivity and detection levels, or about the vast range of analytical applications to which the ion trap is now being applied. These aspects are considered in detail in the chapters that follow, and in Volumes II and III.

REFERENCES

1. Paul, W.; Steinwedel, H. *U.S. Patent* 1960, 2,939,952.
2. Paul, W. *Angew. Chem.* 1990, *29*, 739.
3. Dawson, P.H.; Whetten, N.R. *Adv. Electr. Electron. Phys.* 1969, *27*, 59.
4. Dawson, P.H. *Quadrupole Mass Spectrometry and Its Applications*; Elsevier: Amsterdam, 1976.
5. March, R.E.; Hughes, R.J. *Quadrupole Storage Mass Spectrometry*; Wiley Interscience: New York, 1989.
6. Todd, J.F.J. *Mass Spectrom. Rev.* 1991, *10*, 3.
7. Nourse, B.D.; Cooks, R.G. *Anal. Chem. Acta.* 1990, *228*, 1.
8. Cooks, R.G.; Kaiser, R.E., Jr., *Accounts Chem. Res.* 1990, *23*, 213.
9. Glish, G.; McLuckey, S.A. *Int. J. Mass Spectrom. Ion Processes.* 1991, *106*, 1–20.
10. March, R.E. *Int. J. Mass Spectrom. Ion Processes.* 1992, *118/119*, 71.
11. Fischer, E. Z. *Phys.* 1959, *156*, 1.
12. Dawson, P.H.; Whetten, N.R. *J. Vac. Sci. Technol.* 1968, *5*, 1.
13. Dawson, P.H.; Whetten, N.R. *J. Vac. Sci. Technol.* 1968, *5*, 11.
14. Bonner, R.F.; Lawson, G; Todd, J.F.J. *Int. J. Mass Spectrom. Ion Phys.* 1972/73, *10*, 197.
15. Lawson, G; Bonner, R.F.; Todd, J.F.J. *J. Phys. E: Sci. Instrum.* 1973, *6*, 357.
16. Mather, R.E.; Waldren, R.M.; Todd, J.F.J. in *Dynamic Mass Spectrometry, Vol. 5*; D. Price, J.F.J. Todd, Eds.; Heyden & Son: London, 1978, p. 71.
17. Stafford, Jr., G.C.; Kelley, P.E.; Syka, J.E.P.; Reynolds, W.E.; Todd, J.F.J., *Int. J. Mass Spectrom. Ion Processes.* 1984, *60*, 85.
18. Kelley, P.E.; Stafford, Jr., G.C.; Stephens, D.R., *U.S. Patent*, 1985, 4,540,884.
19. Finnigan MAT, 355 River Oaks Parkway, San Jose, CA.
20. Bexon, J.J.; Todd, J.F.J. unpublished results, 1987.
21. Louris, J; Schwartz, J.; Stafford, G.; Syka, J.; Taylor, D. *Proc. 40th ASMS Conf. Mass Spectrometry and Allied Topics.* Washington, May/June 1992, p. 1003.
22. Todd, J.F.J.; Waldren R.M.; Mather, R.E. *Int. J. Mass Spectrom. Ion Phys.* 1980, *34*, 325.
23. Stafford, Jr., G.C.; Taylor, D.M.; Bradshaw, S.C.; Syka, J.E.P. *Proc. 35th ASMS Conf. Mass Spectrometry and Allied Topics.* Denver, CO, May 1987, p. 775.

24. Yost, R.A.; McClennen, W.; Snyder, A.P. *Proc. 35th ASMS Conf. Mass Spectrometry and Allied Topics.* Denver, CO, May 1987, p. 789.
25. Weber-Grabau, M.; Todd, J.F.J. unpublished results, 1987.
26. Tucker, D.B.; Hameister, C.H.; Bradshaw, S.C.; Hoekman, D.J.; Weber-Grabau, M. *Proc. 36th ASMS Conf. Mass Spectrometry and Allied Topics.* San Francisco, June 1988, p. 628.
27. Schwartz, J.C.; Syka, J.E.P.; Jardine, I. *J. Am. Soc. Mass Spectrom.* 1991, 2, 198.
28. Bonner, R.F.; Lawson, G.; Todd, J.F.J. *J. C. S. Chem. Commun.* 1972, 1179.
29. Brodbelt, J.S.; Louris, J.N.; Cooks, R.G. *Anal. Chem.* 1987, 59, 1278.
30. Boswell, S.M.; Mather, R.E.; Todd, J.F.J. *Int. J. Mass Spectrom. Ion Processes.* 1990, 99, 139.
31. Dorey, R.C. *Org. Mass Spectrom.* 1989, 24, 973.
32. Glish, G.; Van Berkel, J.G.; McLuckey, S.A. in *Advances in Mass Spectrometry,* Volume 11B, P. Longevialle, Ed.: Heyden & Son: London, 1989, 1656.
33. Bonner, R.F. *Int. J. Mass Spectrom. Ion Phys.* 1977, 23, 249.
34. Fulford J.E.; March, R.E. *Int. J. Mass Spectrom. Ion Phys.* 1978, 26, 155.
35. Strife, R.J.; Kelley, P.E.; Weber-Grabau, M. *Rapid Commun. Mass Spectrom.* 1988, 2, 105.
36. Ardanaz, C.E; Traldi, P.; Vettori, U.; Kavka, J.; Guidugli, F. *Rapid Commun. Mass Spectrom.* 1991, 5, 5.
37. McLuckey, S.A.; Goeringer, D.E.; Glish, G.L. *J. Am. Soc. Mass Spectrom.* 1991, 2, 11.
38. Kelley, P.E. *U.S. Patent,* 1992, 5,134,286.
39. Strife R.J.; Keller, P.R. *Org. Mass Spectrom.* 1989, 24, 201.
40. Berberich, D.W.; Hail, M.E.; Johnson, J.V.; Yost, R.A. *Int. J. Mass Spectrom. Ion Processes.* 1089, 94, 115.
41. Creaser, C.S.; Mitchell, D.S.; O'Neill, K.E. *Int. J. Mass Spectrom. Ion Processes.* 1991, 106, 21.
42. Syka, J.E.P.; Fies, W.J., Jr.; *Proc. 35th ASMS Conf. Mass Spectrometry and Allied Topics.* Denver, CO, May 1987, p. 767.
43. Syka, J.E.P.; Fies, W.J., Jr.; *U.S. Patent,* 1988, 4,755,670.
44. Griffiths I.W.; Heesterman, P.J.L. *Int. J. Mass Spectrom. Ion Processes.* 1990, 99, 79.
45. Louris, J.N.; Cooks, R.G.; Syka, J.E.P.; Kelley, P.E.; Stafford, G.C., Jr.; Todd, J.F.J. *Anal. Chem.* 1987, 59, 1677.
46. Johnson, J.V.; Yost, R.A.; Kelley, P.E.; Bradford, D.C. *Anal. Chem.* 1990, 62, 2162.
47. Brodbelt, J.S.; Kenttämaa, H.I.; Cooks, R.G. *Org. Mass Spectrom.* 1988, 23, 6.
48. Lammert, S.A.; Cooks, R.G. *J. Am. Soc. Mass Spectrom.* 1991, 3, 487.
49. Lammert, S.A.; Cooks, R.G. *Rapid Commun. Mass Spectrom.* 1992, 6, 528.
50. Glish, G.; McLuckey, S.A.; Goeringer, D.J.; Van Berkel, G.J.; Hart, K.J. *Proc. 39th ASMS Conf. Mass Spectrometry and Allied Topics.* Nashville, TN, May 1991, p. 536.
51. Pannell, L.K.; Quan-long, P.; Mason, R.T.; Fales, H.M. *Rapid Commun. Mass Spectrom.* 1990, 4, 103.
52. Todd, J.F.J.; Penman, A.D.; Thorner, D.A.; Smith, R.D. *Rapid Commun. Mass Spectrom.* 1990, 4, 108.
53. Johnson, J.V.; Pedder, R.D.; Yost, R.A. *Int. J. Mass Spectrom. Ion Processes.* 1991, 106, 197.
54. Louris, J.N.; Brodbelt, J.S.; Cooks, R.G. *Int. J. Mass Spectrom. Ion Processes.* 1987, 75, 345.
55. Creaser, C.S.; McCoustra, M.R.S.; Mitchell, D.S.; O'Neill, K.E. *Proc. 18th Meet. Br. Mass Spectrom. Soc.,* London, September 1990, p. 49.
56. Penman, A.D.; Todd, J.F.J.; Thorner, D.A.; Smith, R.D. *Rapid Commun. Mass Spectrom.* 1990, 4, 415.
57. Yates, N.A.; Yost, R.A.; Bradshaw, S.C.; Tucker, D.B. *Proc. 39th ASMS Conf. Mass Spectrometry and Allied Topics.* Nashville, TN, May 1991, p. 132.
58. Vedel, F.; Vedel, M.; March, R.E. *Int. J. Mass Spectrom. Ion Processes.* 1991, 108, R11.

59. McLuckey, S.A.; Goeringer, D.E.; Glish, G.L. *Anal. Chem.* 1992, *64*, 1455.
60. March, R.E.; McMahon, A.W.; Londry, F.A.; Alfred, R.L.; Todd, J.F.J.; Vedel, F. *Int. J. Mass Spectrom. Ion Processes.* 1989, *95*, 119.
61. March, R.E.; McMahon, A.W.; Allinson, E.T.; Londry, F.A.; Alfred, R.L.; Todd, J.F.J.; Vedel, F. *Int. J. Mass Spectrom. Ion Processes.* 1990, *99*, 109.
62. Fulford, J.E.; Hoa, D.-N.; Hughes, R.J.; March, R.E.; Bonner, R.F.; Wong, G.J. *J. Vac. Sci. Technol.* 1990, *17*, 829.
63. Vedel, F.; Vedel, M.; March, R.E. *Int. J. Mass Spectrom. Ion Processes.* 1990, *99*, 125.
64. Penman, A.D.; Todd, J.F.J.; Thorner, D.A.; Smith, R.D. *Proc. 38th ASMS Conf. Mass Spectrometry and Allied Topics.* Tucson, AZ, May 1990, p. 888.
65. Franklin, A.M.; Todd, J.F.J.; Penman, A.D.; March, R.E. *19th Annual Meeting, British Mass Spectrometry Society.* St. Andrews, Scotland, September 1992, p. 261.
66. Todd, J.F.J.; Bexon, J.J.; Weber-Grabau, M.; Kelley, P.E.; Stafford, G.C. Jr., *Proc. 35th ASMS Conf. Mass Spectrometry and Allied Topics.* Denver, CO, May 1987, p. 263.
67. Todd, J.F.J.; Penman, A.D.; Smith, R.D. *Int. J. Mass Spectrom. Ion Processes.* 1991, *106*, 117.
68. Kaiser, R.E., Jr.; Cooks, R.G.; Moss, J.; Hemberger, P.H. *Rapid Commun. Mass Spectrom.* 1989, *3*, 50.
69. Kaiser, R.E., Jr.; Louris, J.N.; Amy, J.W.; Cooks, R.G. *Rapid Commun. Mass Spectrom.* 1989, *3*, 225.
70. Kaiser, R.E., Jr.; Cooks, R.G.; Stafford, G.C. Jr.,; Syka, J.E.P.; Hemberger, P.H. *Int. J. Mass Spectrom. Ion Processes.* 1991, *106*, 79.
71. Heller, D.N.; Lys, I.; Cotter, R.J.; Uy, O.M. *Anal. Chem.* 1989, *61*, 1083.
72. Glish, G.L.; Goeringer, D.E.; Asano, K.G.; McLuckey, S.A. *Int. J. Mass Spectrom. Ion Processes.* 1989, *94*, 15.
73. Kishore, M. N.; Ghosh, P.K. *Int. J. Mass Spectrom. Ion Phys.* 1979, *29*, 345.
74. Todd, J.F.J.; Freer, D.A.; Waldren, R.M. *Int. J. Mass Spectrom. Ion Phys.* 1980, *36*, 371.
75. O, C.-S.; Schuessler, H.A. *Int. J. Mass Spectrom. Ion Phys.* 1981, *40*, 67.
76. Ho, M.; Hughes, R.J.; Kazdan, E.; Matthews, P.J.; Young, A.B.; March, R.E. *Proc. 32nd ASMS Conf. Mass Spectrometry and Allied Topics.* San Antonio, May/June 1984, p. 513.
77. Louris, J.N.; Amy, J.W.; Ridley, T.Y.; Cooks, R.G. *Int. J. Mass Spectrom. Ion Processes.* 1989, *88*, 97.
78. Dehmelt, H.G. in *Advances in Atomic and Molecular Physics*, Vol. 3; D.R. Bates and I. Estermann, Eds.; Academic Press: New York, 1967, p. 53.
79. Pedder, R.E.; Yost, R.A.; Weber-Grabau, M. *Proc. 37th ASMS Conf. Mass Spectrometry and Allied Topics.* Miami Beach, FL, May 1989, p. 468.
80. McLuckey, S.A.; Glish, G.L.; Kelley, P.E. *Anal. Chem.* 1987, *59*, 1670.
81. Suter, M.J.-F.; Gfeller, H.; Schlunegger, U.P. *Rapid Commun. Mass Spectrom.* 1989, *3*, 62.
82. Schwartz, J.C.; Kaiser, R.E., Jr.; Cooks, R.G.; Savickas, P.J. *Int. J. Mass Spectrom. Ion Processes.* 1990, *98*, 209.
83. McLuckey, S.A.; Glish, G.L.; Asano, K.G. *Anal. Chim. Acta.* 1989, *225*, 25.
84. Langford, M.L.; Todd, J.F.J. *Org. Mass Spectrom.* 1993, *28*, 773.
85. Eckenrode, B.A.; Glish, G.L.; McLuckey, S.A. *Int. J. Mass Spectrom. Ion Processes.* 1990, *99*, 151.
86. Flurer, R.A.; Glish, G.L.; McLuckey, S.A. *J. Am. Soc. Mass Spectrom.* 1990, *1*, 217.
87. Fenn, J.B.; Mann, M.; Meng, C.K.; Wong, S.F.; Whitehouse, C.M. *Mass Spectrom. Rev.* 1990, *9*, 37.
88. Van Berkel, G.J.; Glish, G.L.; McLuckey, S.A. *Anal. Chem.* 1990, *62*, 1284.
89. Covey, T.R.; Bonner, R.F.; Shushan, B.I.; Henion, J.D. *Rapid Commun. Mass Spectrom.* 1988, *2*, 249.
90. McLuckey, S.A.; Van Berkel, G.J.; Glish, G.L. *J. Am. Chem. Soc.* 1990, *112*, 5668.
91. Mordenhai, A.V.; Hopfgartner, G.; Huggins, T.G.; Henion, J.D. *Rapid Commun. Mass Spectrom.* 1992, *6*, 508.

92. Franzen, J.; Gabling, R.-H.; Heinen, G.; Weiss, G. *Proc. 38th ASMS Conf. Mass Spectrometry and Allied Topics.* Tucson, AZ, May 1990, p. 417.
93. Franzen, J. *Int. J. Mass Spectrom. Ion Processes.* 1991, *106*, 63.
94. Eichelberger, J.W.; Budde, W.L.; Slivon, L.E. *Anal. Chem.* 1987, *59*, 2731.
95. McLuckey, S.A.; Glish, G.L.; Asano, K.G.; Van Berkel, G.J. *Anal. Chem.* 1988, *60*, 2312.
96. Lawson, G.; Bonner, R.F.; Mather, R.E.; Todd, J.F.J.; March, R.E. *J. Chem. Soc., Faraday Trans. I.* 1976, *73*, 545.
97. Nourse, B.D.; Kenttämaa, H.I. *J. Phys. Chem.* 1990, *94*, 5809.
98. Nourse, B.D.; Cooks, R.G. *Int. J. Mass Spectrom. Ion Processes.* 1991, *106*, 249.
99. Todd, J.F.J.; Mylchreest, I.C.; Berry, A.J.; Games, D.E.; Smith, R.D. *Rapid Commun. Mass Spectrom.* 1988, *2*, 55.
100. Penman, A.D.; Todd, J.F.J.; Berry, A.J.; Mylchreest, I.C.; Games, D.E.; Smith, R.D. in *Advances in Mass Spectrometry*, Vol. 11A; P. Longevialle, Ed.; Heyden & Son: London, 1989, p. 240.
101. Kaiser, R.E., Jr.; Williams, J.D.; Schwartz, J.C.; Lammert, S.A.; Cooks, R.G. *Proc. 37th ASMS Conf. Mass Spectrometry and Allied Topics.* Miami Beach, FL, May 1989, p. 369.

Chapter 2

THEORY OF QUADRUPOLE MASS SPECTROMETRY

Raymond E. March and Frank A. Londry

CONTENTS

0-8493-4452-2/95/$0.00+$.50
© 1995 by CRC Press, Inc.

I. INTRODUCTION

Undoubtedly the greatest barrier to the understanding of quadrupole devices is the theory of operation of such devices which has penetrated little into undergraduate physics programs, and virtually not at all in undergraduate chemistry programs. However, there is increasing interest in treating quadrupole field theory at the graduate level. It is hoped that this chapter will encourage the reader to make the investment in time and energy to become more familiar with the rapidly growing utilization of quadrupolar devices, both quadrupole mass filters and quadrupole ion traps.

This chapter seeks to explain the basic theory of quadrupolar devices, so as to present to the reader a compact treatment of the subject matter and to provide a basis in theory for the accompanying chapters. The treatment present here is not exhaustive, and the reader who seeks to learn of nonpure quadrupole ion traps is directed to Chapter 3, where a detailed examination of such traps is presented.

II. QUADRUPOLE ION TRAP

The three-dimensional quadrupole ion trap (Paul trap or QUISTOR) is but one of a family of devices that utilize path stability as a means of separating ions according to their mass/charge (m/z) ratio.

A. Origin

The original public disclosure of the quadrupole ion trap was made in the very same patent[1] as that in which Paul and Steinwedel first described the operating principles of the quadrupole mass filter; while not quite an afterthought of the inventors, the ion trap was nevertheless described as "still another electrode arrangement". In the same year in which Paul and Steinwedel filed their quadrupole patent (1953), similar ideas were put forward by Post and Heinrich[2] for a "mass spectrograph using strong focusing principles" and by Good[3] for "a particle containment device". It appears that these proposals were stimulated through the 1952 publication by Courant et al.[4] of the theory of strong focusing of charged particle beams using alternating gradient quadrupole magnetic fields. Yet the strong-focusing technique had been discovered 2 years earlier by Christofilos,[5] an electrical engineer in Athens, Greece. His work was overlooked at that time, despite the submission of a report on his work to the University of California Radiation Laboratory and his applications for patents. The principle of using strong focusing fields for mass analysis was recognized by Paul and colleagues[6] at the University of Bonn

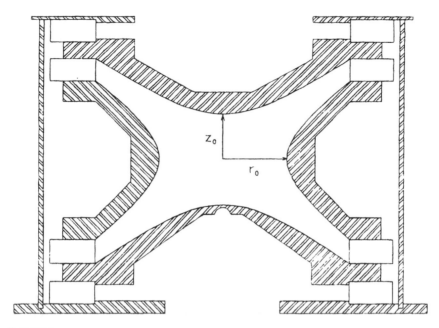

FIGURE 1
Schematic diagram of the electrode structure of the three-dimensional quadrupole ion trap.

(Germany), and the first detailed account of the operation of the quadrupole ion trap appeared in the thesis by Berkling aus Leipzig[7] in 1956.

B. Description

The quadrupole ion trap is the three-dimensional analogue of the two-dimensional quadrupole mass filter. The quadrupole ion trap consists of an arrangement of three electrodes wherein gaseous ions can be held by application of electric fields only. The device is composed of a central electrode having the form of a hyperboloid of two sheets (or ring electrode) located between two hyperboloids of one sheet (or end-cap electrodes), as shown in Figure 1.

The device is commonly referred to by physicists as the "Paul trap", while chemists have followed Dawson[8] by describing it as the "quadrupole ion trap" (which is somewhat confusing to students who expect to find a fourth electrode!), or the school of Todd, which coined the name QUISTOR[9] (from *quadrupole ion store*). Today, while the device is known widely and simply as "the ion trap," there are additionally a variety of commercial instruments* known as the ITD, ITMS, and Saturn. The first

* The ITD (ion trap detector) and the ITMS (ion trap mass spectrometer) are manufactured by Finnigan MAT of San Jose, CA; the Saturn I and II models of the gas chromatograph/mass spectrometer are manufactured by Varian Associates, Walnut Creek, CA.

major review[8] of the quadrupole ion trap appeared in 1976, while the second[10] appeared in 1989; since the latter review, some 250 publications concerning the quadrupole ion trap have appeared which is a measure of the quite considerable activity in this field. In Table 1 are shown the various "ages" of the quadrupole ion trap.

C. Recognition of the Quadrupole Ion Trap Pioneers

In the autumn of 1989, the nascent Beaujolais Nouveau appeared to our great joy once again and, to the greater joy of the ion trapping fraternity, the Nobel Prize in Physics[11] was shared by three men, the first two being Wolfgang Paul[12] of the University of Bonn and Hans Dehmelt[13] of the University of Washington, Seattle. These particular awards were made "for the development of the ion trap technique . . . which has made it possible to study a single electron or single ion with extreme precision". The Nobel was also awarded[11] to Norman Ramsay for "invention of the separated oscillatory fields method and its use in the hydrogen maser and other atomic clocks". Physicists and chemists of the ion trapping fraternity bathed in the reflected glory of the Nobel awards. All of the contributors to this book take this opportunity to offer their collective congratulations to these Nobelists. The award brought recognition to an area in which activity had become intense since the advent of the commercial version of the ion trap as a mass spectrometer[14] in 1983.

III. QUADRUPOLE DEVICES

Quadrupole devices belong to the type of mass analyzers that are *dynamic* as opposed to *static*. In static devices, electric and magnetic fields are applied so as to produce a constant force on the ions under study. Under these circumstances of constant acceleration, the ion trajectories are well defined and easily calculated. In the case of dynamic instruments, the ion trajectories are influenced by a set of time-dependent forces which render their trajectories somewhat more difficult to predict. Ions in such quadrupole fields experience strong focusing in which the restoring force, which drives the ions back toward the center of the device, increases linearly with displacement from the origin. Instruments in which dynamic and static devices are combined in any constructive order are described as *hybrid* instruments.

The motion of ions in quadrupole fields is described mathematically by the solutions to a second-order linear differential equation described by Mathieu in 1868.[15] The equations governing the motion of ions in quadrupole fields have been treated in an introductory fashion[16-18] and

TABLE 1

Ages of the Quadrupole Ion Trap

Mass-Selective Detection

1953	First disclosure (Paul and Steinwedel)
1959	Storage of microparticles (Wuerker, Shelton, and Langmuir)
	Use as a mass spectrometer (Fischer)

Mass-Selective Storage

1962	Storage of ions for RF spectroscopy (Dehmelt and Major)
1968	Ejection of ions into an external detector (Dawson and Whetten)
	Use as a mass storage spectrometer (Dawson and Whetten)
1972	QUISTOR combined with quadrupole mass filter for analysis of ejected ions (Todd, Lawson, and Bonner)
	Trap characterization, CI, ion/molecule kinetics, etc. (Todd et al.)
1976	Collisional focusing of ions (Bonner, March, and Durup)
1978	Selective ion reactor (Fulford and March)
1979	Resonant ejection of ions (Armitage, Fulford, Hoa, Hughes, March, Bonner, and Wong)
1980	Use as a GC detector (Armitage and March)
1982	Multiphoton (IR) dissociation of ions (Hughes, March, and Young)

Mass-Selective Ejection

1984	Disclosure of ion trap detector, ITD™ (Stafford, Kelley, Syka, Reynolds, and Todd)
1985	Ion trap mass spectrometer, ITMS™ (Kelley, Stafford, Syka, Reynolds, Louris, and Todd)
1987	MS/MS, CI, photodissociation, injection of ions, mass-range extension, etc.
	Fourier transform quadrupole ion trap (Syka and Fies)
	Automatic gain control (Stafford, Taylor, Bradshaw, and Syka)
	Reverse scans (Todd et al.)
1988	Axial modulation (Tucker, Hameister, Bradshaw, Hoekman, and Weber-Grabau)
	Ion isolation (Strife, Kelley, and Weber-Grabau)
	Quantitative internal excitation (Brodbelt, Kenttämaa, and Cooks)
	Supercritical fluid chromatography (Todd et al.)

Recent Advances and Developments

1989	Nobel Prize in Physics to W. Paul and H. G. Dehmelt
	Mass range extension (Kaiser, Cooks, Moss, and Hemberger; Kaiser, Louris, Amy, and Cooks)
	Laser desorption (Heller, Lys, Cotter, and Uy; Glish, Goeringer, Asano, and McLuckey)
	Hybrid instrument (Suter, Gfeller, and Schlunegger; Schwartz, Kaiser, Cooks, and Savickas)
	Atmospheric sampling glow discharge ionization (McLuckey, Glish, and Asano)
	Thermospray LC (Kaiser, Williams, Schwartz, Lammert, and Cooks)
1990	$(MS)^n$ (Johnson, Yost, Kelley, and Bradford)
	Dynamically programmed scans (Todd, Penner, Thorner, and Smith)

TABLE 1 (*Continued*)

	Observation and simulation of resonance absorptions (Vedel et al.)
	Electrospray ionization (Van Berkel, Glish, and McLuckey)
	New trap designs (Franzen)
	Pulsed valve operation (Emary, Kaiser, Kenttämaa, and Cooks)
1991	Multiple ion resonance (Johnson, Pedder, and Yost)
	Enhanced mass resolution (Syka et al.; Cooks et al.)
	Black holes (Traldi et al.)
	Black canyons (Todd et al.)
	Boundary activation (Creaser; Traldi et al.)
	Mass defects (Cooks et al.; Traldi et al.)
	Consecutive isolation (Cooks et al.; Traldi et al.)
1992	Broadband waveform ejection (Buttrill et al.)
	"Stretched" ion trap (Louris et al.)
	Enhanced mass resolution (Londry, Wells, and March)
1993	Precooling and boundary activation (Londry, Catinella, March, and Traldi)

in detail.[8,10,19] The relevant mathematics have been examined by McLachlan.[20] The following treatment is intended to provide a common base for the descriptions of recent research that follow. As an understanding of the mathematics and physical theory of the device is essential to an appreciation of the operation of the three-dimensional quadrupole ion trap, it is hoped that this treatment will serve as a guide to the novitiate and as an aide memoire to the researcher.

We commence by developing an expression for the electric potential within a quadrupole ion trap, and from which the quadrupole field can be derived. From the field, the differential equation of motion is obtained; when this equation is cast in the canonical form of the Mathieu equation, the stability parameters can be obtained. We proceed to examine stable solutions to the Mathieu equation. Sample calculations are presented. By examination of stability criteria, we proceed to a discussion of regions of stability and instability in which we develop stability diagrams. Finally, we examine the quadrupole ion trap stability diagram closest to the origin in stability parameter space.

A. Theory of the Quadrupole Ion Trap

From the theory of differential equations, it is well known that a solution of Laplace's equation in spherical polar coordinates (ρ, θ, ϕ) for a system with axial symmetry about the z-axis, has the general form

$$\Phi(\rho, \theta, \phi) = \sum_{n=0}^{\infty} A_n^0 \rho^n P_n(\cos \theta) \tag{1}$$

where A_n^0 are arbitrary coefficients, and P_n (cos θ) denotes a Legendre polynomial. Rewriting Equation 1 with $\rho^n P_n$ (cos θ) expressed in cylindrical polar coordinates (r, ϕ, z), one obtains

$$\Phi(r, \phi, z) = A_0^0 + A_1^0 z + A_2^0(\tfrac{1}{2}r^2 - z^2) + A_3^0 z(\tfrac{3}{2}r^2 - z^2)$$
$$+ A_4^0(\tfrac{3}{8}r^4 - 3r^2 z^2 + z^4) + \cdots \quad (2)$$

The terms with $n = 0, 1, 2, 3$, and 4 in Equation 2 correspond to the monopole, dipole, quadrupole, hexapole, and octopole components, respectively, of the potential field Φ. Considered in this chapter is the case of a pure quadrupole ion trap, for which only the coefficients corresponding to $n = 0$ and $n = 2$ in Equations 1 and 2 are nonzero. A discussion of ion traps in which higher order multipole components are present in the potential is given in Chapter 3.

Setting $A_0^0 = B$ and $A_2^0 = 2A$ for convenience, the potential in a pure quadrupole ion trap can be written

$$\Phi(r, z) = A(r^2 - 2z^2) + B \quad (3)$$

where A and B are constants to be determined from the boundary conditions on the trap electrodes.

The trap electrodes must be surfaces that correspond to equipotential surfaces in the trapping field. For a potential field given by Equation 3, the equipotential surface corresponding to the ring electrode will have the form

$$r^2 - 2z^2 = \rho_0^2 \lambda \quad (4)$$

where λ is a positive dimensionless number and ρ_0 is an arbitrary displacement. Note that λ and ρ_0 are not independent parameters. The end-cap electrodes will have the form

$$r^2 - 2z^2 = -\rho_0^2 \mu \quad (5)$$

where μ is an arbitrary positive dimensionless number. As above, μ and ρ_0 are not independent.

For the electrode surfaces given by Equations 4 and 5, the equatorial radius, r_0, is given by

$$r_0^2 = \lambda \rho_0^2 \quad (6)$$

and the polar distance, the displacement of the end-cap electrodes along the line $r = 0$ from the origin at the center of the trap, z_0, by

$$z_0^2 = \tfrac{1}{2}\mu\rho_0^2 \tag{7}$$

In order to fix the values of μ and λ, the arbitrary parameter ρ_0 is assigned the value of the equatorial radius:

$$\rho_0 \equiv r_0 \tag{8}$$

Adopting these conventions, fixed values for μ and λ can be calculated from Equations 6 and 7 as

$$\lambda = 1 \tag{9}$$

and

$$\mu = \frac{2z_0^2}{r_0^2} \tag{10}$$

respectively.

Now, expressions for the coefficients A and B can be obtained from the boundary conditions on the electrode surfaces. When the potential is $\Phi = \Phi_0^R$ on the ring electrode, Equations 3, 4, and 9 can be combined to obtain

$$\Phi(r, z) = \Phi(r_0, 0) = Ar_0^2 + B = \Phi_0^R \tag{11}$$

Similarly, when the potential is $\Phi = \Phi_0^E$ on the end-cap electrodes, Equations 3, 5, and 10 can be combined to obtain

$$\Phi(r, z) = \Phi(0, z_0) = A(-2z_0^2) + B = \Phi_0^E \tag{12}$$

Solving Equations 11 and 12 for A and B, one obtains

$$A = \frac{\Phi_0^R - \Phi_0^E}{r_0^2 + 2z_0^2} \tag{13}$$

and

$$B = \frac{2z_0^2\Phi_0^R + r_0^2\Phi_0^E}{r_0^2 + 2z_0^2} \tag{14}$$

respectively. Finally, substituting Equations 13 and 14 in Equation 3 yields the potential

$$\Phi(r, z) = \frac{\Phi_0^R - \Phi_0^E}{r_0^2 + 2z_0^2}\left[r^2 - 2z^2\right] + \frac{2z_0^2\Phi_0^R + r_0^2\Phi_0^E}{r_0^2 + 2z_0^2} \tag{15}$$

Equation 15 is identical in form to the general expression for the potential inside a quadrupolar ion trap given by Knight.[21] Furthermore, when the ratio of r_0^2 to z_0^2 is set equal to 2 in Equation 15 the more commonly used expression

$$\Phi(r, z) = \frac{\Phi_0^R - \Phi_0^E}{2r_0^2}\left[r^2 - 2z^2\right] + \tfrac{1}{2}\left(\Phi_0^R + \Phi_0^E\right) \tag{16}$$

is obtained.

The differential equation of motion for a singly charged positive ion, subject to the potential of Equation 15, is readily obtained from

$$\frac{d^2\vec{r}}{dt^2} = -\frac{e}{m}\nabla\Phi \tag{17}$$

where m is the mass of the ion and e is the electronic charge. In this discussion it is assumed that both end-cap electrodes are grounded and that an electric potential of

$$\Phi_0^R = U + V\cos(\Omega t + \gamma) \tag{18}$$

is applied to the ring electrode (mode II[22]), where U and V are, respectively, the amplitudes of the direct and alternating components of Φ_0^R, and Ω is the angular frequency of the alternating component with an initial phase angle of γ. Then, Equation 17 can be written as

$$\frac{d^2u}{dt^2} = \frac{\alpha e}{m\left(r_0^2 + 2z_0^2\right)}\left[U + V\cos(\Omega t + \gamma)\right]u \tag{19}$$

where $\alpha = -2$ for $u = r$ and $\alpha = 4$ for $u = z$. With γ set to zero for simplicity, Equation 19 can be cast in the canonical form of the Mathieu equation[15] developed some 150 years ago while investigating the motion of vibrating membranes:

$$\frac{d^2u_0}{d\xi^2} + \left[a_u - 2q_u\cos(2\xi)\right]u_0 = 0 \tag{20}$$

where $\xi = \Omega t/2$, and the stability parameters a_u and q_u are given by

$$a_r = \frac{8eU}{m\left(r_0^2 + 2z_0^2\right)\Omega^2} \tag{21}$$

$$a_z = \frac{-16eU}{m\left(r_0^2 + 2z_0^2\right)\Omega^2} \tag{22}$$

$$q_r = \frac{-4eV}{m\left(r_0^2 + 2z_0^2\right)\Omega^2} \tag{23}$$

and

$$q_z = \frac{8eV}{m\left(r_0^2 + 2z_0^2\right)\Omega^2} \tag{24}$$

Solutions of Equation 20 and applications thereof have been studied in detail by McLachlan.[20] Stable solutions are known to have the general form

$$u_0(\xi) = A_u u_0^C(\xi) + B_u u_0^S(\xi) \tag{25}$$

where A_u and B_u are arbitrary constants and

$$u_0^C(\xi) = \sum_{n=-\infty}^{+\infty} C_{2n,u} \cos(2n + \beta_u)\xi \tag{26}$$

and

$$u_0^S(\xi) = \sum_{n=-\infty}^{+\infty} C_{2n,u} \sin(2n + \beta_u)\xi \tag{27}$$

with β_u determining the spectrum of frequencies corresponding to the parameter pair (a_u, q_u). The $C_{2n,u}$ coefficients give the amplitudes of the allowed modes in the spectral analysis of u_0.

Defining $\omega_{u,n}$ as the angular frequency of order n for motion in the direction $\hat{u}$ ($\hat{u} = \hat{r}, \hat{z}$) and recalling that $\xi = \frac{1}{2}\Omega t$ it can be seen that

$$\omega_{u,n} = \left(n + \tfrac{1}{2}\beta_u\right)\Omega, \quad 0 \le n < \infty \tag{28}$$

and

$$\omega_{u,n} = -\left(n + \tfrac{1}{2}\beta_u\right)\Omega, \quad -\infty < n < 0 \tag{29}$$

When $n = 0$, the fundamental frequency $\omega_{u,0}$ in either the $\hat{r}$ or $\hat{z}$ direction, is given by

$$\omega_{u,0} = \tfrac{1}{2}\beta_u\Omega \tag{30}$$

Therefore, ions with the same β_u values will have identical frequencies. The form of an ion trajectory in the r-z plane has the general appearance of a Lissajous curve composed of two fundamental frequency components $\omega_{r,0}$ and $\omega_{z,0}$ of the secular motion, with a superimposed micromotion of frequency $\Omega/2\pi$ Hz. Because the magnitude of the $C_{2n,u}$ coefficients fall off very rapidly as n increases, only the higher-order frequencies corresponding to $n = \pm1, \pm2$ are of practical significance. Specifically, these are

$$\omega_{u,1} = \left(1 + \tfrac{1}{2}\beta_u\right)\Omega \qquad \omega_{u,-1} = \left(1 - \tfrac{1}{2}\beta_u\right)\Omega$$
$$\omega_{u,2} = \left(2 + \tfrac{1}{2}\beta_u\right)\Omega \qquad \omega_{u,-2} = \left(2 - \tfrac{1}{2}\beta_u\right)\Omega \tag{31}$$

Substituting Equations 26 and 27 in Equation 25 and expressing the displacement as a function of time one obtains

$$u_0(t) = A_u \sum_{n=-\infty}^{+\infty} C_{2n,u} \cos\left((2n + \beta_u)\frac{\Omega}{2}t\right)$$

$$+ B_u \sum_{n=-\infty}^{+\infty} C_{2n,u} \sin\left((2n + \beta_u)\frac{\Omega}{2}t\right) \tag{32}$$

The velocity of an ion can be obtained by differentiating Equation 32 with respect to time as

$$\dot{u}_0(t) = -A_u \frac{\Omega}{2} \sum_{n=-\infty}^{+\infty} C_{2n,u} (2n + \beta_u) \sin\left[(2n + \beta_u)\frac{\Omega}{2}t\right]$$

$$+ B_u \frac{\Omega}{2} \sum_{n=-\infty}^{+\infty} C_{2n,u} (2n + \beta_u) \cos\left[(2n + \beta_u)\frac{\Omega}{2}t\right] \tag{33}$$

Recursion formulas for the $C_{2n,u}$ coefficients can be obtained by evaluating the second time derivative of the displacement, $\ddot{u}(t)$, and equating coefficients with Equation 20 to obtain

$$C_{2n,u} = \frac{-C_{2n-2,u}}{q_u} \cfrac{q_u^2}{(2n+\beta_u)^2 - a_u - \cfrac{q_u^2}{(2n+2+\beta_u)^2 - a_u - \cfrac{q_u^2}{(2n+4+\beta_u)^2 - a_u - \cdots}}}$$

$$\tag{34}$$

and

$$C_{2n,u} = \frac{-C_{2n+2,u}}{q_u} \cfrac{q_u^2}{(2n+\beta_u)^2 - a_u - \cfrac{q_u^2}{(2n-2+\beta_u)^2 - a_u - \cfrac{q_u^2}{(2n-4+\beta_u)^2 - a_u - \cdots}}}$$

$$\tag{35}$$

Replacing n in Equation 35 by $n - 1$ yields another relationship between $C_{2n,u}$ and $C_{2n-2,u}$. Elimination of these coefficients between the new expression and Equation 34 leads to a continued fraction expression for β_u in terms of a_u and q_u as

$$\beta_u^2 = a_u + \cfrac{q_u^2}{(\beta_u+2)^2 - a_u - \cfrac{q_u^2}{(\beta_u+4)^2 - a_u - \cfrac{q_u^2}{(\beta_u+6)^2 - a_u - \cdots}}}$$

$$+ \cfrac{q_u^2}{(\beta_u-2)^2 - a_u - \cfrac{q_u^2}{(\beta_u-4)^2 - a_u - \cfrac{q_u^2}{(\beta_u-6)^2 - a_u - \cdots}}} \qquad (36)$$

In our work we have used this relationship, which is given by McLachlan,[14] in order to calculate β_u for a given a_u and q_u or to determine the required a_u and q_u to achieve a specific β_u. For example, the equation is rewritten in terms of the desired quantity; a trial value of the variable is substituted into the equation to obtain a first approximation. This new value is substituted into the equation to obtain a better approximation; i.e., $\beta_{u,n+1} = f(\beta_{u,n}, a_u, q_u)$. This iterative process is repeated until the difference between $\beta_{u,n+1}$ and $\beta_{u,n}$ becomes negligible (usually <10 times).

Knowing β_u and assuming $C_{0,u} = 1$, the $C_{2n,u}$ coefficients can be calculated from Equations 34 and 35, and the resonant frequencies $\omega_{u,n}$ can be calculated from Equations 28 and 29. Finally, if the initial conditions given by u and $\dot{u}$ are known, Equations 32 and 33 can be solved simultaneously to obtain values for the coefficients A_u and B_u.

B.   A Note on the Applied Potential

We perceive that some confusion arises from the literature as to what exactly constitutes an applied potential, particularly the value of the applied RF voltage V_{RF}. The value of the applied RF potential will depend on whether the measurement is made from the zero potential (giving $V_{zero-to-peak}$, $V_{(0-p)}$) or from the maximum of the wave to the minimum of the wave (giving $V_{peak-to-peak}$, $V_{(p-p)}$). Normally we employ $V_{(0-p)}$ and, when giving experimental details, will state explicitly in which fashion V_{RF} was measured. It is imperative that V_{RF} be properly defined; when comparing calculated values with literature or experimental values, differences of a factor(s) of 2 can arise readily. A further complication is apparent when considering the mode in which the device is operated. The ion trap may be operated in one of several different modes which are described elsewhere.[10] In the most commonly used mode, wherein the end-cap electrodes are grounded and the potential Φ_0^R is applied to the ring electrode, the applied RF is measured with respect to ground (and, in this case, with respect to the end-cap electrodes) as $V_{(0-p)}$.

In summary, the magnitudes of the terms involving the potential will depend on the following three factors: how V_{RF} is measured, how Φ_0^R is defined, and the operational mode of the device.

C. Regions of Stability

The primary concerns in the utilization of quadrupole devices are the criteria that govern the stability (and instability) of an ion trajectory within the field; i.e., the experimental conditions that determine whether an ion is stored within the device or is ejected from the device, and either lost to the environment or detected externally. The boundaries between stable and unstable regions of the stability diagram correspond to those values of a_u and q_u for which β_u is an integer $(0,1,2,\ldots)$. For such values the solutions to the Mathieu equation are periodic but unbounded, and they represent, in practical terms, the point at which the trajectory of an ion becomes unbounded. Of particular interest here is the portion of the diagram for which $0 \le \beta_u \le 1$. These limits have been shown to correspond to combinations of cosine and sine elliptic series.[19]

Thus, the stable solutions can be plotted using a_u and q_u as coordinate axes. Such graphical representations of stable solutions to the Mathieu equation are called stability diagrams, and one such diagram is depicted in Figure 2. Figure 2 is shown in a deliberately simplified fashion; only the positive

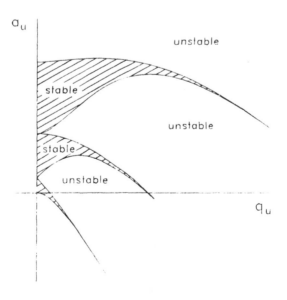

FIGURE 2
Graphical representation of stable solutions of the Mathieu equation plotted in (a_u, q_u) space.[20]

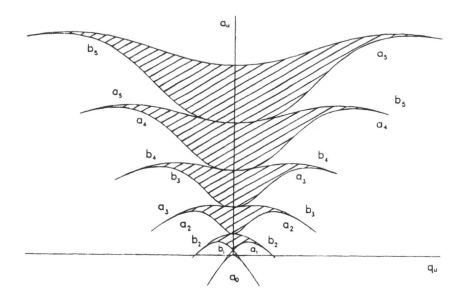

FIGURE 3

Mathieu stability diagram in one dimension of (a_u, q_u) space. The characteristic curves a_0, $b_1, a_1, b_2, \ldots$ divide the plane into regions of stability and instability. The even-order curves are symmetric about the a_u-axis, but the odd-order curves are not. The diagram itself, however, appears as being symmetric about the a_u-axis.

values are shown along the q_u axis because symmetry exists about the a_u axis. Regions in which the values of a_u and q_u represent stable solutions to the Mathieu equation are shaded; those unshaded are unstable. Such diagrams may be constructed for ion motion in both the $\hat{r}$ and $\hat{z}$ directions.

Presented in Figure 3 is the Mathieu stability diagram in one dimension of (a_u, q_u) space showing the regions delineated by characteristic numbers of a cosine-type function (a_m) of order m and a sine-type function (b_m) of order m. This diagram is labeled in the terminology used by McLachlan.[14] The boundaries of even order are symmetric about the a_u axis, but the boundaries of odd order are not; however, the diagrams themselves appear as being symmetric about the a_u axis.

IV. QUADRUPOLE ION TRAP STABILITY DIAGRAMS

From Equations 21 to 24, it is seen that $a_z = -2a_r$ and $q_z = -2q_r$, and thus the stability parameters for the $\hat{r}$ and $\hat{z}$ directions differ by a factor of -2. The stability regions defining the (a_u, q_u) values corresponding to solutions of the Mathieu equation that are stable in the $\hat{z}$ direction are labeled z-stable in Figure 4. The regions corresponding to solutions that

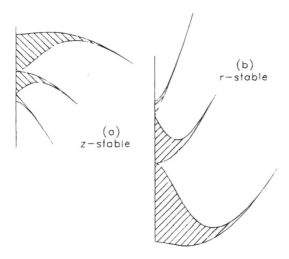

FIGURE 4
Several Mathieu stability regions for the three-dimensional quadrupole field: (a) diagrams
for the $\hat{z}$ direction of (a_z, q_z) space; (b) diagrams for the $\hat{r}$ direction of (a_r, q_r) space where the
q_r dimension has been multiplied by -1.

are stable in the $\hat{r}$ direction, and are labeled r-stable in Figure 4, are twice
the size of the $\hat{z}$ direction regions rotated about the q_z-axis.

Ions can be stored in the ion trap provided they are stable in the $\hat{r}$ and
$\hat{z}$ directions simultaneously. The region closest to the origin, i.e., region A
in Figure 5 is of the greatest importance at this time and this region is
shown in greater detail in Figure 6. From the known solutions to the Mathieu
equation one can generate a stability diagram in parametric form. The coordinates of this stability diagram are the Mathieu equation parameters a_z
and q_z as shown in Figure 6. This stability diagram shows the common region in (a_z, q_z) space for which the radial and axial components of the ion
trajectory are stable simultaneously. For practical purposes, one may evaluate the a_u, q_u coordinates for constant β_u (the so-called iso-β lines). These
lines are depicted in Figure 6 between the limits $\beta_u = 0$ and $\beta_u = 1$.

A. Commercial "Stretched" Ion Traps

The values of a_z and q_z as defined by Equations 22 and 24, respectively, pertain to a single ion isolated in an ideal quadrupole ion trap for
which the electrodes extend to infinity. In practice, the presence of other
ions is ignored, provided the ion number does not exceed ca. 10^5; at this
point, space-charge effects, discussed in Chapter 8, become evident.
Truncation of the electrodes, which is necessary in order to produce a
practical working instrument, introduces higher-order multipole components to the potential, as is discussed in detail in Chapter 3.

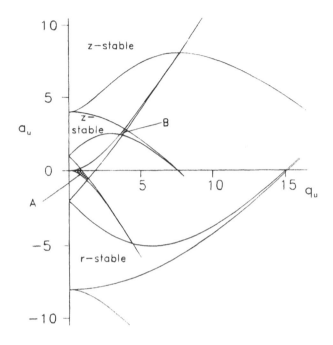

FIGURE 5

The Mathieu stability diagram in (a_u, q_u) space for the quadrupole ion trap in both the $\hat{r}$ and $\hat{z}$ directions. Regions of simultaneous overlap are labeled A and B. While the axes are labeled a_u and q_u, the diagrammatic representation shown here of simultaneous stability in both the $\hat{r}$ and $\hat{z}$ directions shows the ordinate and the abscissa scales in units of a_z and q_z, respectively.

Until recently, the effects of higher-order multipoles have been ignored. However, following the recent announcement[23] that the ion trap used in the Finnigan MAT Ion Trap Mass Spectrometer™ (ITMS™) and Ion Trap Detector™ (ITD™), and in the Varian Associates Saturn models of gas chromatograph/mass spectrometer differs significantly from the pure quadrupole geometry, a great deal of interest has arisen in how the stability diagram is affected by the presence of higher-order multipoles. A full account of the origin of the altered geometry of the commercial trap is presented in Chapter 4.

The electrode surfaces for the commercial devices cited above were formed according to Equations 4 and 5, but the electrodes are assembled in such a way that the distance between the end-cap electrodes has been increased or "stretched"; the value of z_0 has been increased by 10.6%, with no corresponding modification to the shapes of the electrodes, which would be required in order to maintain a purely quadrupolar geometry. Despite the change in commercial ion trap geometry, the stability diagram should

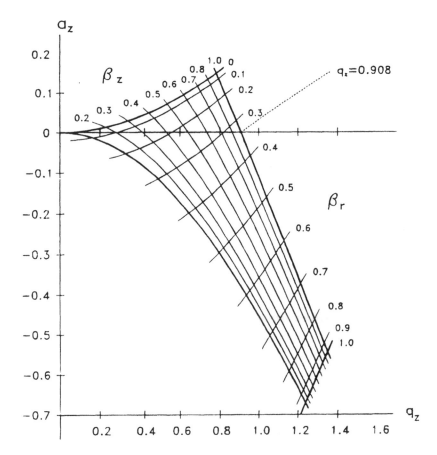

FIGURE 6

Stability diagram in (a_u, q_u) space for the region of simultaneous stability in both $\hat{r}$ and $\hat{z}$ directions near the origin for the three-dimensional ion trap; the iso-β_r and iso-β_z lines are shown in the diagram. The q_z-axis intersects the $\beta_z = 1$ boundary at $q_z = 0.908$, which corresponds to q_{max} in the mass-selective instability mode.

be little changed (see Chapter 7), if β_u is calculated in the usual manner (i.e., using Equation 36) as a function of a_u and q_u provided these parameters are calculated using Equations 21 and 24, respectively, and remembering that $r_0^2 \neq 2z_0^2$ as had been assumed prior to the announcement.[23]

It is noteworthy that q_z is not calculated directly in the operation of commercial ion traps; hence, reported values of this parameter should be treated with caution. Commercial instruments employ the lowest m/z ratio to be stored in setting the value of the RF drive voltage amplitude; thus, in order to calculate the q_z value for a given ion species, the working point of mass ejection should be calculated from Equation 24.

B. Sample Calculations of Potentials for Apex Isolation and Mass Limit

When working with a quadrupole ion trap it occasionally becomes necessary to calculate some of the relevant parameters, and so we present calculations using values of a_z and q_z for two commonly used working points on the stability diagram. The a_z and q_z values, or coordinates, for a given ion species define the working point for that species as a location on a stability diagram. It should be noted that it is incorrect to refer to an ion or ion species as "approaching a boundary of a stability diagram", when it is the working point which is made to approach a boundary; under such conditions, the ion or ion species experiences a change in the field due to the change in working point. The conditions have been chosen to reflect common instrumentation and operating conditions for commercial quadrupole ion traps. The composite potential Φ_0^R is applied to the ring electrode, and the end-cap electrodes are grounded; under these conditions (mode II) the relevant parameters a_z and q_z are given by Equations 22 and 24, respectively.

1. Calculation 1—Apex Isolation

Let us consider a singly charged ion of 217 u in an ITMS™ that has a ring electrode of equatorial radius $r_0 = 10^{-2}$ m; the end-cap electrode distance from the origin at the center of the trap is $z_0 = \dfrac{1.106}{\sqrt{2}} r_0$ i.e., $z_0 = 0.782\ r_0$. The task is to calculate the values of U_{DC} and V_{RF} which will place the working point of the subject ion species in the upper apex of the stability diagram at $a_z = 0.148$, $q_z = 0.771$, for ion isolation. At 1.1 MHz, $\Omega = 2\pi f = 2\pi \times 1.1 \times 10^6$ rad s^{-1}; m = 217 u = 217×10^{-3} kg mol^{-1}/6.022 $\times$ 10^{23} mol^{-1}. As $r_0^2 = 1.635 z_0^2$, $(r_0^2 + 2z_0^2) = 2.223\ r_0^2$. Thus, from Equation 22 we obtain

$$U_{DC} = \frac{(0.148)(217 \times 10^{-3}\ \text{kg mol}^{-1})(2.223 \times 10^{-4}\ \text{m}^2)(2\pi \times 1.1 \times 10^6\ \text{rad s}^{-1})^2}{(-16)(1.602 \times 10^{-19}\ \text{C})(6.022 \times 10^{23}\ \text{mol}^{-1})}$$

$$= -220.9\ \text{kg m}^2\ \text{s}^{-2}\ \text{C}^{-1}$$

$$= -221\,V$$

Similarly, from Equation 24, we obtain $V_{RF} = 2300\ V_{(0-p)}$.

2. Calculation 2—Mass Limit

The intersection of the q_z-axis with the $\beta_z = 1$ boundary of the stability diagram is found at $a_z = 0$, $q_z = 0.908$. The stated mass range of the ITD™ is 650 u. The task is to calculate the RF drive voltage amplitude, V_{RF}, at which the ion of 650 u is ejected at the $\beta_z = 1$ stability boundary.

At 1.0 MHz, $\Omega = 2\pi f = 2\pi \times 1.0 \times 10^6$ rad s^{-1} (for the ITD$^{\text{™}}$), $m = 650$ u,

$$V_{RF} = \frac{(0.908)(650\times10^{-3} \text{ kg mol}^{-1})(2.223\times10^{-4} \text{ m}^2)(2\pi\times1.0\times10^6 \text{ rad s}^{-1})^2}{(8)(1.602\times10^{-19} \text{ C})(6.022\times10^{23} \text{ mol}^{-1})}$$

$$= 6710\,V_{(0-p)}$$

Thus, the RF voltage required to achieve the mass limit of 650 u in the commercial ITD$^{\text{™}}$ is ca. 6710 $V_{(0-p)}$, when the instrument is operated in the mass-selective axial instability mode.

C. Pseudopotential Well

The pseudopotential well model, derived originally by Major and Dehmelt[24] and developed by Todd and co-workers [Chapter 4 in Reference 8], approaches the motion of the ion with the view that said motion in the direction $\hat{u}$ can be approximated by the sum of two components: a micromotion, δ_u, at the frequency of the RF drive superimposed on the much larger amplitude fundamental secular motion, U, such that

$$u(\xi) = U(\xi) + \delta_u(\xi) \tag{37}$$

If q_u, on which the driving force depends, is sufficiently small, then $\delta_u \ll U$ and $d^2\delta_u/dt^2 \gg d^2U/dt^2$, and the Mathieu equation can be written

$$\frac{d^2\delta_u}{d\xi^2} + [a_u - 2q_u \cos(2\xi)]U = 0 \tag{38}$$

Assuming $a_u \ll q_u$ and U to be constant over one RF cycle, Equation 38 can be integrated to obtain

$$\delta_u = -\tfrac{1}{2}q_u U \cos 2\xi \tag{39}$$

Using this approximation to δ_u in Equation 37 yields

$$u(\xi) = U(\xi) - \tfrac{1}{2}q_u U(\xi) \cos 2\xi \tag{40}$$

This expression for $u(\xi)$ can be used in Equation 38 to calculate the average acceleration over one period of the RF drive. Expressing the result as a function of time, one obtains

$$\left\langle\frac{d^2u}{dt^2}\right\rangle_{1 \text{ rf cycle}} = \left\langle\frac{d^2U}{dt^2}\right\rangle_{1 \text{ rf cycle}} = -\frac{\Omega^2}{4}\left(a_u + \tfrac{1}{2}q_u^2\right)U \tag{41}$$

This is an equation for simple harmonic motion at the fundamental secular frequency which also can be expressed as

$$\frac{d^2U}{dt^2} = -\omega_{u,0}^2 U \qquad (42)$$

Using the expression for $\omega_{u,0}$ given by Equation 30, it can be seen from Equation 41 that β_u has been approximated by

$$\beta_u = \left(a_u + \tfrac{1}{2}q_u^2\right)^{\frac{1}{2}} \qquad (43)$$

Equation 43 is known as the Dehmelt or adiabatic approximation. Wuerker et al.[25] have validated this result by direct observation as being accurate to within 1% for $q_z \leq 0.4$.

Following Equation 17, the average force on an ion of mass m and charge e is

$$m\left(\frac{d^2U}{dt^2}\right)_{1 \text{ rf cycle}} = -e\frac{\partial \overline{D}_u}{\partial U} \qquad (44)$$

where $\overline{D}_u$ is the depth of the pseudopotential well in which ions oscillate in the $\hat{u}$ direction.

Substituting in Equation 44 the expression for acceleration given in Equation 42 gives

$$\frac{\partial \overline{D}_u}{\partial U} = \frac{m\,\omega_{u,0}^2}{e}U \qquad (45)$$

Upon integrating this equation one obtains an expression for the depth of the parabolic potential well in which the ions oscillate at the fundamental secular frequency $\omega_{u,0}$ as

$$\overline{D}_u = \frac{m\,\omega_{u,0}^2}{2e}U_0^2 \qquad (46)$$

where U_0 corresponds to either r_0 or z_0.

The solution to Equation 41 is that for a simple harmonic oscillator whose displacement is given by

$$U(t) = U_0 \sin \omega_{u,0}t \qquad (47)$$

with velocity

$$\dot{U}(t) = U_0 \omega_{u,0} \cos \omega_{u,0} t \tag{48}$$

An expression for a mean kinetic energy can be obtained by calculating the expectation value of the square of the velocity given by Equation 48 over one period of the secular motion as

$$\langle E \rangle_{1 \text{ secular cycle}} = \tfrac{1}{2} m \langle \dot{U}^2(t) \rangle_{1 \text{ secular cycle}} = \frac{m \omega_{u,0}^2 U_0^2}{4\pi} \int_{t=0}^{\frac{2\pi}{\omega_{u,0}}} \cos^2 \omega_{u,0} t \, dt \tag{49}$$

Performing this integration yields the result

$$\langle E \rangle_{1 \text{ secular cycle}} = \tfrac{1}{4} m \, \omega_{u,0}^2 U_0^2 = \frac{e}{2} \overline{D}_u \tag{50}$$

Recalling that $\omega_{u,0} = 1/2 \, \beta_u \Omega$ and using the approximate expression for β_u given by Equation 43 with $a_u = 0$ yields

$$\langle E \rangle_{1 \text{ secular cycle}} = \frac{m q_u^2 \Omega_{u,0}^2 U_0^2}{32} \tag{51}$$

This result is larger by a factor of $\pi^2/8$ than that given in Reference 8, p. 221, where the expectation value for the energy was calculated according to $\langle E \rangle = \tfrac{1}{2} m \langle \dot{U} \rangle^2$ rather than $\langle E \rangle = \tfrac{1}{2} m \langle \dot{U}^2 \rangle$.

Using Equations 50 and 51, the average depth of the potential well $\overline{D}_u$ can be expressed as a function of q_u as

$$\overline{D}_u = \frac{m q_u^2 \Omega^2 u_0^2}{16e} \tag{52}$$

Because the magnitude of q_r is only half that of q_z, the depth of the potential well for motion in the direction $\hat{r}$, $\overline{D}_r$, would be only one fourth that of $\overline{D}_z$ provided that $r_0 = z_0$. However, in most practical instruments $r_0^2 \approx 2z_0^2$ and consequently, $\overline{D}_z \approx 2\overline{D}_y$ as shown schematically in Figure 7.

D. Negative Ions

The stability diagram for negative ions is the mirror image of that for positive ions, and is obtained by reflecting the stability diagram for positive ions about the q_z-axis, as is shown in Figure 8. A region of stability is common to both negative and positive ions due to the overlap of the two stability diagrams. The shape of this region of simultaneous stability is determined by the $\beta_r = 0$, $\beta_z = 0$, and $\beta_z = 1$ boundaries of the respective diagrams, and not by the $\beta_r = 1$ boundaries of either; as the asymmetry of the ion trap stability diagram is due mainly to the relative position

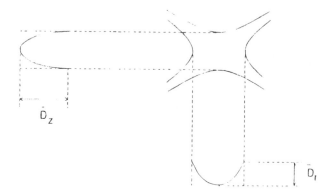

FIGURE 7

Representation of the parabolic pseudopotential wells $\overline{D}_r$ and $\overline{D}_z$. The respective secular frequencies of oscillation of the ion within the wells are $\omega_r, 0$ and $\omega_z, 0$.

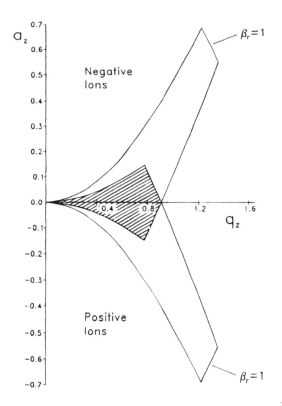

FIGURE 8

Superimposed stability diagrams in (a_u, q_u) space for the quadrupole ion trap for positive and negative ions showing a common area of overlap (simultaneous stability).

of the $\beta_r = 1$ boundary, the stability region common to both anions and cations resembles the symmetrical stability diagram for the quadrupole mass filter. Negative and positive ions are stored simultaneously when the ion trap is operated in the "RF only" mode, and ions of each polarity are ejected in the mass-selective axial instability scan at $q_z = 0.908$ or by axial modulation (see Chapter 1) at a lower value of q_z. Thus, mass spectra should be interpreted with care; one should bear in mind the possible inclusion of ion signals in the mass spectrum due to adventitious negative ions.

E.   Stability Diagrams in U/V Space

An alternative representation of the stability diagram can be given in U/V space as shown in Figure 9. In this form of representation, there is a continuum of stability diagrams in that a stability diagram exists for each m/z ratio. Two such stability diagrams for two ion species are shown in Figure 9, with that for the ion of greater m/z ratio lying to the right in the figure. The U/V space representation of the stability diagram is helpful in appreciating an early method of *in situ* ion detection used by Rettinghaus.[26] In this method, the detection frequency was held constant and the RF voltage amplitude was scanned so as to bring each mass into resonance at the point where the $\beta_z = 0.5$ line intersects the V-axis of the stability diagram. V was scanned slowly from a value corresponding to a little less than that for the ion of lowest m/z ratio to a higher value such that ions of higher m/z ratio were brought successively into resonance.

The U/V space representation of the stability diagram is helpful also for illustrating resonance ejection in rapid succession of two ion species

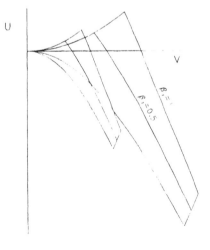

FIGURE 9

Stability diagrams in U, V space for two ion species of different m/z ratio, with that for the ion of greater m/z ratio located at the right-hand side of the figure. The figure illustrates the scanning of the mass spectrum by increasing the RF potential. As V is increased each species becomes resonant with an external field at a unique value of the applied potential, shown here as the intersection of the $\beta_z = 0.5$ line in the stability diagram with the V-axis. Alternately, each species becomes unstable and leaves the ion trap at a unique value of the applied potential, shown here as the intersection of the $\beta_z = 1$ boundary with the V-axis.

which differ appreciably in m/z ratio. For example, the ion of greater m/z ratio can be ejected resonantly at a value of V corresponding to the intersection of the $\beta_z = 0.5$ line for the ion of lower m/z ratio with the V-axis, while the ion of lower m/z ratio is ejected at a slightly greater value of V but at a different frequency. Two-color resonant ejection of ions in this manner, with ejection of the ion of greater m/z ratio preceding that of the ion of lower m/z ratio, may be applied to the problem of accurate mass assignment.

REFERENCES

1. Paul, W.; Steinwedel, H., German Patent 944,900, 1956; U.S. Patent 2,939,952, 7 June 1960.
2. Post, R.F.; Heinrich, L., Univ. Calif. Rad. Lab. Rep. (S. Shewchuk) UCRL-2209, Berkeley, CA, 1953.
3. Good, M.L., Univ. Calif. Rad. Lab. Rep. UCRL-4146, Berkeley, CA, 1953.
4. Courant, E.L.; Livingston, M.S.; Snyder, H.S., *Phys. Rev.* 1952, *88*, 1190.
5. Dawson, P.H.; Whetten, N.R., *Res/Dev.* 1969, *19*, 46.
6. Paul, W.; Steinwedel, H., *Z. Naturforsch.* 1953, *8*, 448.
7. Berkling aus Leipzig, K. Physikalisches Institut der Universitat, Bonn, West Germany, 1956.
8. Dawson, P.H., *Quadrupole Mass Spectrometry and Its Applications.* Elsevier: Amsterdam, 1976.
9. Lawson, G.; Todd, J.F.J., Mass Spectrometry Group Meeting, Bristol, 1971, Abstr. 44.
10. March, R.E.; Hughes. R.J., *Quadrupole Storage Mass Spectometry.* Chemical Analysis Series, Vol. 102, John Wiley & Sons: New York, 1989.
11. Royal Swedish Academy of Sciences, Press release, 12 October 1989, p. 5.
12. Paul, W., *Rev. Mod. Phys.* 1990, *62*, 531.
13. Dehmelt, H.G., *Rev. Mod. Phys.* 1990 *62*, 525.
14. Kelley, P.E.; Stafford, G.C., Jr.; Stephens, D.R., U.S. Patent 4,540,884, 10 September 1985; Can. Patent 1,207,918, 15 July 1986.
15. Mathieu, E., *J. Math. Pure Appl.* 1868, *13*, 137.
16. Dawson, P.H.; Whetten, N.R., *J. Vac. Sci. Technol.* 1968, *5*, 1.
17. Dawson, P.H.; Whetten, N.R., *J. Vac. Sci. Technol.* 1968, *5*, 11.
18. Campana, J.E., *Int. J. Mass Spectrom. Ion Phys.* 1980, *33*, 101.
19. Lawson, G.; Todd, J.F.J.; Bonner, R.F., *Dyn. Mass Spectrom.* 1975, *4*, 39.
20. McLachlan, N.W., *Theory and Applications of Mathieu Functions.* Clarendon Press: Oxford, 1947. See also: Campbell, R., *Théorie Générale de l'Equation de Mathieu.* Masson: Paris, 1955.
21. Knight, R.D., *Int. J. Mass Spectrom. Ion Processes*, 1991, *106*, 63.
22. Bonner, R. F., *Int. J. Mass Spectrom. Ion Phys.* 1977, *23*, 249.
23. Louris, J., Presentation WOB 10:45 at the 40th ASMS Conf. Mass Spectrometry and Allied Topics, Washington, D.C., June 1–5, 1992.
24. Major, F.G.; Dehmelt, H.G., *Phys. Rev.* 1968, *179*, 91.
25. Wuerker, R.F.; Shelton, H.; Langmuir, R.V., *J. Appl. Phys.* 1959, *30*, 342.
26. Rettinghaus, V. von G., *Z. Angew. Phys.* 1967, *22*, 321.

Chapter 3

NONLINEAR ION TRAPS

J. Franzen, R.-H. Gabling, M. Schubert, and Y. Wang

CONTENTS

0-8493-4452-2/95/$0.00+$.50
© 1995 by CRC Press, Inc.
49

I. INTRODUCTION

In a mathematically pure quadrupole (or "linear") ion trap, the radiofrequency (RF) field increases linearly in both r- and z-directions. Nonlinear ion traps are characterized by a field for which the increase deviates slightly from linearity. Such deviations sometimes have been called field faults. From another point of view, these field faults may be called superposition of higher multipole fields, because all rotationally symmetric fields without space charge can be described by a weighted sum of such multipoles.

In the past, such field faults caused by deviations from the ideal quadrupole field were regarded as bad throughout the literature, in that they were held to be responsible for ion losses by so-called nonlinear resonances. The nonlinear resonance effects induced by these field faults, first detected and explained by Fritz von Busch and Wolfgang Paul[1] in two-dimensional quadrupole fields, were studied mainly with the announced intention of avoiding such ion losses. Dawson and Whetten[2,3] have investigated ion losses and the resulting peak splitting in great detail. The influences of any nonlinear field contributions were such as to deteriorate ion storage and to cause undesirable ion losses.

A.   Why Does Interest in Nonlinear Ion Traps Exist?

In practice, it is not really possible to create an ion trap with a pure quadrupole field in the interior of the trap. Even with precisely shaped electrodes, the truncation of each electrode to a finite size immediately induces some superimposed multipole fields. As seen below, the influence on modern mass scanning methods of these weak multipole fields generated by truncation is surprisingly strong and by no means can be neglected. Assembly tolerances and end-cap electrode perforations add more multipole fields.

During recent years, however, the beneficial influence of purposefully generated superimposed weak multipoles on mass resolution, scan speed, and ion storage stability of ion trap mass spectrometers has become more widely known.[4-6]

The knowledge of the influence of superimposed multipole fields has become of even greater interest since it was made public[7] that all commercial Finnigan MAT ion traps (ITD™, ITS40™, ITMS™) are by no means pure quadrupole-type ion traps. The separation between the end-cap electrodes in all Finnigan MAT ion traps is intentionally "stretched", with the effect that the pure quadrupole field originally intended for the unstretched device is now superimposed with weak, even multipole fields. It was shown that the mass-selective instability scan[8] works only sufficiently well when the quadrupole trap is somewhat distorted, and unit mass resolution has only been achieved by this geometrical stretching. This route is not, however, the sole way to improve mass resolution for this type of scan; other kinds of distortions yield even better results.[5]

In the light of these contradictions in the literature and in practice, it is the aim of this chapter to describe clearly the different types of effects that can be caused by superimposed weak multipole fields and to illustrate the predominantly positive, though in some cases negative, influences of these so-called nonlinear fields on the behavior of ions inside the traps.

B. What is a NonLinear Ion Trap?

In the pure (or mathematically ideal) three-dimensional RF quadrupole field, the amplitudes of the RF drive field increase strictly linearly in both z- and r-directions. From these RF fields in the z- and r-directions, the ions see time-averaged forces, retracting them in the z-direction toward the plane $z = 0$ and in the r-direction toward the z-axis characterized by $r = 0$. These time-independent forces are sometimes called harmonic pseudoforces, the expression "harmonic" coming from the harmonic oscillator, which is characterized by centripetal forces that increase linearly with the distance from the center. The harmonic pseudoforces shown in Figure 1 give rise to the picture of the pseudopotential well explained in Chapter 2. Because of the linearity of both the RF field amplitude and the harmonic pseudoforce field, the pure quadrupole trap is sometimes called a linear ion trap.

It is possible to produce rotationally symmetric three-dimensional RF fields which increase in higher powers of the distance from the center. A strictly quadratic increase of the RF field amplitude without any linear, cubic, or higher terms defines the pure three-dimensional hexapole field (see Section I.C.2); a strictly cubic increase produces the pure three-

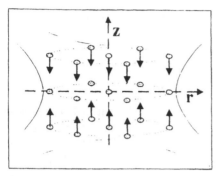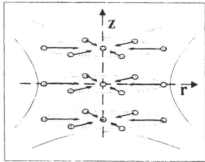

FIGURE 1

The two harmonic oscillators of the pure quadrupole filled with corresponding pseudo-force fields. The ions are trapped in the r-direction by the cylindrical oscillator, and in the z-direction by the planar oscillator. Both oscillators are completely independent, i.e., the movements in the r-direction do not influence the movements in the z-direction.

dimensional octopole field. These fields are not considered here as our main topics, but a short description is given and a few investigations are visited briefly below.

Nonlinear ion traps are understood to consist basically of a three-dimensional RF (and optional direct current, DC) quadrupole field, but the linear increase of the RF (and DC) field amplitude in the r- or z-direction is altered slightly by the addition (or subtraction) of weak quadratic, cubic, or higher power functions (Figure 2). The production of these nonlinear fields by shaping the electrodes is described below. Any rotationally symmetric nonlinearity may be described, in a very comprehensive and mathematically correct way, by the superposition of weak higher-order multipole fields on top of the basic quadrupole field. Thus, the term nonlinear ion trap can be interpreted as an ion trap with superimposed weak multipole fields. Both terms are identical in meaning.

C.   Classification of Pure Multipole Fields

Before we can consider the superposition of weak multipole fields, we should know a little more about pure multipole fields, their features, and classification.

1.   Multipole Fields

Multipole fields are defined mathematically as spatial distribution functions of the electrical potential. In theory, these multipole fields should be source-free; there must not be any sources or sinks of field lines, i.e.,

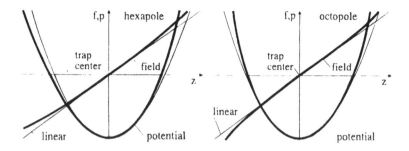

FIGURE 2

The field increases linearly in both z- and r-directions in a pure ion trap, hence the name "linear ion trap". The superimposed weak hexapole field adds a quadratic term, therefore, a small parabolic contribution is added to the linear increase. The octopole field adds a cubic term, as can be seen in the figure. It is from these deviations from a linear increase that the name "nonlinear ion traps" is derived.

any electrical charges in the described field region. The poles, however, form sources and sinks for the electrical field. As a consequence, theory requires the poles to be outside the region under consideration. Only the field inside the electrode structure can be described by superimposed multipoles.

2. Two- and Three-Dimensional Multipole Fields

Two-dimensional multipole fields can be imagined as being produced by long poles arranged in equally spaced parallel distances at the wall of an imaginary cylinder and connected with suitable potentials, alternating in sign from pole to pole. The familiar quadrupole mass filter consisting of four parallel rods is a fair example; the quadrupole mass filter with four hyperbolic surfaces is a better example because it produces a purer quadrupole field. Two-dimensional multipole fields do not represent the topics of this book; they are only mentioned for reasons of clarity.

Rotationally symmetric three-dimensional multipole fields are produced by two end-cap electrodes and a defined number of ring electrodes, arranged around an axis of rotational symmetry (named the z-axis) at the walls of an imaginary sphere. Voltages alternating in sign are applied to the electrodes from the first end-cap over the rings to the second end-cap. As can be derived from a cross-section through this structure, the end-caps represent one pole each, whereas the rings represent two poles each. The quadrupole ion trap with its hyperbolic end-caps and its hyperbolic ring is a good example. When we speak of multipole fields in this book, we always think of these rotationally symmetric three-dimensional multipole fields.

3. Naming Multipole Fields

Both two- and three-dimensional multipole fields are named after the number of poles, using Greek names for numbers above four. The poles are arranged around the respective horizon in a sequence with alternating polarity (plus and minus). Thus, the number of poles with positive polarity must equal the number of poles with negative polarity. It follows, then, that the total number of poles always must be even: the series starts with the dipole field having 2 poles, and continues with the quadrupole field (4 poles), hexapole field (6 poles), octopole field (8 poles), decapole field (10 poles), dodecapole field (12 poles), and so on.

The electrode structures of multipole devices can be supplied with DC, RF, or mixed DC plus RF voltages. Two-dimensional multipoles are often used with DC voltages as ion lenses with special characteristics. Ion traps, however, can store ions only if they are supplied with RF voltages, and they require at least four poles.

Figure 3 exhibits some electrode structures for multipole fields: the dipole field generated by a plate condenser (not forming an ion trap), the quadrupole ion trap structure, the pure hexapole ion trap, and the pure octopole ion trap.

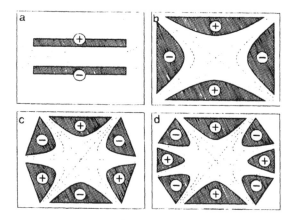

FIGURE 3

Cross-sections of three-dimensional electrode structures for the four lowest multipole fields: (a) the plate condenser generates the dipole field which is spatially constant within the region of interest; (b) two end-caps and one ring electrode (if correctly shaped) produce the quadrupole field; (c) two end-caps and two rings span the hexapole field; (d) the octopole field is made from two end-caps and three ring electrodes. All fields of order higher than the dipole field have field centers with vanishing field strength; when driven by a high frequency voltage, all such traps can store ions in the region around the field center.

Mathematical Characterization of Multipole Fields

So far, the multipoles are mainly characterized by the number of poles. mathematical characterization of the multipole fields, however, is :n by the power with which the field (or the potential) increases along z-axis (in fact, the change along the r-axis follows a different func-), as shown in Table 1.

The mathematical formulation of the dipole field is simple: it is a ipletely homogeneous field. In spite of its simplicity, the dipole field fully into the mathematical scheme of multipole fields, but it has some queness because it is the only multipole field without a field center. field center of all higher multipoles is characterized by vanishing l strength.

Mathematically, the multitude of three-dimensional multipole fields ns an orthogonal class of potential distributions, equivalent to the or-ʒonal class of the power functions. This identity means that any rce-free rotationally symmetric three-dimensional multipole field can :xpressed mathematically as a weighted superposition of such pure tipole fields. The weights may be imagined as the percentages of the tipole fields which are all superimposed with the same z-axis, and same field center. Such is the mathematical reasoning for the fact that linear ion traps can be defined as quadrupole fields with superim-ed multipole fields.

Even and Odd Multipoles; Order of the Multipole Field

The multipoles are classified further into even and odd multipoles. ontrast to the naming of the multipoles, the count for this classifica-₊ is not the total number of poles, but the number of pole *pairs*, n, in-d. Dipole (2 poles = 1 pair), hexapole (6 poles = 3 pairs), decapole (10 ?s = 5 pairs), and tetradecapole (14 poles = 7 pairs) are examples of l multipoles, whereas quadrupole (4 poles = 2 pairs), octopole (8 poles pairs), dodecapole (12 poles = 6 pairs), and sedecapole (16 poles = 8

TABLE 1

Characteristics of Pure Multipole Fields

Multipole name	Field increases	Potential increases
Dipole (2 poles)	Constant, no increase	Linearly
Quadrupole (4 poles)	Linearly	Quadratically
Hexapole (6 poles)	Quadratically	Cubically
Octopole (8 poles)	Cubically	With fourth power
Decapole (10 poles)	With fourth power	With fifth power
Dodecapole (12)	With fifth power	With sixth power
Hexadecapole (16)	With seventh power	With eighth power

pairs, sometimes called hexadecapole) belong to the even multipoles. The number of pole pairs n is called the order n of the multipole field. The field in the z-direction increases with $(n - 1)$th power; the electrical potential increases with nth power.

Odd multipoles are characterized by the fact that opposite poles have different polarity, whereas opposite poles of even multipoles carry the same sign or polarity. As a result, three-dimensional ion traps with even multipoles are symmetrical with respect to the $z = 0$ plane. The same is true for any mixture of even multipoles. The $z = 0$ plane may be called the "equator" for these types of symmetric ion traps. Contributions of odd multipoles, however, destroy the symmetry with respect to this plane.

D. Three Simple Basic Effects in Nonlinear Ion Traps

In spite of the complicated mathematics necessary to obtain some useful information from the equations of motion in nonlinear ion traps, the basic effects of superimposed weak multipole fields are easy to comprehend. In fact, only three simple basic principles explain qualitatively (or almost so) all of the effects of nonlinear ion traps. All effects in nonlinear ion traps can be explained by these three basic principles.

1. Pure Quadrupole Field at the Center

When the ions are cooled down to rest in the center of a nonlinear ion trap, they do not see any nonlinear effects, they see only the trapping quadrupole field. As can be seen from Figure 2, all the nonlinear deviations of the field vanish in the direct neighborhood of the field center.

2. Nonlinear Resonances

For each kind of superimposed multipole field, some nonlinear resonance conditions of varying strength exist inside the stability region. The oscillator for the ions is no longer pseudoharmonic; it is distorted by the multipole superposition. The nonharmonic distortions generate multipole-specific overtones (or higher harmonics) of the secular oscillation. These overtones, in turn, produce sideband frequencies of overtones in addition to the normal Mathieu sideband frequencies of the basic secular frequency. Resonance occurs when these overtones, or combinations of these overtones in the r- and z-directions, match with the sideband frequencies. There are different types of nonlinear resonances. Most cause the ions to take up a certain amount of energy from the RF drive field, thereby increasing the amplitude of the secular oscillations.

3. Frequency Dependence on Amplitude

In cases of superposition with even multipoles, the frequency of the secular ion oscillation becomes amplitude dependent. With odd multipoles, this dependence is weaker but is nevertheless present; therefore, no resonance condition can put an ion into a truly unstable condition. When the ion takes up energy and increases its amplitude, the frequency shifts out of resonance and the amplitude of the oscillation decreases again, ending up in a beat. This statement is true for the above-mentioned nonlinear resonances caused by the multipole field, for resonances with an additional dipolar alternating current (AC) voltage applied across the end-cap electrodes, and for resonances with frequencies superimposed on the RF drive voltage such as hum frequencies. Thus, weak, even multipole fields stabilize ions inside the trap, and only strong effects can eject ions out of the trap. On the other hand, the change in frequency with increasing amplitude can affect strongly the quality of scan procedures. Both deterioration and improvements are possible.

E. Pure Multipole Ion Traps without a Quadrupole Field

Quadrupole ion traps with superimposed weak multipole fields show some favorable features for mass spectrometric applications. The quadrupole part of the potential is used for the trapping of the ions. The multipole part can be designed to optimize the resolution and peak shape of the mass spectrum, and the scan speed, quite independently from the quadrupole part.

In a pure multipole ion trap of order higher than quadrupole, the behavior of ions in the trap is different. The ions "feel" the nonlinearity of the trapping field everywhere in the trapping volume. The characteristics of the motion of an ion, including the stability of its trajectory, can depend critically on its initial conditions. Hence, pure multipole ion traps have not been used as mass-selective devices.

However, it is worthwhile to investigate the properties of these traps. A pure multipole field can be generated by electrodes with appropriate surfaces (see Figure 3). For a multipole with n pairs of poles, two end-cap electrodes and $(n - 1)$ ring electrodes are necessary. As an example, for an octopole, the two end-cap electrodes and the middle ring electrode would be connected to ground, whereas the two intermediate ring electrodes would be connected to the RF drive circuit.

When the RF amplitude is not too large, then the adiabatic (pseudopotential) approximation[9] is applicable and gives some information on the dynamics of the ions. The term "pseudopotential" arises from the fact that the motion of a trapped ion in the oscillating potential can be approximated by the motion of an ion in the (time independent) pseudopotential. The pseudopotential of a pure multipole field with n pairs of poles depends on the $(2n - 2)$th power of the coordinates. For a quadrupole, the pseudopotential is a three-dimensional parabolic potential well, and the motion of a trapped ion is a harmonic oscillation at the secular frequencies. Compared to a quadrupole with the same pseudopotential well depth, the pseudopotential for a multipole with n pairs of poles ($n \geq 3$) is less steep in the center of the trap and steeper at the edges of the trap. Thus, the volume accessible to ions of definite energy increases with n, and the use of higher-order multipole traps for large storage capacity has been discussed.[10]

The trapping properties of a pure octopole trap have been investigated experimentally by use of laser-induced fluorescence of Ba+ ions (m/z 138).[11] At low RF amplitudes, the stability diagram has sharp edges, whereas at high RF amplitudes, the edges are very shallow. Due to the space charge repulsion, the spatial density distribution of the ions has a minimum at the trap center and a maximum at its edges. At low RF amplitudes, the ions near the trap center behave like quasi-free particles, while the ions at the edges of the trap feel the strong restoring force of the trapping field.

II. ION MOTION IN ION TRAPS

A. Ion Motion in Pure Quadrupole Ion Traps

The RF quadrupole ion trap with its rotationally symmetrical hyperbolic end-cap and ring electrodes, invented by Paul and Steinwedel,[12] utilizes an ideal three-dimensional quadrupole field. As described in Chapter 2, the movements of ions inside the field of such a trap and the stability of their trajectories may be derived from the Mathieu equation, which can be solved exactly using analytical methods. The following fundamental properties of these ion movements in mathematically ideal quadrupole traps will be changed by the superposition of higher multipole fields:

1. The ion movements in the axial or z-direction in mathematically ideal quadrupole fields are completely independent and decoupled from the movements perpendicular to the z-axis (normally called the r-direction). Thus, movements in the r-direction do not influence in any way the movements in the z-direction, and vice versa. (The independence of axial and radial movement, and the linear field increase, allow the application of the Mathieu differential equations to investigate the stability of ion movement.)

2. The RF field amplitudes in the r- and z-directions increase linearly with the distance from the field center. The resulting harmonic pseudoforces onto the ions, driving them back to the central z-axis and r-plane, respectively, also increase linearly.

3. The stability of ions of a given m/z ratio in an infinitely large quadrupole field does not depend on the initial movement conditions; it depends only on the field parameters a_u and q_u where $u = r$ or z (see Chapter 2 for definitions of a_u and q_u). In other words, only the two parameters a_u and q_u for the RF and DC fields, respectively, determine whether the solutions for the ion movement are stable or whether the oscillations of the ions increase their amplitude without limit to infinity. This condition is described by the well-known stability diagrams for quadrupole ion traps. (In practice, however, with fields confined by the electrode structure, the practical stability may deviate somewhat from the theoretical stability.)

4. When the a_u and q_u coordinates remain inside the stability region of the stability diagram, the ions perform stable oscillations in both the r- and z-directions, the so-called secular oscillations with frequencies ω_r and ω_z, which are less than half the RF drive frequency Ω. The frequencies of these secular oscillations can be described by so-called iso-β_r and iso-β_z lines, which subdivide the ion stability region (see Chapter 2 for definitions).

5. The frequency ω of the secular oscillation is independent from the oscillation's amplitude; it is strictly a harmonic oscillation. The oscillator is called pseudoharmonic because the harmonic force field is achieved only by integration over time.

6. When the secular oscillations of the ions in the z-direction become excited by an axial dipolar field of matching frequency (for example, by the application of an additional AC voltage across the end-caps), the ion oscillation becomes unstable. The amplitude would increase linearly with time and without limit, if there were no end-cap electrodes to stop the motion. No threshold exists for the AC voltage. If there is enough time, and if the oscillations are not damped by a cooling gas inside the trap, even the smallest voltages cause the resonating ions to leave the ion trap.

B. Ion Motion in Nonlinear Ion Traps

When we superimpose weak multipole fields (such as hexapole, octopole, decapole, dodecapole, and higher-order fields), the resulting nonlinear ion traps will exhibit effects differing considerably from those of a linear trap. The main characteristics of nonlinear traps comprise the following features:

1. The RF field increases nonlinearly in the r- and z-directions with distance from the field center. The same is true for the back-driving forces of the pseudopotential wells. The deviation from linearity is the stronger, the farther the location of the ion from the center. (Ions in the center

of the nonlinear ion trap do not see the nonlinear increase of the fields caused by the superimposed multipole fields; they see only the linear part, i.e., the quadrupole field. The field components of higher power equal zero around the center. Once the ions are cooled by a damping gas into a small cloud around the center, they are trapped exactly as they would be in a purely quadrupolar trap.)

2. The ion motions in the r- and z-directions are no longer independent; they are coupled. Kinetic energy can even be exchanged between radial and axial movements. The significance of this effect for practical ion trap experiments, however, remains to be investigated. (In the pure coupling resonance explained below, this exchange of energy between motion in the two directions is the only effect of this type of nonlinear resonance.)

3. Several numbers and types of nonlinear resonance conditions exist for each kind of higher multipole field, forming curves that cut through the first stability region of the stability diagram. Even when the multipole fields are weak, the effects of the nonlinear resonances may be dramatic because they may enable the ions to take up energy from the driving field. However, the resonances are not effective in the center of the trap. The ions can see those resonances only as their amplitudes of oscillation increase away from the center. Ion losses by nonlinear resonances thus are restricted to phases of strong oscillations, for example, during the ionization phase, the introduction phase for externally generated ions, or the fragmentation phase when ions are excited resonantly by an additional AC voltage across the end-caps.

4. Each of the nonlinear resonances belongs to one of the following five categories:

- z-direction resonances (energy is taken up in the z-direction only)
- r-direction resonances (energy is taken up in the r-direction only)
- Sum resonances (energy uptake in both directions at the same time)
- Difference resonances: (besides a major exchange of energy between the z- and r-oscillations, energy may be taken up from the RF drive field)
- Pure coupling resonances (no energy is taken up from the RF drive field, only an exchange of energy between the r- and z-directions)

C. Amplitude Dependence of Secular Frequencies

For even multipole superpositions, the secular frequencies ω_z and ω_r become strongly amplitude dependent because the back-driving forces are no longer proportional to the distance from the center. No longer do we have a pseudoharmonic oscillation, rather the oscillation becomes inharmonic. For the addition of an even multipole field (the z-direction term of the even multipole has the same sign as the quadrupole z-direction term), the frequency ω_z in the z-direction increases with increasing amplitude; for subtraction, the frequency ω_z decreases.

1. Even Multipole Superpositions

In the presence of weak even multipoles such as octopoles or dodecapoles, and in the presence of stronger odd multipoles such as hexapoles and decapoles, the trajectories of ions encountering any type of resonance do not become unstable, as they do at the boundary of the stability region. The ions may take up a limited amount of energy from the RF drive field, and thus increase their secular oscillation amplitude. Because of the amplitude dependence of the secular frequency, this frequency now shifts out of resonance and the ion amplitude decreases again. With decreasing amplitude, the frequency shifts back, and the process starts again, resulting in oscillations with a beat. The maximum amplitude of the secular oscillation, in this case, depends on the initial oscillations of the ions at the beginning of the resonance. Nevertheless, this limited enlargement of the secular oscillation amplitude can be very fast, and thus can be used for fast scanning or ion elimination processes.

This effect of even multipoles to stabilize ions and to produce a beat is also effective for resonances with additional AC voltages. These AC voltages may be applied across the end-caps of the trap, forming a kind of AC dipole field, or they may be superimposed on the RF drive voltage, forming a quadrupolar excitation field. A rather high threshold exists for ion ejection by the dipole AC voltage, quite in contrast to pure quadrupole fields. This mode of behavior can lead to substantial improvements in fragmentation procedures. The collision energy of parent ions may be increased by resonant excitation with a dipolar AC voltage without ion losses.

Whereas in pure quadrupole fields the ion ejection by dipolar resonances is strongly dependent on the density of the damping gas, this effect is much weaker in ion traps with superimposed octopole fields.

2. Odd Multipole Superpositions

For odd multipole superpositions, the frequency of the secular oscillation is not dependent, to a first-order approximation, on the amplitude. In the z-direction, the stronger back-driving force at one side of the plane $z = 0$ is compensated by a weaker back-driving force at the other side. In higher orders, odd multipole superpositions also cause a shift of the secular frequency. The secular frequency ω_z in the z-direction always decreases with increasing amplitude, regardless of the sign of the multipole superposition.

III. SUPERPOSITION OF MULTIPOLE FIELDS

The following parts of this section seek to give a comprehensive description of all the effects with verbal explanations but without deluging the reader with too much mathematics. In Section IV the effects are illustrated by pictures derived from computer simulations.

A. How to Produce Superpositions

According to the laws of electrodynamics, the potential distribution inside an enclosed volume is determined completely by the potential along the enclosing border surfaces. Neglecting tiny potential drops caused by the resistance effect of possible surface currents, and neglecting possible charge-up effects of unclean surfaces, metal electrodes have the same electrical potential over all the surface, thus showing equipotential surfaces. If we know the exact shape of some equipotential surfaces embracing our cell volume, we can form metal electrodes of exactly that shape using numerically controlled lathes and mills. When we then apply the correct voltages to the correctly positioned electrodes, we have created the desired potential distribution inside the structure. This simple method can be easily used to create any physically possible potential distribution inside a limited cell provided edge problems, where the surfaces approach but never completely close, do not play an essential role. The surfaces necessarily do not form a completely closed cell; they must leave open several edge regions, otherwise they would cause breakdowns of the applied voltages.

1. Dipole Field

A simple example for such a cell is the plate condenser, consisting of two parallel metal plates of sufficient size. The plate condenser forms a dipole field between the plates. The word "sufficient" already signals the edge problem: the field between the plate becomes distorted near the edges of the plates, and the true, completely homogeneous dipole field of the plate condenser is only formed far from the edges, deep inside the plates.

2. Quadrupole Field

As already shown by Paul and Steinwedel[12] (Figure 4), the quadrupole field of the pure quadrupole ion trap can be formed by three electrodes instead of the two for the dipole field: two rotationally symmetric hyperbolic end-cap electrodes and one rotationally symmetric hyperbolic torus electrode (or ring electrode).

In a pure quadrupole field, the angle of the asymptotic double cone with respect to the $z = 0$ plane is

$$\tan \alpha = 1/\sqrt{2} \tag{1}$$

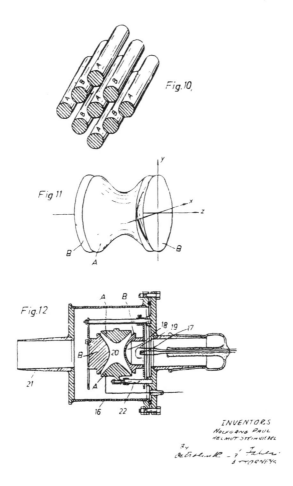

June 7, 1960   W. PAUL ET AL   2,939,952
                    APPARATUS FOR SEPARATING CHARGED PARTICLES
                          OF DIFFERENT SPECIFIC CHARGES
Filed Dec. 21, 1954   4 Sheets-Sheet 4

FIGURE 4
The quadrupole structure as designed by the inventors, Wolfgang Paul and Helmut Steinwedel, and taken from the U.S. patent 2,939,952, applied for (in Germany) on December 24, 1953. We see a cross-section through quite a modern ion trap, with ring and end-cap electrodes, and with an electron gun for internal ionization. The sketch above the electrode structure demonstrates the coordinate system with (x, y)-plane and z-direction, still used today to describe ion motion inside the trap.

It is often assumed that the ratio z_0/r_0 of the distances from the trap center to an end-cap electrode and to the ring electrode must have the same value of $1/\sqrt{2}$. This is not the case.[13] Figure 5 shows three different geometries for ion traps, and all three geometries generate pure quadrupole fields (if we neglect the influence of the edge fields), but have quite different ratios of z_0/r_0.

3.   Hexapole, Octopole Fields, etc.

The pure hexapole field needs four electrodes (two end-cap electrodes and two rings), whereas the pure octopole field needs three rings in addition to the two end-caps for a total of five electrodes. In general, the *n*-tupole needs $(n-2)/2$ ring electrodes in addition to the two end-caps.

4.   Equipotential Surfaces

The exact equations for the potential distribution of quadrupoles with superimposed multipoles are presented in Section VII. From the equations given there, we can easily calculate the equations for equipotential surfaces representing the potential distribution for superimposed weak hexapole and octopole fields.

In Figure 6 through 8, are demonstrated the superpositions of hexapole and (positive) octopole fields. None of these examples contain other multipoles such as dodecapoles or the like.

In general, the quadrupole field with superimposed weak multipole fields needs only three electrodes, as does the pure quadrupole field, in spite of the many poles taking part. This statement, however, is true only for the superposition of weak multipoles. Surprisingly, these superpositions of mathematically pure hexapoles or octopoles on the basic quadru-

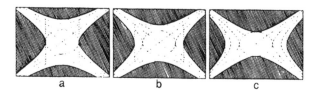

a   b   c

FIGURE 5

Three-electrode structures with different z_0/r_0 ratios; these structures generate the same pure quadrupole field, neglecting the influence of the edges. The surfaces of the electrodes follow exact equipotential surfaces of the pure quadrupole field in all three cases, but with different potential values. The useful space for ion storage is marked by the rectangles. The effects of such structural differences on the resolution of modern scan methods with ion ejection through the end-caps have been little investigated experimentally. There are indications that for structures having relatively high z_0/r_0 ratios, such as is shown in (a), resolution as well as signal-to-noise ratios are improved.

FIGURE 6
Shapes of the electrodes needed to create an ion trap
with 8% superimposed hexapole field. The trap is no
longer symmetrical with respect to the z = 0 plane.

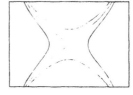

FIGURE 7
Electrode structure for an ion trap with 4% octopole field.

FIGURE 8
Structure of the electrodes needed to superimpose both
hexapole (8%) and octopole (2%) fields.

pole fields have not been investigated experimentally up to the present time. Hitherto, workers in the field either built rather coarse cylindrical ion traps, in most cases not really interested in the potential distribution inside as long as the structure was able to store ions, or they somewhat distorted the normal hyperbolic quadrupole ion trap in order to achieve certain performance goals.

The main interest of the effects of these superimposed weak multipole fields must be directed to the behavior of the ions in the z-direction because all ion trap mass spectrometers eject ions in this direction. Considering the z-direction, the multipole fields can be either added or subtracted. When even multipole fields are added, the field strength increases more strongly than linearly in the outer region of the trap, and the back-driving forces become stronger. In the case of subtracted even multipole fields, the field increases less strongly than linearly in the outer region, and the back-driving forces become weaker. In the case of even multipole fields, the direction of the frequency shift with amplitude growth is influenced.

5. Stretched Ion Traps

Ion traps having mixtures of superimposed higher even multipoles can be produced in various ways. Commercial ion traps[14] utilize a symmetrical stretching of the separation between the two end-cap electrodes of an ideal quadrupole ion trap by 10.8%, as shown in Figure 9a (left-

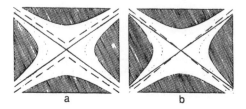

FIGURE 9

Two nonlinear ion trap structures that have been investigated experimentally. (a) The left-hand structure is the stretched end-cap distance design (Finnigan MAT) with its enlarged distance between the two end-cap electrodes, without any further geometrical change of the shape of the end-caps or the ring. The dashed lines are the asymptotes to the end-caps, whereas the solid lines form the asymptotes to the pure quadrupole design. They still form the asymptotes to the hyperbolic ring, so the ring and end-cap electrodes are no longer asymptotic to each other. Calculated percentages of multipoles induced by this type of stretching are given in Table 2. (b) The right-hand structure is the "stretched angle" design (Bruker-Franzen Analytik) with a corresponding change of all electrode shapes. The dashed lines indicate the new double cone asymptotic to the electrode structure, the solid lines show the asymptotic cone of the pure quadrupole ion trap. Calculated percentages of multipoles are presented in Table 2.

hand side). This stretched end-cap separation design was selected to achieve unit mass resolution with the sequential mass-instability scan which was not possible with a truncated ideal quadrupole electrode geometry. A detailed account of the events leading up to this change in ion trap design is given in Chapter 4. An explanation of the poor mass resolution of the truncated ideal ion trap for this type of scan is presented below. The stretching of the separation of the end-caps results in a superposition of not only quadrupole fields, but also dodecapole and higher even multipole fields. The potential distribution inside this type of ion trap is calculated below (Table 2), and the relative strengths of the different higher multipoles are shown graphically (Figure 10).

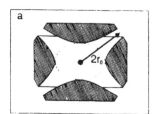
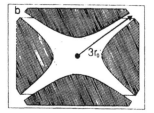

FIGURE 10

Electrode truncation; electrode structures where the electrodes are truncated at distances of (a) $2r_0$ and (b) $3r_0$ from the field center. Multipole fields are induced by the necessary truncature of the quadrupole ion trap electrode structure in order to realize a practical trap. Table 3 presents the percentages of higher-multipole fields for the traps shown in a and b.

TABLE 2

Comparison of Multipole Weights in Two Types of Ion Traps

Multipole	Stretched end-cap distance ion trap	Modified hyperbolic angle ion trap
4-pole (quadrupole)	89.4034%	97.7557%
8-pole (octopole)	1.4390%	1.4083%
12-pole (dodecapole)	0.6280%	0.1593%
16-pole (hexadecapole)	0.0830%	0.0792%
20-pole (ikosipole)	0.0304%	0.0062%
24-pole	0.0034%	0.0019%

6. Modified Hyperbolic Angle Ion Traps

Another type of ion trap development[15] resulted in a different design. The half-angle of the double cone which is asymptotic to the ideal hyperbolic quadrupole surfaces was changed from $\tan \alpha = 1/\sqrt{2}$ to $\tan \alpha = 1/\sqrt{1.9}$. This type of modified hyperbolic angle ion trap (see Figure 9b, right-hand side) also leads to the superposition of different types of higher even multipoles. The slit between the electrodes, however, is narrower in this design, and the dodecapole superposition is significantly lower. The ion storage behavior is much better than that of the stretched end-cap distance design. The performance goal, in this case, was the utilization of a nonlinear octopole resonance for a fast ion ejection scan.

The mathematical equation of the potential distribution for the latter type of ion trap, as well as graphical representations of the relative strengths of the different types of higher-order multipole fields, are presented in Section VII. Some experimental results with this type of ion trap are described in Section VI.

Even if the electrode structure is shaped precisely so as to generate pure quadrupole fields, multipoles are introduced when the edges of the electrodes are formed. Figure 10 and Table 3 present the results of calculations with two different electrode structures of the truncated Paul traps, cut to finite size.

7. Other Production Method for Quadrupole Field

It should be noted here that an ion trap need not be formed by equipotential surfaces. Wang[16] has proposed a modification to Arnold's technique for the construction of an ion trap, the electrodes of which do not have the same potential at every point. If we are able to produce a mathematically correct potential gradient across the surface, we can build very interesting types of ion traps. Such surfaces can be produced by different methods: for example, by a composite structure of isolating and non-isolating sheets of material and resistors as connecting elements between

TABLE 3

Relative Weights $A_n/A_{2,\,ideal}$ of Multipoles in Truncated Paul Ion Traps

Multipole	Truncation at $r = 2r_0$	Truncation at $r = 3r_0$
4-pole (quadrupole)	99.9592%	99.9974%
8-pole (octopole)	−0.1363%	−0.0366%
12-pole (dodecapole)	−0.2014%	−0.0190%
16-pole (hexadecapole)	−0.1514%	−0.0106%
20-pole (ikosipole)	−0.0844%	−0.0026%
24-pole (tetraikosipole)	−0.0313%	−0.0006%
28-pole	−0.0087%	−0.0001%

the conducting sheets. With a special quadrupole ion trap of this type, it is possible to superimpose the quadrupole field with an ideal, completely homogeneous dipole field. This superposition cannot be achieved exactly with ion traps made from equipotential electrodes.

IV. NONLINEAR RESONANCES

The conditions for nonlinear resonances have been derived[17] (see Section VIII for details) using the Hamilton formalism of theoretical mechanics. Here, a short extract of the results is given.

A. Resonance Conditions

The resonance conditions in the first stability region can be expressed by the general equation

$$n_r\omega_r + n_z\omega_z = v\Omega \tag{2}$$

ω_r and ω_z are the secular frequencies in the r- and z-directions, Ω is the RF drive frequency, and n_r, n_z, and v are integer values. For these integers, the following equalities can be formulated:

$$n_r = \text{even, for all types of multipoles} \tag{3}$$

$$n_z = \text{even, for superposition of even multipoles only} \tag{4}$$

$$n_z = \text{any integer for odd multipoles} \tag{5}$$

The connection of the secular frequencies with the parameters β_r and β_z of the Mathieu diagram is given by

$$\omega_r = \beta_r \Omega/2 \qquad \omega_z = \beta_z \Omega/2 \qquad\qquad (6)$$

allowing Equation 2 to be expressed as

$$n_r\beta_r + n_z\beta_z = 2v, \quad \text{with } 0 \le \beta_r, \beta_z \le 1 \qquad (7)$$

describing the resonances in the first stability region. The strengths of the nonlinear resonances decrease strongly with higher numbers of n_r, n_z, and v.

B. Nature

Equation 2 can be easily interpreted. It is well known from the Mathieu theory that the oscillations of the ions in a pure quadrupole ion trap, when analyzed with respect to the frequencies, consist of the basic secular frequency ω and a set of sideband frequencies of the form $i\Omega + \omega$ and $i\Omega - \omega$ ($i > 0$ is an arbitrary integer). In nonlinear quistors, overtone frequencies $k\omega$ (or so-called higher harmonics) are produced by the non-harmonic oscillator, $k > 1$ being an integer. The overtones $k\omega$ in turn produce sidebands of the form $i\Omega - k\omega$ and $i\Omega + k\omega$. Resonances simply take place when overtone frequencies match sideband frequencies. This simple phenomenon is shown below by Fourier analysis of simulated ion oscillations.

A closer examination reveals that there is not just a single match of one overtone frequency with one sideband. When such matches take place, groups of overtone frequencies match groups of sideband frequencies. The superimposed multipoles produce multipole-specific overtones. Even multipoles produce odd overtones only (3., 5., 7. harmonics, etc.), which give rise to the condition in Equation 4. Odd multipoles produce even and odd overtones for the secular oscillations (2., 3., 4., 5. harmonics, etc.). The weight factors for overtone generation depend on the kind of multipole, and on the amplitude of the oscillation. In general, the higher the order of the multipole, the higher the relative weights for the higher harmonics.

C. Types

By computer simulations in the z-direction, resonances have been found so far for $v = 0$ and $v = 1$ only, v being the factor of the RF drive frequency Ω in Equation 2. By different constellations for v, n_r, and n_z, the nonlinear resonances can be classified into five categories:

1. *Pure coupling resonance.* For $v = 0$, the term with the RF drive frequency Ω vanishes, and n_r and n_z must have different signs. Energy cannot be taken up from the RF drive field. There is only energy exchange between the oscillations in the r- and z-directions.[18] The amplitudes of the secular oscillations in both directions fluctuate anticyclicly between minima and maxima.

For $v = 1$, the following categories exist:

2. *Sum resonance condition.* n_r and n_z have the same sign, and energy is taken up from the quadrupole Rf drive field in both directions.

3. *Difference resonance condition.* n_r and n_z have different signs. Ion motion in the r- and z-directions is coupled and energy is exchanged between both directions. A smaller amount of energy may be taken up from the RF drive field.

4. *z-direction resonance condition.* n_r equals zero, and energy is taken up from the RF drive field in the z-direction only. An ion cloud can leave the ion trap, under certain circumstances, through one or more small holes in the center of one of the end-cap electrodes.

5. *r-direction resonance condition.* n_z equals zero, and energy is taken up in the r-direction. In an ion trap, the ions may hit the equator of the ring electrode.

D. Major Nonlinear Resonances for Hexapole and Octopole

The effects of nonlinear resonances on ion motion are mainly produced by low-order multipole fields, i.e., hexapole (three pairs of poles, order $n = 3$) and octopole (four pairs of poles, order $n = 4$). Resonances with higher-order multipoles are less effective. We assume here that operation of the ion trap is confined to the first stability region:

$$0 < \beta_r, \beta_z < 1 \tag{8}$$

The nonlinear resonances are given by Equation 7. For $v = 0$, only pure coupling resonances exist, which do not cause the ions to take up energy from the RF drive field. Only the resonances with $v = 1$ take up energy. Taking into account the rules given by Equations 3 to 5, and limiting the sum $(n_r + n_z)$ to values not greater than the multipole order n, which is a reasonable but not strictly correct assumption, the following resonances can be found in the first stability region:

$$\text{hexapole} : 3\beta_z = 2 \qquad 2\beta_r + \beta_z = 2 \tag{9}$$

$$\text{octopole} : 4\beta_r = 2 \qquad 4\beta_z = 2 \qquad 2\beta_r + 2\beta_z = 2 \tag{10}$$

These nonlinear resonances are presented in Figure 11 for an ion trap with rotationally symmetrical superpositions of hexapole and octopole fields. For the hexapole within the stability region, there are only two resonances by which energy may be taken up, while for the octopole there are three.

It should be mentioned, however, that more resonances can be found in simulation experiments. The additional resonances generally are extremely weak.

E. Graphic Form of Resonance Structure for Multipoles n = 3 to 10

This section presents in graphic form the resonance structure for superposition with multipole fields from 6-pole fields (hexapole, n = 3) to 20-pole fields (ikosipoles, n = 10), restricted to $v \leq 1$. We include here the pure coupling resonances (v = 0). The sum ($|n_r| + |n_z|$) will be restricted to

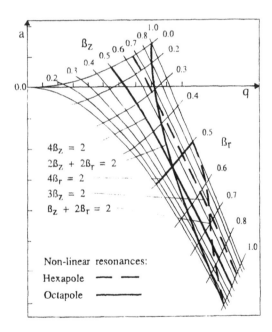

FIGURE 11
The nonlinear resonance conditions for superimposed weak hexapole and octopole fields form curved lines in the a, q plane. Only the lines inside the first stability region are shown; some of the lines (not all) continue into the surrounding instability area. The equations for the 2 hexapole and 3 octopole resonance conditions are presented in a somewhat unusual form so as to emphasize the symmetry within the equations. For a rotationally symmetric trap structure, the equations given here form the full set of hexapole and octopole resonance conditions.

values not greater than $n + 1$, n being the multipole order. This superposition introduces a few more resonances for odd multipole fields, e.g., see the above examples of the hexapole field. The additional resonances are very weak. For the octopole and other even multipole fields, there is no change by the increase of the sum ($|n_r| + |n_z|$) from n to $n + 1$.

The resonance conditions are not presented in the usual a_z, q_z diagram. If we transform the a_z vs. q_z stability diagram into a β_z vs. β_r stability diagram, the stability region becomes a rectangle, and all resonance conditions become straight lines (Figure 12). The shift of the stability borders by the superposition of the hexapole field and a corresponding distortion of the stability region in the a_z vs. q_z diagram is neglected. In the β_z vs. β_r diagram, the shifts do not exist because the shifts are frequency changes already reflected by the β_u values.

The β_z vs. β_r stability diagrams are much easier to read (and to draw) than are the a_z vs. q_z stability diagrams, and the full sets of nonlinear resonance conditions can be presented in this type of diagram. However, some disadvantage is associated with the β_z vs. β_r stability diagram: the operation line of the trap used without DC voltages, the so-called RF-

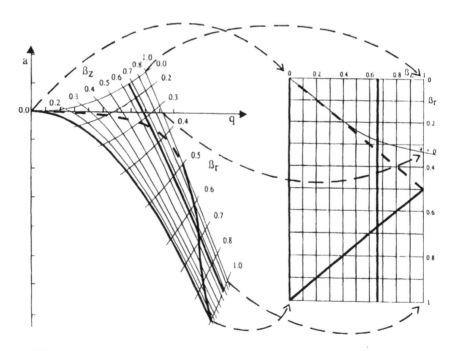

FIGURE 12
The first stability region of the a/q diagram becomes a rectangle when transformed into the β_z/β_r plane, and the nonlinear resonance conditions become straight lines. The transformation is shown here for the hexapole resonances.

only mode, cannot easily be seen. Therefore, the line $a_z = 0$ is drawn on the β_z vs. β_r stability diagrams to represent this mode.

In Figure 13, the working points of ions with relative mass values from 1 to 10 u are drawn on the stability region for the RF-only mode, showing the distorted mass scale along the $a_z = 0$ line. The points may reflect the masses 10 to 100 u, or 50 to 500 u.

It should be noted that the working points for all masses greater than 3.6 times the cutoff mass at the $\beta_z = 1$ boundary are located in the region $\beta_z < 0.2$. It will be shown that this region is completely undisturbed by resonance conditions of multipoles up to $n = 10$.

The resulting resonance conditions are presented in Figures 14a through h. Because the pure coupling resonances do not take up energy from the RF drive field, they are marked in the graphic representations by dashed lines. All conditions that cause energy to be taken up from the RF drive field are represented by solid lines. The resonances at the boundary lines and within the corners of the stability area are not shown here. All odd multipole fields (hexapole, decapole, etc.) exhibit nonlinear resonances only at the $\beta_z = 0$ boundary, whereas even multipoles exhibit resonances at all four boundaries. Furthermore, all odd multipoles have singular resonance points in the two corners $\beta_z = 0$, $\beta_r = 0$ and $\beta_z = 0$, $\beta_r = 1$, but they are not shown in the graphic representations. Even multipoles exhibit additional singular resonance points in all four corners. The singular resonance points within the corners contribute to the poor performance of ion isolation by mass-selective ion storage in the corners. The corners were often used for the isolation of ions of single masses as is discussed elsewhere.[19]

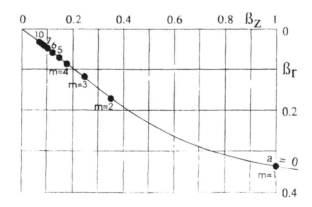

FIGURE 13
The upper part of the first stability region with the curve $a = 0$, corresponding to the operating line (or q-axis) of the "RF-only" mode. Ten working points for relative masses from 1 to 10 are marked, showing that masses higher than 3.6-fold the cutoff mass have their working points below $\beta_z < 0.2$.

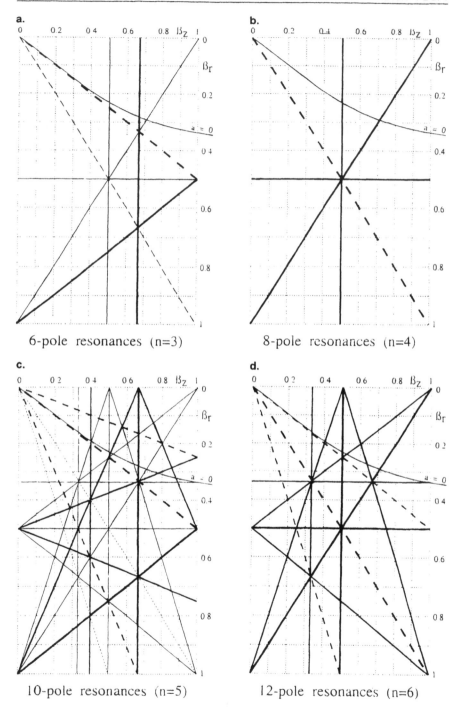

a.

6-pole resonances (n=3)

b.

8-pole resonances (n=4)

c.

10-pole resonances (n=5)

d.

12-pole resonances (n=6)

FIGURE 14

(a to h) Nonlinear resonance conditions for multipoles with orders $n = 3$ to $n = 10$, respectively.

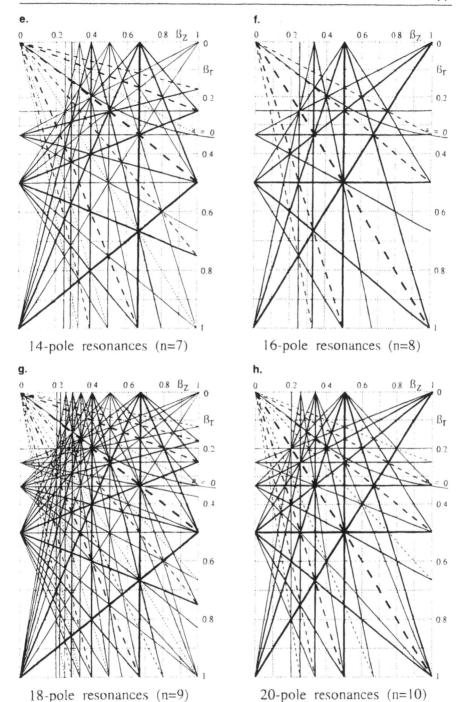

e.
14-pole resonances (n=7)

f.
16-pole resonances (n=8)

g.
18-pole resonances (n=9)

h.
20-pole resonances (n=10)

FIGURE 14
(*Continued*)

F.   Avoidance of Nonlinear Resonance Conditions

The discussion is now directed toward the avoidance of effects induced by the nonlinear resonances. Several cases may be discussed here:

1. If no resonances at all are to appear inside the stability region, then all multipole field superpositions must be strictly avoided. As shown above, this condition cannot be achieved simply by a truncated Paul ion trap. Some hitherto unknown corrections must be applied to the shapes of the electrodes.

2. The pure octopole superposition (with no other higher multipoles present) has a resonance-free storage region for all masses > 1.7 times the cutoff mass, corresponding to the resonance $\beta_z = 1/2$. On the other hand, the octopole superposition stabilizes ion storage and improves the performance of the mass-selective instability scan. If this effect is to be utilized, and no other resonances are to appear, then the hexapole field and all higher multipole fields must be avoided.

3. If the coupling of the motions of high m/z ratio ions between the r- and z-directions proves to be poor for an operation of the ion trap in the RF-only mode (this effect has yet to be investigated), all multipole fields other than the octopole field should be avoided. All the other multipoles cause such a coupling. The stronger the coupling, the higher the m/z ratio of the ions.

4. If resonances in the r-direction are to be avoided along the $a_z = 0$ operation line, all multipoles higher than the octopole must be avoided. The resonance at $\beta_r = 1/3$ may have the worst influence because the r-direction resonance is very near the mass-selective instability ejection working point.

5. Even in ion traps of poor design, with considerable amounts of higher multipoles up to $n = 10$, ions with masses > 3.7 times the cutoff mass ($\beta_z < 0.2$) are relatively undisturbed by nonlinear resonances. Such ions can be produced inside the trap or injected into the trap from outside with high efficiency. Ion losses then appear only during collisional fragmentation with increased oscillation amplitudes, and only for daughter ions which are produced and have oscillations matching such resonance conditions. The loss of ions by this effect has been described as the "black hole effect" or the "black canyon effect" (see Section V).

6. When the ion trap is used to isolate ions of a single m/z ratio in the upper corner $\beta_z = 1$, $\beta_r = 0$, or in the lower corner $\beta_z = 0$, $\beta_r = 1$, all multipole fields must be avoided. They all contribute resonance conditions within both corners.

It should be noted, however, that none of the resonance conditions represents a condition of complete trajectory instability, except those at the boundaries of the stability region. The even multipole fields (octopole, dodecapole, etc.) exert a braking effect upon their own resonances in the first order, the odd multipoles in second order, both by the effect of shifting the secular frequency out of resonance as the secular oscillation amplitude increases.

G. Black Holes, Black Canyons, and Other Ion Losses

There are three cases in which ions pass locations far from the center:

1. During ion generation by electrons which are directed into the trap from an external electron gun, a major fraction of the ions is produced far away from the center. These ions oscillate widely and may encounter resonances.

2. During fragmentation of parent ions, either parent or daughter ions may encounter nonlinear resonances in the main body of the trap away from the center. The resonance effect will be stronger the higher the amplitude of the dipolar excitation voltage used for the fragmentation.

3. When the ions are produced in an ion source separate from the trap, and induced into the trap from outside, all ions will see the maximum strength of the nonlinear resonances. For resonating ions, black holes and canyons will have maximum depth in this case.

1. Black Holes

Black holes and black canyons denote a phenomenon first encountered[20] in tandem mass spectrometry (MS/MS) experiments in ion traps. In these experiments, the axial oscillations of prospective parent ions are excited. Following collisions with the background gas, these ions increase in internal energy to the point of fragmentation. The charged fragments, called daughter ions, are trapped and mass analyzed subsequently. It was observed that when daughter ions have a q_z value close to 0.78 ($a_z = 0$), their signal intensity drops to very small values. The intensity drop extends a little to each side of this q_z value; thus, a black hole is formed. The effect is especially severe with strong resonant excitation, i.e., when high amplitudes of the dipolar excitation voltage (sometimes called "tickle voltage") are required for fragmentation. The same effect occurs at a value of $q_z = 0.64$ ($a_z = 0$).[21] These q_z values correspond to $\beta_z = 2/3$ ($q_z = 0.78$) and $\beta_z = 1/2$ ($q_z = 0.64$), and thus satisfy nonlinear resonance conditions for the hexapole and the octopole superposition.

2. Black Canyons

Detailed investigations showed that the loss of daughter ions occurs not only at these black holes located on the $a_z = 0$ line, but also along the nonlinear resonance lines for the $\beta_z = 2/3$ and $\beta_z = 1/2$ resonances and hence have the form of black canyons, crossing the line $a_z = 0$ at a certain angle.[22] Figure 15 shows an example of a black canyon along the $\beta_z = 2/3$ resonance. Up to now, black canyons have been observed for z-type resonances only; it appears that the excitation of the parent ions must be along the axis corresponding to the type of resonance. Experiments with radial excitation of the ions have not been reported thus far.

The black canyons indicate a mass-selective diminution in ion storage capacity, which can be explained as follows. During the fragmentation process, daughter ions are generated with a high initial amplitude of oscillation in the z-direction. The multipole fields are most effective off-center, and if a nonlinear resonance condition involving the z-motion is fulfilled, the amplitude of oscillation increases. Due to the high initial amplitude of the parent ions and the protracted duration of the fragmentation process, even a weak resonance condition by a weak multipole field suffices to drive the ions to the end-caps. In general, the phenomenon of ion losses by nonlinear resonances should be observed in all cases in which ions see a matching nonlinear resonance and when they move around in locations outside the trap center, i.e., when they oscillate strongly.

In routine MS/MS measurements with an ion trap, black canyons should be avoided, and this can be accomplished easily by adjusting the amplitude V of the RF drive field in order to position the black valleys between adjacent masses. However, some authors describe applications of the black canyon phenomenon for the study of the kinetic energy release during the fragmentation process.

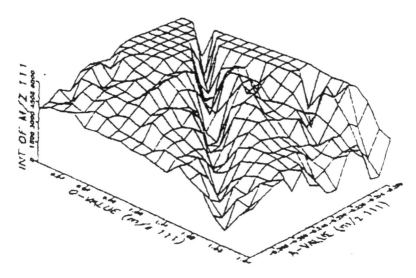

FIGURE 15
Black canyons.[22]

3. Acceptance Mountains

Figure 16 exhibits a wonderful range of mountains within the stability area of the stability diagram.[23] The mountains represent the acceptance profile of the ion trap for ions introduced from outside. The maximum ratios of canyon depth to summit height amount to more than 1:4. By comparison with Figure 14, the canyons for hexapoles, octopoles, decapoles, and dodecapoles can be identified.

4. Electron Impact Ionization

During electron impact (EI) ionization, ions may have large amplitudes in the z-direction. However, the ions are generated with a broad distribution of amplitudes, and the trapping efficiency of the pure quadrupole trap generally increases with decreasing initial amplitude. Therefore, the initial amplitudes for many of the trapped ions are not sufficiently large for the nonlinear resonance to have an effect. This condition explains the relatively weak influence of the effect of black canyons on mass spectra generated by EI ionization. This weak influence, however, does not mean complete absence. There are some indications that distortions of spectral intensities might be due to such effects.

V. SIMULATION STUDIES OF NONLINEAR EFFECTS

Computer simulations of the ion movements inside ion traps are sometimes more confusing than clarifying. It is nice to see, for a few times, the wonderfully entangled oscillations of a single ion in three-dimensional space. However, the influence of the superposition of multipoles can scarcely be seen in this view.

The often irregular behavior of the ions near resonances, for instance, can be studied only by a large number of simulation experiments wherein the initial parameters are slightly changed from experiment to experiment in a stochastic way. Such simulations are tiring, and the statistics often remain poor, with no clear indication of the general behavior. Automatically performed simulations may be helpful in this case; however, often the experimental investigation of real (not computed) ions exiting the ion trapping device is much easier and faster than when done by computer simulations.

A. Can We Gain Insight by Computer Simulation?

There are exceptions to this rule, however. Mastery, whether it be in art or in science, is often gained by the perceptive delineation of a specific line of inquiry, by casting aside superfluities and focusing on the

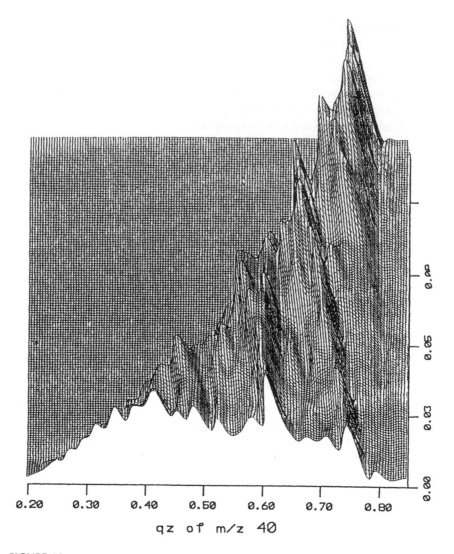

FIGURE 16
Acceptance profile for the storage of external ions in the upper part of the stability region.[23]

essential. Thoughtful selection of some aspects that can be studied without confusion may help, in certain cases, to gain vivid insight into the processes taking place in linear and nonlinear ion traps. A simple example may be enlightening at this juncture. If we wish to study the exchange of energy between oscillations in the r- and z-directions, a view of each of two parallel tracks showing fast r- and z-oscillations vs. time may easily verify this exchange, or it may not. If the two maximum oscillation amplitudes in the r- and z-directions oscillate by themselves in a contrary sense, such an energy exchange takes place and can be easily studied by changing the values of the initial parameters. It is easy to program these two projected views in computer simulations.

B. The Fruits of Computer Simulation of z-Axis Motion

We now introduce a very simple computer simulation method[4] that permits the study of some distinct aspects of nonlinear effects. Some other aspects, however, cannot be seen by this method. It is essential to know which can be seen and which cannot.

We simulate the oscillation of an ion moving only up and down exactly along the z-axis. This simulation is simple; it even may appear primitive and hardly worth trying, but bear with us. The ion in this method does not have any motion component in the r-direction; its speed in the r-direction is zero. Initial parameters for the ion oscillation are restricted to the location and speed on the z-axis. The ion in this case does not represent the behavior of a real ion inside the trap with all its degrees of freedom. The statistical probability is, indeed, zero for a real ion in a trap to move strictly along the z-axis. Yet we can learn a great deal about the effects of superimposed multipole fields on the oscillations of ions by this simple method.

Because this movement is restricted to just one linear dimension, we can easily display, on a two-dimensional computer monitor screen, the movement in the z-direction as a function of time. Thus, we can observe the secular oscillation of the ion, superimposed by the faster oscillation imposed by the RF drive field. We can investigate the influence of the field parameters (RF amplitude or RF frequency) on an ion of given m/z ratio, and the effect of the instability condition outside the stability region. We can investigate the influence of hexapole or octopole superposition on the secular frequency. We can apply an excitation dipole field along the z-axis (simulating an additional alternating current, AC, voltage across the end-caps), and study the influence of the dipole AC field on the oscillation of the ion.

1. Secular Oscillations

As an example, Figure 17 shows the secular frequencies of the oscillation of an ion in a pure quadrupole field for three different values of β_z. For small values of β_z, the oscillation is a neat sine wave with only tiny impregnations of the driving frequency Ω. Near the value $\beta_z = 1$, we see a beat structure of two superimposed sine waves of about the same frequency and amplitude. Section V.B.2 shows that this structure is an interference pattern caused by the secular frequency ω_z and its sideband $\Omega - \omega_z$. Furthermore, we can change some scan parameters with time (e.g.,

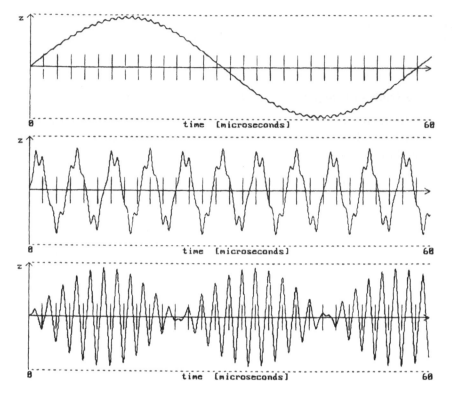

FIGURE 17

The secular oscillations in a pure quadrupole field ion trap, shown for three values of β_z: 0.0345 (top), 0.364 (center), 0.955 (bottom). When the operating point inside the stability area approaches the $\beta_z = 1$ boundary, the oscillation becomes increasingly dominated by an interference with the sideband frequency, $\Omega - \omega_z$. Thus, the sine shape of the secular frequency for small values of β_z (top) changes into a beat structure of two superimposed sine waves with very similar frequency (bottom). The curves were calculated by a z-axis simulation program.

the RF amplitude) and study the scan behavior of the ion. The effect of these measures on the ion oscillations can clearly be seen in the two-dimensional pictures. In spite of this primitive approach to trajectory simulation, the clarity of the effects is often surprising, and understanding is gained easily.

2. Beat Structure

Let us consider a second example. We show here the wonderful beat structure of the oscillations of the ion when the storing field amplitude is scanned. The scan moves the working point in the stability diagram, and thus the frequency of the secular oscillation. The beat is shown in Figure 18 for several outstanding points along the scan. This figure, however, is somewhat distorted by the pixel structure of the presentation which diminishes the beauty of the presentation.

A more detailed picture is given in Figure 19, which shows the fine structure of some short sections of the full scan. The envelopes of the beat structure form twisted sine curves of changing frequencies. Around the point with one third of the RF drive frequency, three sine curves are twisted, forming maxima which alternate on the positive and negative side of the

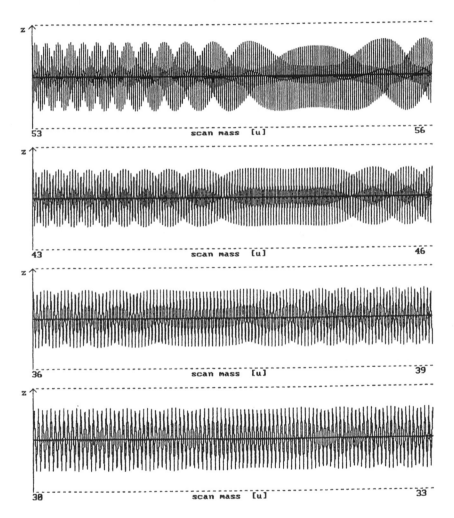

FIGURE 18
Structure of the beat during a scan. Even in a pure quadrupole field as shown here, a beat appears during the scan. The shape of this beat is shown around mass scale locations where the secular frequency equals one third (top), one fourth, one fifth, and one sixth (bottom) of the RF drive frequency. This scan beat does not change very much by superposition of weak multipole fields.

time axis. Around one fourth, four interlaced sine curves can be seen, with the maxima having symmetrical structure.

3. Secular Oscillation Distorted by Multipole Fields

In Figure 20, we see ions oscillating freely in two different ion traps, a pure quadrupole trap (top), and a nonlinear ion trap with a superimposed weak octopole field (bottom). The strength of the octopole field is 2% (defined by the ratio of the expansion weights A_n).

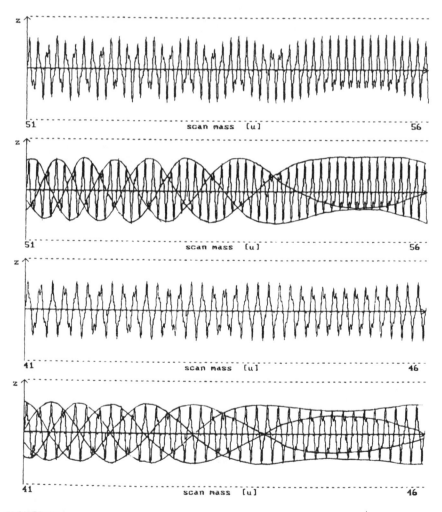

FIGURE 19

Fine structure of the scan beat around one third (upper two scans) and one fourth (lower two scans) of the RF drive frequency. Each scan beat is shown with and without beat envelopes in the form of three or four twisted sine waves.

In each of the ion traps, oscillations are shown with narrow and wide amplitudes to show the amplitude dependence of the frequency. Just a little more than a full secular period is displayed in each of the tracks, and the period of the oscillations is marked. The secular oscillation is superimposed by the smaller and faster oscillations caused by the RF drive

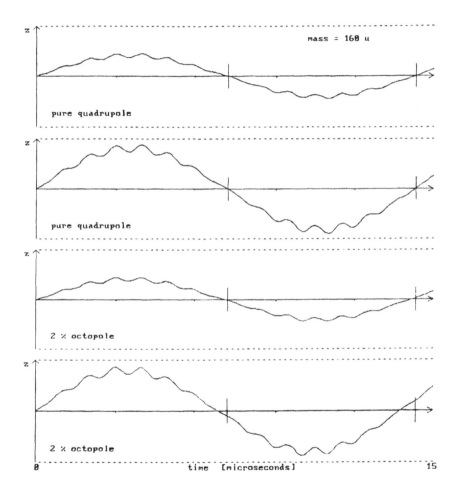

FIGURE 20
Secular oscillation of an ion in the z-axis, in a pure quadrupole field (top), and in a superimposed 2% octopole field (bottom). The vertical axis represents the time; only 15 μs are shown here. The wide secular oscillation is by itself slightly modulated by the RF drive frequency of the ion trap which amounts, in this case and in all the following simulations, to 1 MHz. The secular frequency becomes slightly higher in the case of octopole superposition because the back-driving forces are slightly higher. The frequency now is dependent on its own amplitude. The small difference in frequency may not seem essential, but it will prove to be extremely so in all resonance cases.

field. The superimposed oscillations are larger in amplitude, the further the ion is away from the center. The oscillations shown here will go on and on if the movement is not disturbed by collisions with residual gas molecules.

The difference between the ion oscillations in both kinds of traps is, at first sight, scarcely noticeable. Upon closer examination, we discover that the oscillation period of the secular frequency in the nonlinear trap (bottom) is just a little shorter, and depends clearly on the amplitude. However, the basic storage mechanism seems not to be changed by the superposition of a weak octopole field.

Figure 21 presents the wave forms of the secular frequencies in the z-direction for the pure quadrupole ion trap and for superposition traps with 4% of each hexapole, octopole, decapole, and dodecapole. The secular frequencies are shown for $\beta_z = 1/7$ ($\omega_z = \Omega/14$). For small amplitudes, the frequencies for all the ion traps shown are exactly the same, and the visible changes in frequency stem from the amplitude shifts of the frequencies. The dashed lines represent the distance to the end-caps for a normal Paul ion trap with $r_0 = 1$ cm, $z_0 = 1/\sqrt{2}$ cm = 0.707 cm.

We recognize that the rather large superposition of a 4% hexapole field (second from top) changes the secular frequency by only a very small amount, even with large oscillation amplitudes. However, the waveform becomes asymmetric, and the positive halfwave is shorter in length and amplitude than the negative halfwave. A superposition of a negative hexapole field reverses these relations, but the slight elongation of the wavelength (a slight decrease of the frequency) is independent of the sign of the hexapole field. The asymmetric waveform indicates the presence of even overtones (i.e., even higher harmonics of the form $\cos^2 \omega_z$, $\cos^4 \omega_z$, etc.).

The even multipoles (octopole and dodecapole, third and fifth tracks in Figure 21) produce much larger changes in the secular frequency. With positive multipoles, the frequency increases, whereas negative multipoles decrease the frequency. The frequency shift is dependent on the oscillation amplitude and the order of the multipole. The waveform remains symmetric, and deviations from the sine shape cannot be observed visually. They can be analyzed only by Fourier analysis, showing the existence of odd higher harmonics ($\cos^3 \omega_z$, $\cos^5 \omega_z$, etc.). With higher odd multipoles (decapole), frequency shifts and asymmetry are stronger than with the hexapole.

4. Oscillation Frequencies of the Pure Quadrupole Field

In the pure quadrupole ion trap, the secular oscillation (top track of Figure 21) is a sine wave of frequency ω_z, upon which is superimposed weaker oscillations of the RF drive frequency Ω. The superimposed

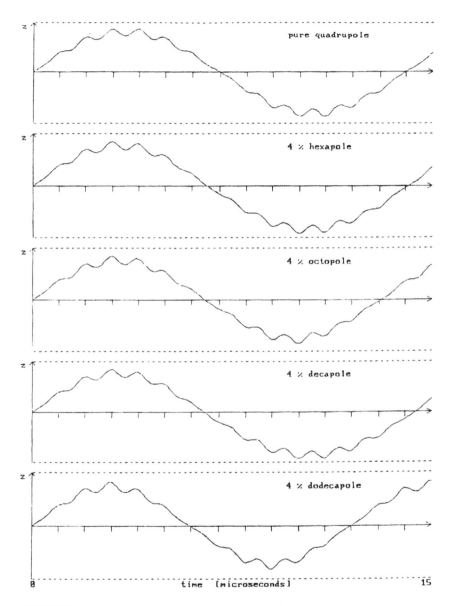

FIGURE 21

Change of waveform and frequency of the secular oscillations in the z-direction by the superposition of higher multipole fields. The dashed lines represent the distance of the endcaps from the trap center. The asymmetric wave form created by odd multipoles is clearly visible, as is the marked frequency change (depending on the amplitude) for even multipoles. More information can be obtained by Fourier analysis of these computed oscillations.

oscillations are proportional in amplitude to z (because the RF drive field is proportional to z), so that the impregnation has the form $\sin \omega_z \times \cos \Omega$, and the full secular oscillation $s_z(t)$ can be written

$$s_z(t) = a\sin \omega_z + b\sin \omega_z \cos \Omega \qquad (11)$$

With the trigonometric relation

$$\sin \omega_z \cos \Omega = 1/2 \sin(\Omega - \omega_z) + 1/2 \sin(\Omega + \omega_z) \qquad (12)$$

the superposition must contain the sidebands $\Omega - \omega_z$ and $\Omega + \omega_z$:

$$s_z(t) = a\sin \omega_z + b\sin(\Omega - \omega_z)/2 + b\sin(\Omega + \omega_z)/2 \qquad (13)$$

In other words, if we analyze the frequency spectrum of the oscillations, we should not find the RF drive frequency Ω because the symmetric impregnations on the positive and negative branches of the sine wave compensate each other completely. Instead, we will see the sidebands.

C. Limitations of This Simple Kind of Simulation

Seeing such nice results, however, we must not forget that these simulation experiments are severely limited. Because the motion components in the r-direction are nonexistent, coupling effects between the r- and z-directions cannot be studied. Only the z-type resonances can be investigated, the method is completely blind for all other types of nonlinear resonances. This blindness refers to sum-type resonances as well as to difference types and to r-type resonances. No focusing or defocusing effects of ejected ions can be investigated. The study of the effects at the boundaries of the stability region is restricted to the $\beta_z = 0$ and $\beta_z = 1$ boundaries; the influence of the $\beta_r = 0$ and $\beta_r = 1$ boundaries remains hidden. The mass-selective ion instability scan can be studied easily, and the influence of axial modulation can be demonstrated neatly. Even the vastly deteriorating influence of odd or negative even multipole field superposition on the resolution of sequential mass-instability scans can be studied.

In the hexapole superposition, only the resonance at $\beta_z = 2/3$ can be studied; the resonance at $\beta_r + \beta_z/2 = 1$ will not be seen. For the octopole, only the resonance $\beta_r = 1/2$ is visible; neither of the resonances $\beta_z + \beta_r = 1$ and $\beta_r = 1/2$ can be investigated.

Nevertheless, this method has increased our understanding of the effects of nonlinear fields. The simulation experiments demonstrate clearly the way in which the beat is connected with even multipoles when the ions are excited by axial AC voltages. The exponential increase of the secular amplitude by the hexapole z-type resonance can be seen.

In the following sections, we have outlined some of the results obtained which, on occasion, show surprising effects. Even more insight can be gained by a Fourier analysis of the computed secular oscillations, showing the sideband and harmonic frequencies induced by the multipole superposition. All of the following z-axis simulations are performed in gas-free traps. The effects of damping gas inside the trap are not investigated here, although the effects of such a gas, as discussed in Chapter 5, can be included easily.

D. Fourier Analysis of the Secular Oscillation

A simple Fourier transform program for 1024 points in the time domain was used. It was applied to calculated secular oscillations in the z-direction stored as digital values in a data file by the simulation program. The Fourier program used for this purpose assumes a periodic repetition of the selected time interval. This presupposition has some influence on the resulting frequency spectrum. The frequency peaks become very narrow if there is an integral number of frequency periods inside the 1024-point time interval. Otherwise, a phase jump exists at the borders of the interval, and the frequency lines show pronounced broadening at the base. In the following discussion, therefore, the sharpness of the frequency lines does not have any significance. It tells only whether the wavelength is an integral fraction of the 1024 point time interval.

1. Pure Quadrupole Ion Trap

The Fourier analysis of the oscillations in a pure quadrupole ion trap, as presented in Figure 22, shows the complete absence of the driving frequency Ω, which was set to 1 MHz in the calculations of the secular oscillation. In addition to the secular frequency ω_z, only the sidebands $\Omega - \omega_z$, $\Omega + \omega_z$, and $2\Omega - \omega_z$ can be seen. About 25 secular oscillations were stored in 1024 points for the Fourier analysis of Figure 22, and $\beta_z = 0.54$ was chosen. The existence of these sidebands is already well known from the Mathieu theory.

2. Overtones of the Superimposed Hexapole Field

Figure 23 exhibits a completely different frequency spectrum of the z-direction oscillations of an ion in a trap with 20% hexapole field superposition. The strong hexapole field was chosen to show the existence of the higher harmonics more clearly.

Besides the basic secular frequency ω_z (designated in Figure 23 as ω), the overtone series $2\omega_z$, $3\omega_z$, $4\omega_z$, and $5\omega_z$ can be seen. Because of the asymmetry of the secular oscillation (see track 2 in Figure 21), the

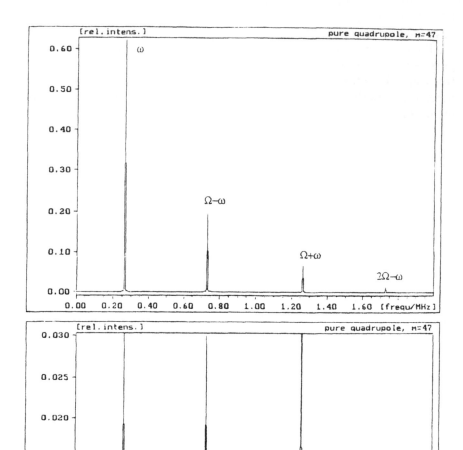

FIGURE 22

Frequency spectrum of the z-direction oscillations in a pure quadrupole ion trap, shown as 1024-point Fourier analysis of the calculated oscillations. About 25 oscillations are analyzed. In addition to the fundamental secular frequency, ω_z = 0.27 MHz, only the sideband frequencies $\Omega - \omega_z$, $\Omega + \omega_z$ and $2\Omega - \omega_z$ are to be seen. The RF drive frequency, $\Omega = 1$ MHz is not contained in the frequency spectrum.

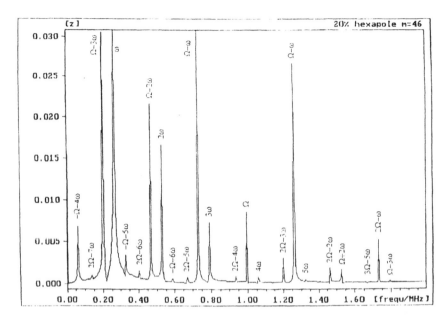

FIGURE 23

Frequency spectrum for 20% hexapole superposition. In addition to the sidebands $\Omega - \omega_z$, $\Omega + \omega_z$, and $2\Omega - \omega_z$ of the pure quadrupole trap, complete series of sidebands of higher harmonics can be observed. See, for instance, the series $2\Omega - \omega_z$, $2\Omega - 2\omega_z$, $2\Omega - 3\omega_z$, $2\Omega - 4\Omega \omega_z$, $2\Omega - 5\omega_z$, $2\Omega - 6\omega_z$, and $2\Omega - 7\omega_z$. The other observable series are $-\Omega + n\omega_z$, $\Omega - n\omega_z$, $\Omega + n\omega_z$, and $3\Omega - n\omega_z$.

superimposed oscillations with the RF drive frequency Ω no longer compensate in the positive and negative halfwave, and the frequency Ω emerges in the frequency spectrum. In addition to the well-known basic sidebands $\Omega - \omega_z$, $\Omega + \omega_z$, and $2\Omega - \omega_z$ of the pure quadrupole trap, a complete series of sidebands of higher harmonics can be observed (see, for instance, the series $2\Omega - \omega_z$, $2\Omega - 2\omega_z$, $2\Omega - 3\omega_z$, $2\Omega - 4\omega_z$, $2\Omega - 5\omega_z$, $2\Omega - 6\omega_z$, and $2\Omega - 7\omega_z$). The other observable series is composed of $-\Omega + n\omega_z$, $\Omega - n\omega_z$, $\Omega + n\omega_z$, and $3\Omega - n\omega_z$.

The frequency lines are generally stronger in the low-frequency part of the spectrum. In addition, line pairs being symmetric in the interval 0 through Ω (for example, the pair $\Omega - 2\omega$ and 2ω, or the pair ω and $\Omega - \omega$) show equal line intensities if they match by changing the operation point β_z within the stability region.

3. Overtones of the Octopole Field

The frequency spectrum of an ion oscillating in a 20% octopole field superposition (Figure 24) is much less rich than that of a hexapole superposition. Only the higher harmonics $3\omega_z$ and $5\omega_z$ exist, and the sideband series are also restricted to odd numbers of higher harmonics. In total, the sideband frequencies of higher harmonics appear to be weaker than in the case of the hexapole.

When the secular frequency ω_z becomes a little larger (for example, for an ion of lower m/z ratio), the line pair $2\Omega - 5\omega_z$ and $-\Omega + 5\omega_z$ will match at $2\Omega - 5\omega_z = -\Omega + 5\omega_z$, possibly giving resonance at $\omega_z = 3\Omega/10$ or $\beta_z = 3/5$. The strongest nonlinear resonance of the octopole superposition is a match of the line pair $\Omega - 3\omega_z$ and ω_z with $\beta_z = 1/2$.

4.   The Rich World of Decapole Overtones

Figure 25 illustrates the richness of frequencies in the oscillation of an ion in the 20% decapole superimposed ion trap. In order to see all of these lines, it is necessary to have a wide oscillation with a large amplitude because the ions see the nonlinear field only far from the trap center.

The decapole as an odd multipole produces higher harmonics of even and odd order, showing extremely high overtones in the sidebands if they only are located in a favorable area of the frequency spectrum. In principle, no difference exists in the type of frequency lines between hexapole and decapole superposition, but the decapole superposition forms stronger lines for the higher harmonics, at least in the sidebands.

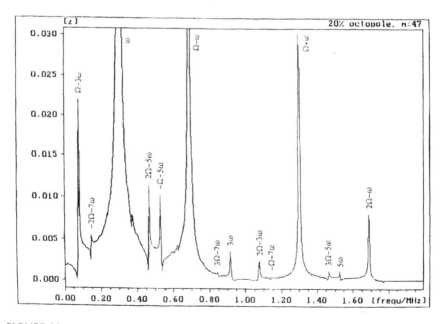

FIGURE 24

The 20% octopole superposition reveals a much less rich frequency spectrum than the hexapole superposition because only odd higher harmonics and their sidebands are generated.

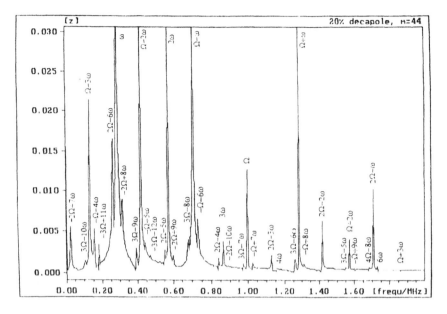

FIGURE 25

The 20% decapole superposition generates overtones of all orders, just like the hexapole superposition, but the higher harmonics appear to be stronger in intensity. The relative intensities of the higher harmonics, however, depend very much on the amplitude of the oscillation, on the frequencies themselves (lower frequencies are preferred), and on the frequency difference to other lines.

5. How Nonlinear Resonances are Formed

Resonances are formed when higher harmonics match sideband frequencies. Such a match does not mean a single harmonic frequency matching a single sideband frequency; normally complete groups of harmonics match with complete groups of sidebands.

In Figure 26 and 27, Fourier analyses of the secular oscillations are shown for masses 35 through 45 u, at a given storage constellation. The ion trap storage parameters are chosen such that mass 40 u encounters a decapole nonlinear resonance at $\beta_z = 2/3$. The Fourier analysis for this mass is omitted because it was not possible to store sufficient frequency periods due to the resonance.

The Fourier analyses for masses 37, 44, and 45 u show near-resonances. In the analysis for mass 37 u, ω_z equals 3/8, corresponding to $\beta_z = 3/4$. For masses 44 and 45 u, the conditions are $\beta_z = 4/7$ and $\beta_z = 6/11$, respectively. The coincident match of many frequencies can be seen best for $\beta_z = 2/3$ at mass 40 u. Both neighboring masses 39 and 41 u show clearly which frequencies are going to match. Tables 4 and 5 give some examples of lines that match for the two main nonlinear resonances of the hexapole ion trap.

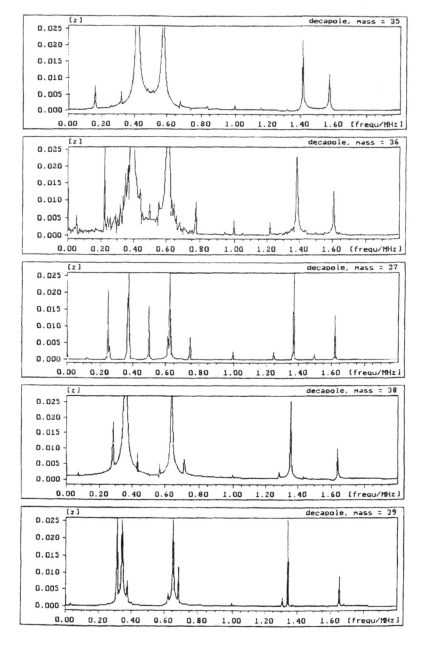

FIGURE 26

Frequency spectra for 20% decapole superposition, for masses 35 to 39 u. The mass scale is calibrated such that mass 40 u experiences the strongest nonlinear resonance at $\beta_z = 2/3$. However, the frequency spectrum for this mass, 40 u, cannot be shown because the simulation oscillation grows so rapidly that the amplitude limit is exceeded immediately. Mass 37 u shows a near-match for a possible nonlinear resonance at $\beta_z = 3/4$ ($\omega_z = 3/8 \; \Omega$).

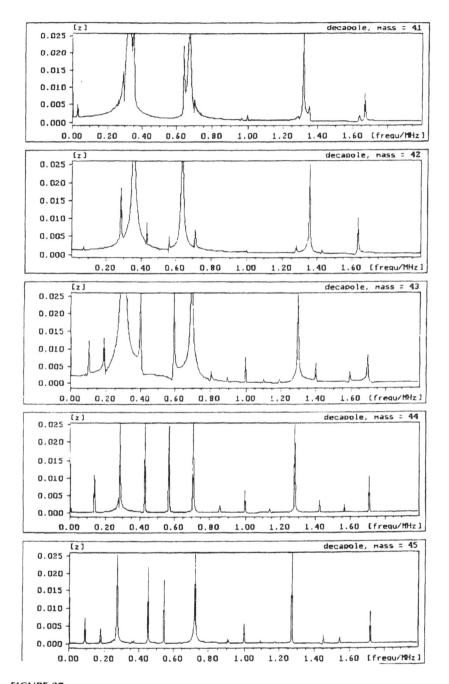

FIGURE 27

Frequency spectra for 20% decapole superposition, for masses 41 to 45 u (continuation of Figure 26). Masses 44 and 45 u represent near-matches for possible resonances at $\beta_z = 4/7$ and $\beta_z = 6/11$.

TABLE 4

Examples for Frequency Pairs that Match Simultaneously for the Resonance $\beta_z = 2/3$ of the Hexapole Field Superposition

$0 = \Omega - 3\,\omega$	$-\Omega + 4\omega = \omega$	$\omega = \Omega - 2\omega$	$-\Omega + 5\omega = 2\omega$
$\Omega - 2\,\omega = 2\Omega - 5\,\omega$	$2\omega = \Omega - \omega$	$-\Omega + 6\omega = 3\omega$	$\Omega - \omega = 2\Omega - 4\,\omega$
$3\omega = \Omega$	$\Omega = 2\Omega - 3\omega$	$4\omega = \Omega + \omega$	$\Omega + \omega = 2\Omega - 2\omega$

TABLE 5

Examples for Frequency Pairs that Match Simultaneously for the Resonance $\beta_z = 1/2$ of the Hexapole Field Superposition

$0 = -\Omega + 4\omega$	$2\Omega - 7\omega = \Omega - 3\omega$	$\Omega - 3\omega = \omega$	$\omega = -\Omega + 5\,\omega$
$2\Omega - 6\omega = \Omega - 2\omega$	$\Omega - 2\omega = 2\omega$	$2\omega = -\Omega + 6\omega$	$2\Omega - 5\omega = \Omega - \omega$
$\Omega - \omega = 3\omega$	$2\Omega - 4\omega = \Omega$	$\Omega = 4\omega$	$2\Omega - 3\omega = \Omega + \omega$

E. Ejection of Ions by Resonances in Stationary Fields

The mathematical analysis of the motion of ions within nonlinear ion traps is difficult to perform because the equation of motion cannot be expressed in the closed form of an analytical function. Below the theory of nonlinear resonances is presented, including a method for studying the relative intensities applied to the two nonlinear resonances of the hexapole. It is, however, much easier to investigate resonances by numerical computer simulations.

1. General Rule for Amplitude Growth by Resonances

The following investigation shows that the amplitude increase induced by an ion trap resonance can be presented, to a good approximation, by the equation

$$dz/dt = C_{n-1} * z^{n-1} \tag{14}$$

n being the order of the multipole. Equation 14 holds true for the dipole ($n = 1$), the quadrupole ($n = 2$) and, approximately, for all odd higher multipoles ($n = 3, 5, 7, \ldots$) such as hexapole, decapole, etc. The higher even multipoles strongly quench their own resonances because of the marked shift of the frequency with increasing amplitude, causing the resonant excitation to abate. The frequency shift is already known from Figure 21.

2. Dipolar Excitation

Figure 28 shows the resonance of an ion with a dipolar AC field, the frequency of which matches the secular frequency of the ion under study. The dipole field can be generated by electrical superposition on the quadrupole field of the ion trap. An RF voltage with frequency $\omega_z < \Omega$

applied across the end-caps produces a good approximation of such a superimposed dipole field; it is of interest to note here that there are quadrupole designs that facilitate the superposition of exact dipole fields.[6]

The simulation shown in Figure 28 reveals the linear increase of the amplitude, in accordance with Equation 14. The constant C_0 is proportional to the AC voltage:

$$dz/dt = C_0 \qquad C_0 = C_0' * V_{AC} \tag{15}$$

The integration of Equation 15 results in a linear increase.

The ions of secular frequency ω_z can also be excited by dipolar frequencies $\omega_d > \Omega/2$ matching their sideband frequencies $\Omega - \omega_z$, $\Omega + \omega_z$, $2\Omega - \omega_z$, $2\Omega + \omega_z$, etc. It can be seen from Figure 29 that the excitation gets increasingly weaker for higher sideband frequencies. The exciting dipolar voltage is held constant in this figure. Figure 30 shows the strictly linear growth of the amplitude with these sideband excitations in another presentation, in which the dipolar voltage is changed so as to obtain about the same slope for the amplitudes.

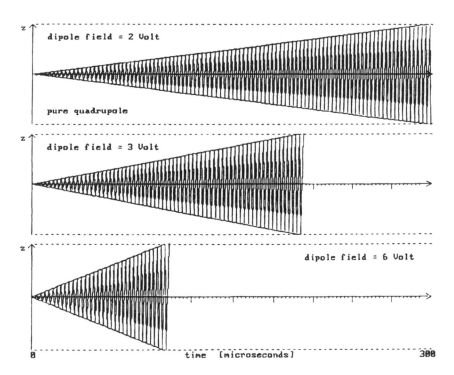

FIGURE 28
Excitation of an ion's oscillations by dipolar fields of three different field strengths results in linearly increasing amplitudes. The constant C_0 (see text) for the slope of the enveloping curve is identical in all three tracks.

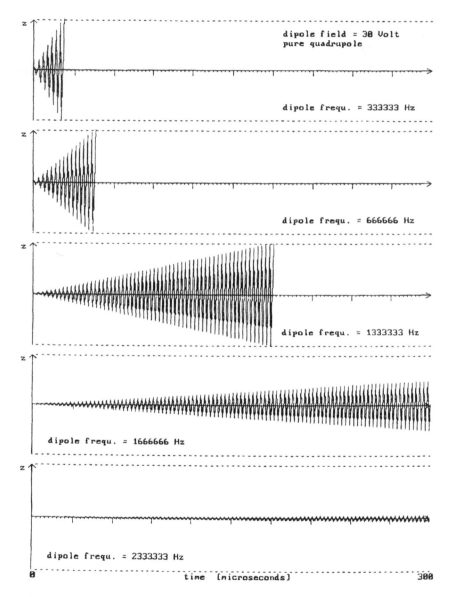

FIGURE 29

Ions with a secular frequency ω_x = 333.333 kHz cannot be excited only by a dipolar RF voltage matching this basic frequency (top), but also by dipolar voltages matching the sideband frequencies $\Omega - \omega_x$ = 666.666 kHz, $\Omega + \omega_x$ = 1333.333 kHz, $2\Omega - \omega_x$ = 1666.666 kHz, etc. The dipolar excitation voltage is kept constant. The relative strengths of the excitation can be seen from the slopes of the growth of the amplitudes.

3. Instability Boundary of the Quadrupole Field

The quadrupole resonance is often called a linear resonance because the strength of the quadrupole field increases linearly. The resonance is very special because it is not a resonance having a sharp mathematical

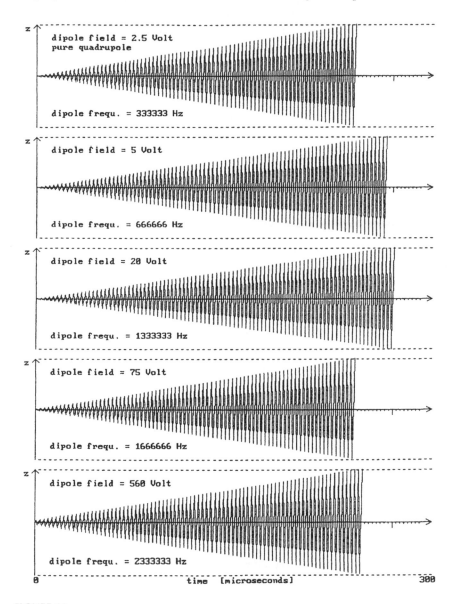

FIGURE 30

Sideband excitation with different dipolar voltages shows the linear amplitude growth more clearly than in Figure 29.

pole for a single and distinct frequency; rather, there is a whole frequency area of resonance outside the stability region of the a_z, q_z diagram. We shall examine first the resonance at an fixed working point in an area of axial instability; i.e., the working point is located outside the stability region, but not far from the point at which the $\beta_z = 1$ stability boundary intersects the q_z-axis ($a_z = 0$, $q_z = 0.908$), which is the point of ion trajectory instability in the mass-selective ion instability scan. In this area, ions are unstable in the z-direction only.

Figure 31 shows the exponential increase of the amplitude, as can be expected from an integration of the equation

$$dz/dt = C_1 * z \qquad (16)$$

The exponential increase is presented for different starting amplitudes for the secular oscillation, and shows the equal shifts of the curves for equal starting amplitude factors. The envelopes of the curves are designed according to Equation 16, using the same constant C_1 for all curves. The enveloping curves fit reasonably well with all of the oscillation amplitudes.

Figure 32 depicts the behavior of the secular oscillation amplitude around the point ($a_z = 0$, $q_z = 0.908$) for different values of q_z. For values of $q_z < 0.908$, we see an oscillation beat of the secular frequency ω_z produced by interference with the sideband frequency $\Omega - \omega_z$ (see also Figure 17 for a better view of the beat). The envelope curve has the form of a sine function. At the point $q_z = 0.908$, the secular oscillation and its amplitude remain constant; the frequency is exactly $\omega_z = \Omega/2$. In the area of axial instability with $q_z > 0.908$, the exponential increase in amplitude occurs. The higher the value of q_z, the more rapidly does the exponential increase.

A more quantitative analysis (by more simulations) reveals, in a rough first-order approximation, the dependence of the constant C_1 on the parameter q_z:

$$dz/dt = C_1' * \sqrt{q - 0.908} * z \qquad (17)$$

which gives upon integration the amplitude function

$$z(t) = C_1'' * e^{\sqrt{q - 0.908} * t} \qquad (18)$$

This function describes the amplitude function on both sides of the stability boundary. Inside the stability region, the square root becomes imaginary, and the exponential function splits (by Euler's equation) into a real

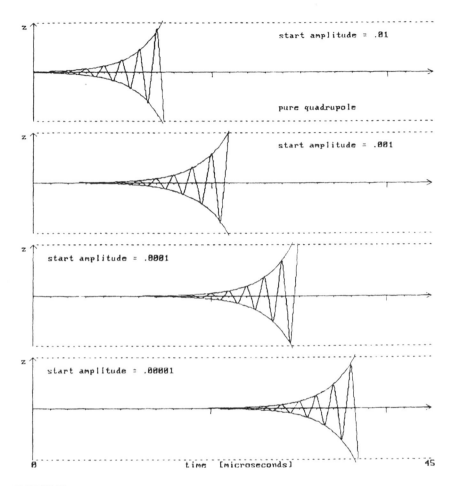

FIGURE 31

The exponential increase of the amplitude at a fixed working point outside the stability area, for different starting amplitudes of the oscillation. The trapping field is purely quadrupolar. The growth constant C_1 of the amplitude envelope is identical for all three simulations. The curves are shifted by equal distances for each tenfold change in the starting amplitude.

cosine function, describing exactly the beat motion seen in Figure 32 and an imaginary sine function.

Equation 18 is interpreted as indicating the absence of a sharp onset of instability at the $\beta_z = 1$ boundary. The storing effect by the back-driving forces of the harmonic oscillator ends smoothly at this boundary, goes through zero, and begins smoothly to defocus the ion oscillation.

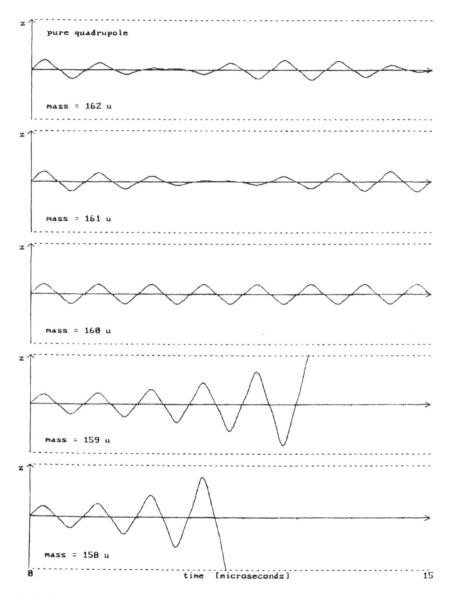

FIGURE 32

The amplitude changes of the secular frequency around the stability boundary is shown. Inside the stability area, the amplitude beats with a sine-shaped envelope, caused by interference of the secular frequency ω_z with its sideband $\Omega - \omega_z$. At the boundary, the secular amplitude remains constant, the frequency being exactly half the RF drive frequency. Beyond the stability area, an exponential increase of the amplitude takes place; the increase becomes more pronounced as the working point is moved further beyond the boundary.

4. *Ejection by Superimposed Hexapole Field*

Figure 33 exhibits the amplitude increase of oscillating ions matching the nonlinear hexapole resonance $\beta_z = 2/3$, for different starting amplitudes of oscillation. The envelope curves are constructed differentially, by the computer program, according to the equation

$$dz/dt = C_2 * z^2 \qquad (19)$$

with the same value of the constant C_2 for all three curves. Upon integration, the envelope function has the form of a hyperbolic function

$$z(t) = C_2'/(z - C_2'') \qquad (20)$$

with a mathematical pole at $z = C_2''$.

For ions near the center, $z = 0$, this function is weaker than the exponential function; however, for ions outside the center, the increase in amplitude is stronger than with the exponential function because of the mathematical pole of this function. The envelope curves in Figure 33 do

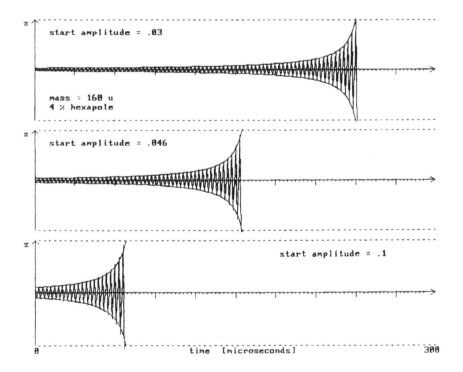

FIGURE 33

The hexapole resonances $\beta_z = 2/3$, for three different starting amplitudes. The envelope curve is drawn by the equation $dz/dt = C_2 * z^2$, with the same constant C_2 for all three curves, and the starting amplitude as parameters.

not fit exactly. A small (second-order) effect makes the frequency of the secular oscillation dependent on the amplitude; the frequency decreases weakly with increasing amplitude. The effect, however, is so small that the deviations can scarcely be seen in Figure 33.

Besides the well-known hexapole resonance at $\beta_z = 2/3$, more resonances can be found at $\beta_z = 1/2$, $\beta_z 2/5$, and $\beta_z = 1/3$, already anticipated from the study of the nature of the resonances. Figure 34 presents these resonances. The resonances also obey Equation 19, but with different C_2 constants. The resonances $\beta_z = 2/5$ and $\beta_z = 1/3$ are extremely weak and are scarcely noticeable. The hyperbolic increase of the amplitude with

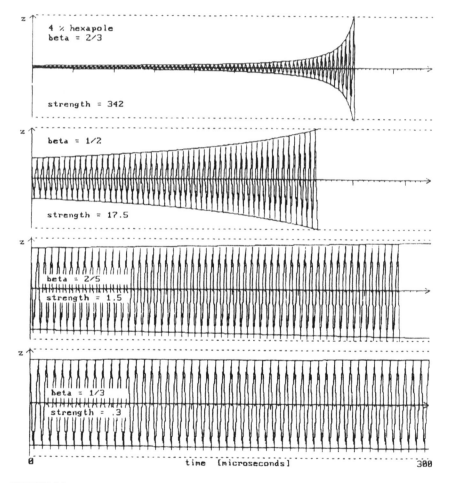

FIGURE 34

The relative strengths of the hexapole resonances $\beta_z = 2/3$, $\beta_z = 1/2$, $\beta_z = 2/5$, and $\beta_z = 1/3$. The latter two resonances are rather weak but still detectable. The resonance $\beta_z = 4/5$ was not found, although it had been anticipated to be expected from the investigation of the resonance's nature.

β_z = 2/3 results in a steep slope, and can be used with advantage for a very fast ejection of ions out of the trap. On the other hand, an ion that rests calmly in the center of the ion trap does not see the nonlinear resonance at all and will not leave the center. It must be pushed out of the center first in order to make it take part in the hyperbolic acceleration.

We know already that this pushing can be performed simply with a dipolar resonance because the dipole field is the only multipolar field that does not vanish in the center. Figure 35 demonstrates how the linear increase of the amplitude by dipolar resonance in a pure quadrupole field (top) merges into the hyperbolic increase when a hexapole is superimposed (bottom). The study of the hexapole nonlinear resonance is especially interesting because there is no end to the oscillation increase inside the ion trap, as observed in the case of even multipoles.

5. Octopole Field Superposition Arrests Its Own Resonances

It is known that the octopole quenches its own resonance.[2] The pseudoforces, which focus the ions to the center, increase more strongly than linearly (with positive octopoles), leading to a higher oscillation frequency with increasing oscillation amplitude. This shift in frequency destroys the resonance, and a beat is formed. Figure 21 demonstrates the frequency shift for different types of multipoles. Thus, it becomes impossible to study Equation 14 for the octopole field.

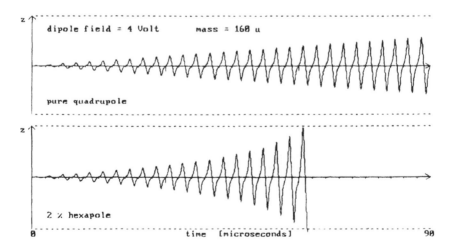

FIGURE 35
Combination of dipolar excitation (linear amplitude growth) and hexapole nonlinear resonance (hyperbolic growth). In the pure quadrupole field (top), only the linear increase of the amplitude is visible. In the hexapole trap (bottom), at first the ion is linearly pushed out of the center, then the hexapole resonance effect takes over and ejects the ion hyperbolically.

The result of the octopole superposition on its own z-type resonance at $\beta_z = 1/2$ is shown in Figure 36. In the upper part, we see the linear increase of the amplitude by resonant excitation in the pure quadrupole field. The conditions are chosen such that the exciting dipole frequency is exactly one fourth of the driving field frequency, identical to the octopole resonance condition $\beta_z = 1/2$. In the lower part, the same experiment is arranged in an octopole-superimposed trap. We see that the ion does not hit the end-cap. Thus, the acquisition of energy is arrested by the phase shift resulting from the nonlinearity.

The octopole superposition arrests its own z-type nonlinear resonance even if it is supported by a matching dipole resonance. It can be shown by full space simulations (not shown here) that the octopole sum resonance

$$\beta_r + \beta_z = 1 \tag{21}$$

is not arrested in the same way. For a positive octopole, the focusing forces in the z-direction increase more than linearly, but the focusing forces in the r-direction increase less than linearly. Thus, the frequency $\omega_z = \beta_z \Omega/2$ increases, and the frequency $\omega_r = \beta_r \Omega/2$ decreases. By Equation 21, there is a first-order compensation of these effects, and the resonance is maintained in spite of the shifting frequencies.

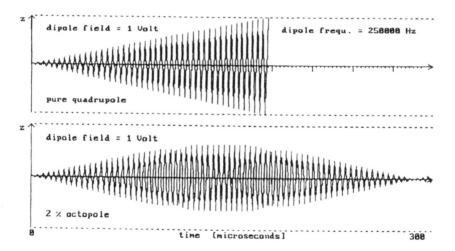

FIGURE 36
The octopole arrests its own z-type nonlinear resonance. The linear amplitude growth of an ion in the pure quadrupole field (top) by a matching dipolar excitation at one fourth of the RF drive frequency is stopped for the case of octopole field superposition, in spite of the fact that the excitation frequency used is the octopole z-resonance frequency. The resulting beat has a symmetric structure, although different from the beat structure at one third of the RF frequency shown in Figure 14.

6. Decapole Field Resonance Ejection

Figure 37 shows the ejection of ions by the four z-type decapole resonances $\beta_z = 2/3$, $\beta_z = 2/5$, $\beta_z = 1/2$, and $\beta_z = 1/3$. The envelope curves are differentially constructed by the equation

$$dz/dt = C_4 * z^4 \tag{22}$$

using different values for the constant C_4. The fit confirms that the amplitude increase is described by the third-order hyperbola

$$z(t) = C_4'/(z - C_4'')^3 \tag{23}$$

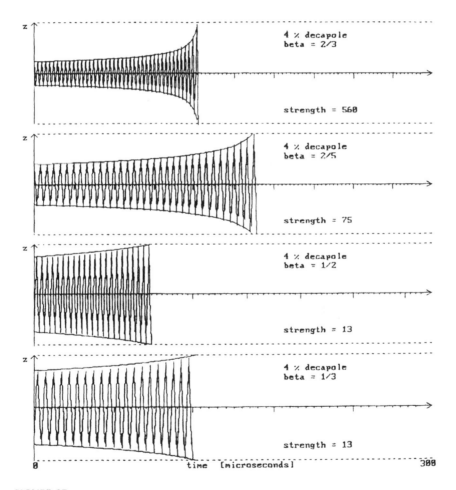

FIGURE 37

The decapole resonances $\beta_z = 2/3$, $\beta_z = 2/5$, $\beta_z = 1/2$, and $\beta_z = 1/3$, with the envelope curves drawn by $dz/dt = C_4 * z^4$.

As in the case of the hexapole resonance, there is a second-order shift of the frequency to smaller values with increasing amplitude, but the effect scarcely can be seen.

7. Nonlinear z-Direction Resonances of the Tetradecapole

Figure 38 presents the ion ejection by the three z-direction resonances $\beta_z = 2/3$, $\beta_z = 2/5$, and $\beta_z = 2/7$ of the multipole of order $n = 7$. Again, the envelope curves are constructed by the equation

$$dz/dt = C_6 * z^6 \tag{24}$$

using different values for C_6. The values for C_6 represent the relative strengths of the resonances, showing the lowest order resonance $\beta_z = 2/3$ to be the strongest. The amplitude function has the form of a fifth-order hyperbola

$$z(t) = C_6'/(z - C_6'')^5 \tag{25}$$

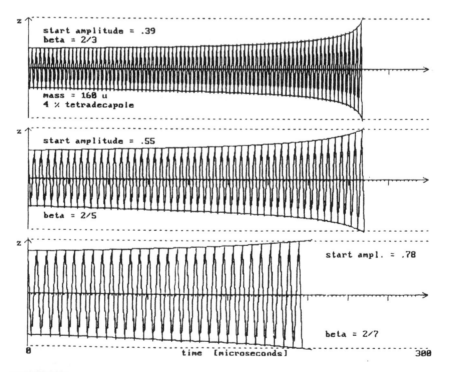

FIGURE 38
The tetradecapole resonances $\beta_z = 2/3$, $\beta_z = 2/5$, and $\beta_z = 2/7$, with the envelope curves drawn by $dz/dt = C_6 * z^n$.

8. Relative Strength of the Resonances

The resonance condition $\beta_z = 2/3$ can be found in three cases: the hexapole, the decapole, and the tetradecapole all show this resonance. In spite of the same β_z for this condition, the resonances for these multipoles differ markedly. As can be seen from Figures 34, 37, and 38, the resonances for the higher multipoles obey the amplitude increase conditions in Equations 22 and 24, respectively, for the decapole and tetradecapole fields. Thus, they can be distinguished clearly from the hexapole resonance.

The values of the constants C_{n-1} used to fit the different envelope curves of a multipole of given order n can be regarded as the relative strengths of the resonances. It turns out that the resonances of low order N are much stronger than those of higher orders, by factors of 6 to 8. Table 6 gives the relative constants found by fitting the simulated ejection processes, with P as the strength of the superimposed multipole.

9. Higher Multipole Resonances Are Not Truly Unstable

It should be noted that all the resonance conditions of higher multipoles are not conditions of complete instability, except those at the boundaries of the stability region. The even multipole fields (octopole, dodecapole, etc.) quench their own z-type resonances in first-order, the odd multipoles (hexapole, decapole, etc.) in second-order, both of them by the effect of shifting the secular frequency out of resonance as the secular oscillation amplitude increases. In the case of odd multipoles, the frequency ω_z decreases with increasing amplitude, regardless of the direction of superposition (addition or subtraction), but the effect is much smaller than that of even multipoles. For even multipoles, the frequency ω_z increases with amplitude for addition, and decreases for subtraction of the multipoles fields. The shift of the frequency can be studied in Figure 21 for the different types of multipole field superpositions.

For each of the z-type resonances of odd multipoles, a threshold exists for the starting oscillation amplitude, below which it is not possible to eject ions from an ion trap under stationary field conditions, even with

TABLE 6

Relative Strengths of the z-Type Resonances for Hexapole, Decapole, and Tetradecapole

$\beta_z =$	2/3	2/5	2/7
	$dz/dt =$	$dz/dt =$	$dz/dt =$
Hexapole	$0.855 * P * z^2$		
Decapole	$1.40 * P * z^4$	$0.185 * P * z^1$	
Tetradecapole	$0.555 * P * z^5$	$0.095 * P * z^n$	$0.015 * P * z^n$

Note: P designates the percentage of the multipole.

unlimited time. Studies of ion ejection from traps such as those shown above, can be performed only with starting amplitudes well above these thresholds. Studies of the equation of the amplitude increase by fitting the envelope curves need even higher starting amplitudes. The deviations between the envelope curves and the peak amplitudes seen in some of the simulations stem from this effect on resonance behavior of a slight change in secular frequency with amplitude.

F.   Multipole Influence on the Quadrupole Mass-Selective Instability Scan

In the above examples of ion ejection from the ion trap, we have not really investigated dynamic scanning conditions; ion behavior was studied only in stationary fields. In modern scanning methods by ion ejection, the parameters of the trapping field are changed such that ions of subsequent masses encounter a resonance condition, take up energy, and leave the trap through small holes in one of the end-caps. The resonance condition may be the instability at the boundary of the stability region, a resonant secular frequency excitation by a dipole field, or a nonlinear resonance condition.

1.   Introducing Scans into z-Axis Simulations

In the simulation method presented here, the behavior of an ion in a scan can be observed easily when we introduce a time-dependent change of the relevant field parameter. We can change either the amplitude V or the frequency Ω of the RF drive voltage, or the frequency ω_d of the dipole field caused by an additional AC voltage across the end-caps. The most common scanning methods use a change in the voltage V of the RF drive field, because a linear change in the RF voltage generates a linear mass scale. Such is not the case for the other parameters.

2.   Octopole Fields and the Mass-Selective Instability Scan

The resolution of the mass-selective instability scan is not sufficiently good when we use a truncated Paul ion trap. This poor mass resolution was found in earlier investigations,[5] and was made public a short time ago by Louris et al.[7] Figure 39 shows experimental results for three different ion trap designs.

When a Paul-type ion trap with ideal hyperbolic electrodes (characterized by $r^2 - 2z^2 = r_0^2$, $r^2 - 2z^2 = -2z_0^2$, and $r_0^2 / z_0^2 = 2$) is cut to finite size, the resulting truncated Paul ion trap contains weak higher multipole fields, the strongest of which is the octopole (see Table 3). The sign of this octopole field is opposite of that of the basic quadrupole field. By

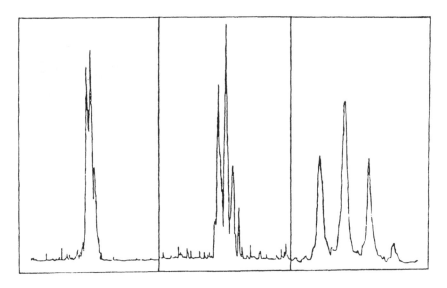

FIGURE 39

Mass resolution for three different types of truncated ion traps with modified hyperbolic angle. The hyperbolic angle can be expressed by the parameter θ, defined by the equations $r^2 - \theta z^2 = r_0^2$ for the ring electrode surface, $r^2 - \theta z^2 = -\theta z_0^2$ for the end-cap electrode surface, and $r_0^2/z_0^2 = \theta$. The scan shows the molecular ions of tetrachloroethene with masses 164, 166, 168, and 170 u, so the lines are two masses apart. The left-hand spectrum is taken with an ion trap with $\theta = 2.1$, corresponding to an octopole field of about -2%. The central spectrum was measured with the truncated Paul ion trap ($\theta = 2$), having about -0.15% octopole and -0.2 dodecapole field. The right-hand spectrum stems from an ion trap with $\theta = 1.9$, corresponding to an octopole field strength of about $+1.5\%$. The scan speed was identical in all three cases, but the scales are somewhat different.

contrast, the ion trap with stretched end-cap separation and the ion trap with modified hyperbolic angle have stronger octopole fields (see Table 2), and the sign of the octopole fields is equal to that of the basic quadrupolar field. It can be assumed that the negative octopole field is responsible for the poor mass resolution of the truncated Paul ion trap.

Figure 40 exhibits the simulated behavior of an ion of mass 185 u encountering the stability boundary during the scan from mass 170 u to mass 190 u, for ion traps having different amounts of superimposed octopole field. With positive octopole field, the ion takes up energy from the RF drive field as soon as its working point crosses the stability boundary, increases its secular oscillation amplitude exponentially, and is almost immediately ejected. With negative octopole fields (compared with the sign of the basic quadrupole field), the ion ejection is delayed.

What happens? With positive octopole field superposition, the pseudopotential forces increase more than linearly. When the amplitude of the secular frequency increases, the repelling pseudoforces cause the ion to oscillate faster; the secular frequency shifts to higher values. As

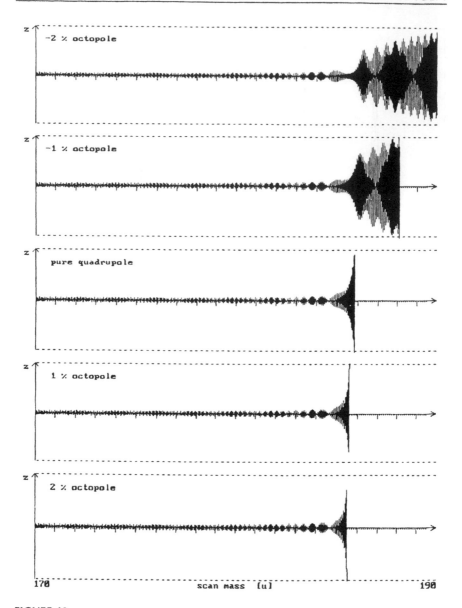

FIGURE 40

Simulated scans of mass 185 u in the scan range 170 to 190 u, for different octopole field superpositions. The octopole field strength ranges from –2% (top) to +2% (bottom). Negative octopole fields delay ion ejection.

soon as the progressing scan shifts the working point of the ion into the region of axial instability, the ion acquires energy and increases its amplitude. Thus, the frequency also increases, and the a_z, q_z working point, which is defined by the secular frequency, is pushed further into the axial instability region. The more the working point is pushed away from the stability boundary into the axial instability region, the stronger the energy uptake, and thus the ejection process is accelerated.

With negative octopole field superposition, the pseudopotential forces increase less than linearly. The secular frequency slows down as the secular amplitude increases. When the ion passes the stability boundary, it acquires energy, increases its amplitude, and the working point is immediately drawn back into the stability region. The energy uptake stops until the scan moves the working point again across the stability boundary. The ion thus can only take up energy while its working point stays in the vicinity of the stability boundary, oscillating a little around it.

The octopole field increases quadratically with the amplitude; therefore, the secular frequency will shift approximately with the square of the amplitude. The condition to hold the frequency constant during a linear scan (equal to that of the boundary) thus leads to an increase of the amplitude which is roughly proportional to the square root of the scan time.

However, the square root condition for constant $\omega/2$ frequency cannot be met by an ion exactly at the time of crossing the stability boundary. The ion must first acquire energy exponentially until the condition for constant $\omega/2$ frequency is met. This exponential acquisition of energy depends on the starting conditions and may take some time when the starting oscillation is extremely small. The working point, therefore, may be transported relatively far into the axial instability region, before the condition for constant $\omega/2$ frequency is reached; at this point, a kind of overshooting takes place, followed by a kind of hunting oscillations.

Figure 41 is a drawing that illustrates this square root condition for constant $\omega/2$ frequency and the oscillation amplitude envelopes of two ions with different starting oscillation amplitudes. A small starting amplitude delays the exponential increase of the amplitude, resulting in a much stronger overshooting effect and a stronger hunting oscillation.

Figure 42 demonstrates that both frequency and phase of the secular oscillation are held constant. The markers designate the exact frequency $\omega/2$ at the boundary $\beta_z = 1$.

Figure 43 shows the simulation of an ion in an ion trap with 1% octopole superposition of negative sign. The starting oscillation amplitude is varied from extremely small values (top) to large values (bottom). With extremely small starting amplitudes, the exponential increase of the amplitude takes more than the scan time of a full mass before it becomes visible; thereafter, the increase is very rapid, and the ion is ejected by the overshooting effect. With larger start amplitudes, the overshooting and the hunting oscillations become smaller, and a delay in ejection is observed.

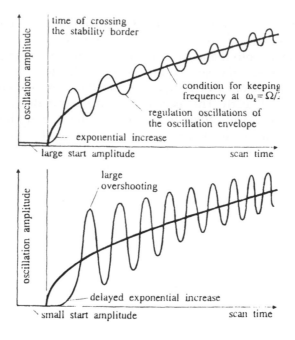

FIGURE 41

Qualitative theoretical curve for the increase of the amplitude with progressing scan for superimposed negative octopole fields. For small starting amplitudes, a longer period of exponential increase is observed, and the increase of the amplitude is overshooting the square root condition for constant $\Omega/2$ frequency. Two oscillation envelopes are shown for two ions with different starting amplitudes.

3. Superposition of Higher Even Multipoles

Superposition of other even multipoles exhibits a similar behavior to that of the octopole. For dodecapoles, the condition for constant $\omega/2$ frequency has the form of a fourth root.

Figure 44 gives two simulation examples for dodecapole fields of opposite signs.

4. Superposition of Odd Multipoles

In ion traps with superimposed hexapole fields, the secular frequency is not shifted to a first-order approximation by an increasing amplitude. The pseudoforce field increases in one z-direction more strongly than directly proportional; however, in the opposite direction the increase in the field is less than directly proportional, and by the same amount. Thus, the effect is compensated. In a higher-order examination, however, this is no longer true because the oscillation shifts out of the center of the trap. Thus, the compensation is not fully valid, and an ejection delay must occur similar to that for even multipoles. The effect here should be weaker

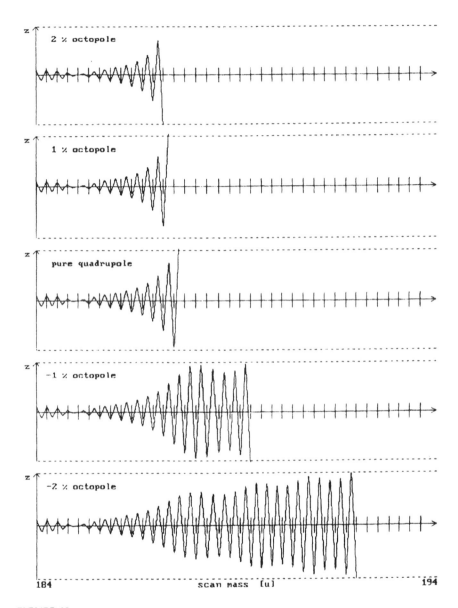

FIGURE 42

Frequency and phase remain strictly constant during the delay period, as can be expected by the delay mechanism. The scan rate is extremely fast (7.5 µs u $^{-1}$, with an RF drive frequency of 1 MHz). The markers designate exactly the $\Omega/2$ frequency at the boundary. We see that frequency and phase are precisely maintained. The scan reaches, for the ion of mass 185 u, from the stability boundary far into the instability area. The frequency is regulated by the increase of the amplitude which has a decreasing effect on the secular frequency. We note the regular oscillations.

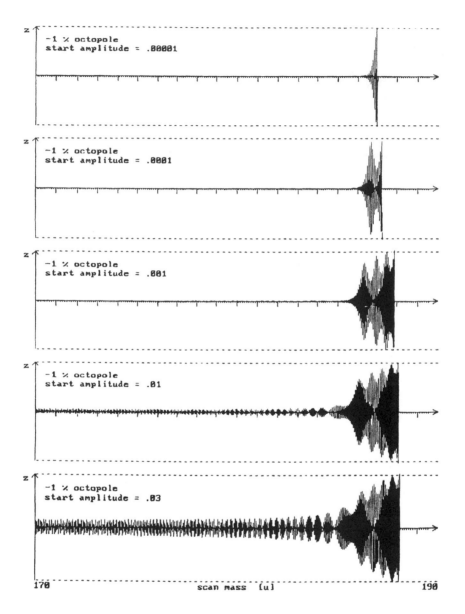

FIGURE 43

Simulated scans of mass 185 u in the scan range 170 to 190 u, for the same negative octopole field of –1%, but different starting amplitudes of the oscillating ion. The starting amplitude ranges from 0.00001 (top) to 0.03 (bottom). For the ion with the extremely small starting amplitude of 0.00001 (top), the exponential increase of the amplitude takes some time, and the observable increase takes place only when the ion is already far away from the stability boundary. Then, the ion is ejected by the overshooting effect, because the amplitude acceleration is strong so far from the boundary.

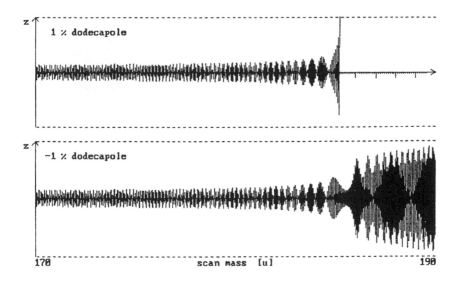

FIGURE 44
The superposition of dodecapole fields has, in principle, the same effect as the superposition of octopole fields. The curve representing the condition for constant frequency looks somewhat different.

than for negative even multipoles. The delay for odd multipoles is not dependent on the sign of the superimposed multipole field; thus a delay must occur with both signs.

Figures 45 and 46 demonstrate the effect for hexapoles and dodecapoles, respectively. The slight asymmetry is caused by the asymmetry of the start conditions; otherwise, the effect would be strongly symmetrical for fields of different sign.

5. Discussion

In these investigations, it was observed that considerable delays occurred in the ejection of ions from ion traps during scans across the stability boundary; these delays were observed whether negative higher multipole fields or odd multipole fields were superimposed on the basic quadrupole field. The delays are dependent on the starting oscillation amplitude with which the ions arrive at the boundary. This somewhat surprising effect was explained by a shift of the secular oscillation frequency with increasing oscillation amplitude; this shift keeps the working point of the ion at the boundary and allows only a controlled acquisition of energy.

The truncated Paul ion trap exhibits a superposition of weak negative octopole and dodecapole fields. The dependency of the ejection upon the starting conditions explains the poor mass resolution obtained with

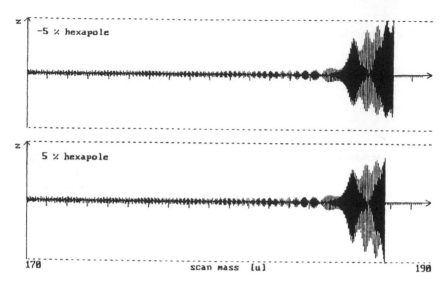

FIGURE 45

The ion ejection delay by superimposed hexapoles is independent of the sign of the hexapole.

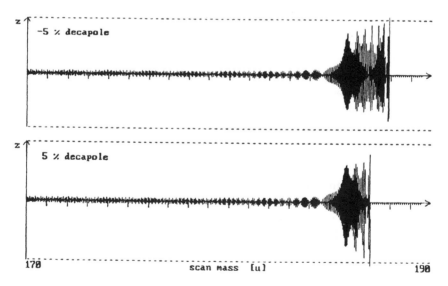

FIGURE 46

The ion ejection delay by decapoles.

the truncated Paul ion trap when the mass-selective instability scan is applied. This deteriorating effect on the mass resolution can be overcome by the artificial superposition of an positive octopole field of sufficient field strength. The stronger the superimposed octopole field, the sharper becomes the ejection by this scan method.

The positive octopole field can be obtained in several ways. Stretching the end-cap separation is one way, but rather strong multipole fields of higher-order are produced by this measure, giving rise to other effects such as fluctuations in the acceptance of external ions. An alternative strategy, which is advantageous compared with stretching the separation of the end-cap electrodes, is to introduce positive octopole fields with low multipole fields by modifying the hyperbolic angle of the trap.

The best way, however, will be an ion trap with specifically shaped electrode surfaces so that the trap contains only the desired amount of octopole field and no other fields. The mathematical and mechanical tools to produce such ion traps are now readily available.

G. Influence of Octopole Superposition on Other Resonances

We shall now investigate the effect of the octopole field on the resonant excitation of ions by a dipole field. The dipole field, in this case, is assumed to be an homogenous ideal dipole field, which is slightly different from the field which we can produce with an additional AC voltage applied across the end-caps.

1. How an Octopole Field Changes Dipolar Resonant Excitation

In Figure 47, we demonstrate the essential effect of even multipoles on ions that are excited resonantly by a dipolar high frequency field that matches their axial secular oscillation frequency. In the pure quadrupole field shown in the top of the figure, the amplitude of the secular oscillations increases linearly, reflecting the constant acquisition of energy from the dipole field. In fact, even the smallest dipole voltage will eject the ions from the trap eventually when the frequencies match (and the motion of the ions is not dampened by a cooling gas). No threshold voltage exists for ion ejection. In the bottom part of the figure, we see the ion excitation with exactly the same parameters, but now in an ion trap with a 2% octopole field. The difference is rather dramatic; the linear increase of the secular amplitude is now arrested, and a beat is formed with an aesthetically pleasing structure. Even with prolonged waiting, the ion is not ejected by the dipole voltage which is identical with that employed with the pure quadrupole in the top part of the figure.

The explanation for this behavior is simple and identical with the reasoning used above, in that the octopole field arrests its own nonlinear resonance. The ions take up energy from the dipole field as in the case of the pure quadrupole field, but the nonlinearity of the field now causes the frequency of the secular oscillation to be dependent on its own

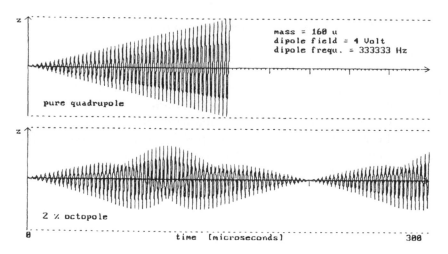

FIGURE 47

Dipole excitation of ions induced by a resonating frequency across the end-caps. The time scale of the figure is 300 μs, showing 100 secular oscillations. In the case of the pure quadrupole field (top), the excitation frequency increases the amplitude of the secular frequency; no threshold voltage exists for ion ejection (no damping gas present). With octopole superposition (bottom), the ion starts to resonate, just as in the case of the pure quadrupole field. Soon, however, the increasing amplitude shifts the secular frequency to higher values, the ion falls off resonance, and a wonderful beat structure is formed in the z vs. time diagram. Thus, in order for ions to be ejected from the trap, a distinct threshold exists for the excitation dipole voltage.

amplitude. With increasing amplitude, the frequency shifts, and falls off resonance with the exciting dipole frequency. The maximum amplitude is reached when the difference in phase equals 180°. Then the amplitude decreases again, and reaches zero when the phase difference equals 360°. In this figure, the ion starts exactly at z = 0 and with zero velocity. For other cases of non-zero minimum amplitude, the beat is only shifted along the time axis.

In order to produce an aesthetic beat structure, we have chosen the frequency of the AC dipole field to be exactly one third of the frequency of the RF drive field. The resulting beat structure is nicely symmetric along the time scale, with five maxima of the envelope changing from positive z values to negative ones, and back. For other simple ratios of the excitation frequency and the drive frequency, we obtain similarly symmetric structures, but the number of maxima on each side differs. When the ratio is not a simple fraction, the beat becomes asymmetric.

In all cases, the envelope of the beat structure is very characteristic. The enveloping curve exhibits a number of maxima which, in most cases, alternate in time from positive z values to negative z values and back. Successive maxima increase one by one, reach a maximum value, and decrease again.

When a dipole frequency is required to eject the ions from the ion trap, a voltage of much higher amplitude is needed compared with that for the pure quadrupole trap. This effect is displayed in Figure 48, which shows the application of four different voltage amplitudes. A distinct minimum voltage amplitude is now required in order to eject the ions by dipolar resonance. Only voltages above this threshold amplitude will eject ions, whereas in the pure quadrupole no threshold is present, and even the smallest voltage will finally eject resonantly excited ions.

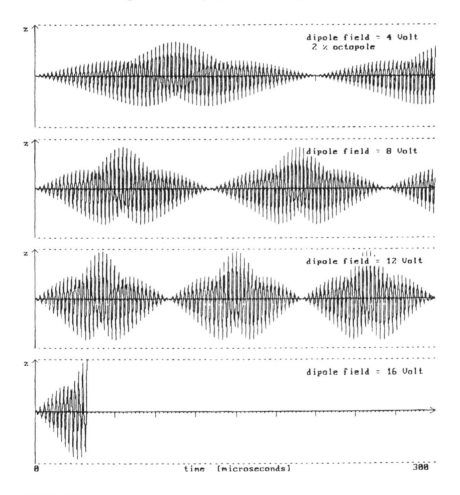

FIGURE 48

Voltage threshold for dipolar resonant ion ejection from superimposed octopole ion traps. The dipolar excitation voltage is increased from 4 V (top) to 8 V, 12 V, and finally 16 V (bottom). The threshold for ion ejection, induced by the octopole superposition, is about 15 V. There exists no threshold at all in pure quadrupole ion traps. The octopole superposition readily facilitates the pumping of kinetic energy into the ion cloud for fragmentation purposes.

This requirement is one of the most essential features of superimposed even multipole fields: ions inside the ion trap are much better protected against losses caused by resonances with small incidental perturbations of the fields, for example, by a small hum voltage across the end-caps. On the other hand, when the central maximum of the beat structure is used for ion ejection, the ejection becomes much sharper and more precise as in the case of pure quadrupoles. The flank of the enveloping curve of the central maximum is much steeper in slope than the slope of the linear increase in quadrupole traps. This effect confines the ejection of ions of the same m/z ratio but having slightly different starting parameters into a smaller time interval, and the mass resolution increases.

2. Improvement of the Dipolar Ejection Scan with Octopole Superposition

In Figure 49, we see the change in the resonant dipole field ion ejection scan when we go from a pure quadrupole trap to traps with increasing octopole field superposition. The scan, from 160 to 180 u in 600 µs with a scan rate of 30 µs/u, is generated by a linear increase of the RF voltage amplitude with time. The dipole voltage is kept the same in all four examples. It can be seen clearly how the ejection is sharpened by increasing strengths of the superimposed octopole. In the case of the pure quadrupole, an ion that already oscillates halfway to an end-cap needs another 16 secular oscillations to reach the point of ejection; for 1% octopole, 10 oscillations are needed; 7 oscillations for 2%; and only 4 oscillations for 3% octopole superposition.

We now investigate the effect of the dipole voltage on the ejection of ions during an RF voltage amplitude scan. Again, the dipole frequency equals one third of the RF drive field frequency. In Figure 50, it is seen that an AC voltage of 16 V is still below the threshold for ion ejection. Upon increasing the amplitude to 17 V, the ion is ejected from the trap; the ejection process takes place on the flank of the highest maximum of the beat structure, and the ion leaves through the lower end-cap. With 24 V, the ejection now takes place on the flank of the second highest maximum, and the ion leaves through the upper end-cap. In the trace at the foot of Figure 50, where ejection took place on the flank of the third highest maximum, the dipolar voltage was 32 V and the ion again left the trap by the lower end-cap.

This behavior is very characteristic. With increasing dipole voltage, the ejection alternates between the two end-caps. This effect also can be seen in experiments with real ion traps; with increasing AC voltage applied across the end-caps, the mass lines in a scan spectrum first reach a maximum and then decrease in height (and shift toward lower masses on the mass scale), then a new set of mass lines appears about half a mass

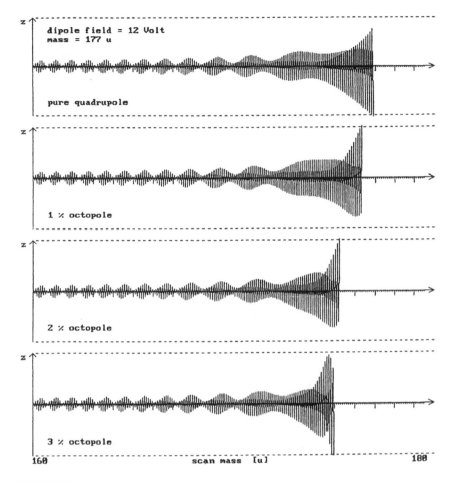

FIGURE 49

Ejection of ions by dipolar excitation, shown for increasing amounts of superimposed octopole. The ejection is "sharpened" by the octopole, essentially caused by the enhanced octopole beat structure.

lower, and these lines increase toward a second maximum as the dipole voltage is increased further. (The experiments were performed with a dipolar frequency equal to one third of the RF drive frequency.) This effect is ambivalent in its value. On the one hand, we can set conditions in which most of the ions are ejected unidirectionally through one end-cap; on the other hand, the effect requires careful adjustment of the exciting dipole voltage throughout the mass scan.

With other frequency ratios, different effects are encountered. When the dipolar frequency is one fourth of the RF drive frequency, unidirectional ejection is replaced by a 50% split of the ions among both end-caps.

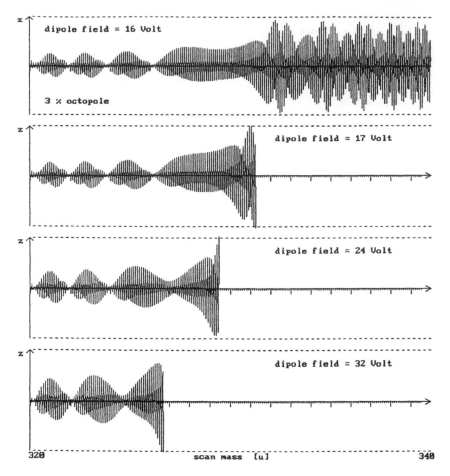

FIGURE 50

The effect of the dipolar excitation voltage on the ejection of ions from the superimposed octopole trap. With 16 V (top), the ions are not yet ejected. With 17 V, the ejection is performed by the highest maximum, the ions hit the lower endcap. With 24 V, the second highest maximum is active for ejection, and the ions are ejected from the upper end-cap. With 32 V, the third highest maximum is used, and the ions again leave through the lower end-cap. This alternating ejection through upper and lower end-cap with increasing dipolar voltage is typical for octopole superpositions.

3. Octopole Influence on the z-Type Resonance of the Hexapole Field

We examine here the effect of octopole superposition on the hexapole z-type resonance $\beta_z = 2/3$, first for the stationary case, and then for the nonlinear resonance ejection scan.

The ejection of ions by the hexapole z-type nonlinear resonance can be suppressed by an additional octopole field applied to the ion trap with 2% hexapole field superposition. Figure 51 demonstrates that a very weak 1% octopole field does not completely arrest the hexapole resonance ejection phenomenon, but a somewhat stronger (2% and more) octopole field is able to suppress ejection. The resulting beat structure in this case appears to be somewhat distorted, compared with that obtained for a pure octopole, but a closer investigation shows all of the features discussed above.

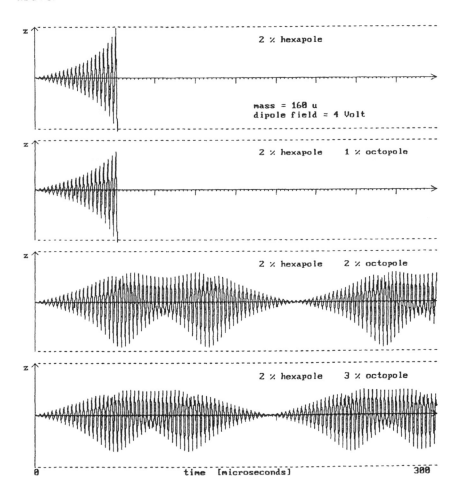

FIGURE 51

The octopole stops the combined 4 V resonant dipole excitation and 2% hexapole resonance ejection. The ejection (top) is only slightly asymmetrically distorted by 1% octopole (second from top). The 2% octopole stops the ejection, with an asymmetrical distortion of the beat (compare with Figure 15). When the octopole is strengthened to 4%, the beat frequency is increased, and the beat amplitude decreased.

Experiments with ion traps purposely designed for the superposition of both pure hexapole and octopole fields (and no other multipole fields) have yet to be carried out. Investigations of this type of ion trap by z-axis simulations look very promising.

As shown in Figure 52, superposition of a favorable mixture of hexapole and octopole fields generates unidirectional ejection independent of the exciting dipole voltage. The frequency of the dipole was selected to match the z-type nonlinear resonance $\beta_z = 2/3$ of the hexapole field superposition. The ejection is extremely sharp, indicating that a very good mass resolution may be possible. The structure of the beat envelope exhibits a single maximum, asymmetric with respect to $z = 0$. The dipole voltages required to eject the ions are very low, compared with those for the pure octopole superposition (compare to the voltages used in Figure 50). The beat structure during the scan, before the resonance is matched, shows only very weak amplitudes. This double resonance phenomenon (dipole excitation and nonlinear hexapole resonance, both at precisely the same frequency) appears to be extremely concise and effective. The secular frequency of the ion matches both the dipole resonance with its linear amplitude amplification and the hexapole nonlinear resonance with its hyperbolic growth of amplitude. The influence of the combined hexapole and octopole field makes the beat asymmetric, causing unidirectional ejection, which is not the case for pure hexapole or octopole field superpositions.

It should be noted here that the scan speed in the above experiments is extremely fast. The scan speed is chosen such that a full mass is scanned in ten oscillations only of the secular movement. Based on an RF drive frequency of 1 MHz, the scan speed is 1 u/30 μs, or 33,333 u s^{-1}.

4. Dipolar Excitation Before Nonlinear Resonance

As has been found experimentally, it is favorable for the ejection process to excite the secular ion oscillation before the ion matches the nonlinear resonance. This effect is demonstrated in Figure 53. The true double resonance case is not the most favorable. When the dipolar frequency matches exactly the nonlinear hexapole resonance at one third of the RF drive frequency, no ejection takes place with the chosen dipolar voltage of 1 V. When the dipolar frequency is reduced slightly, the ion is ejected. The decrease of the frequency can be chosen arbitrarily; a decrease of a few percent is sufficient. In these simulation experiments, the excitation can be much earlier than the ejection process, as shown in Figure 54. In practical ion trap work, however, collisions with the damping gas must be taken into account and they prohibit such long time intervals between excitation and final ejection.

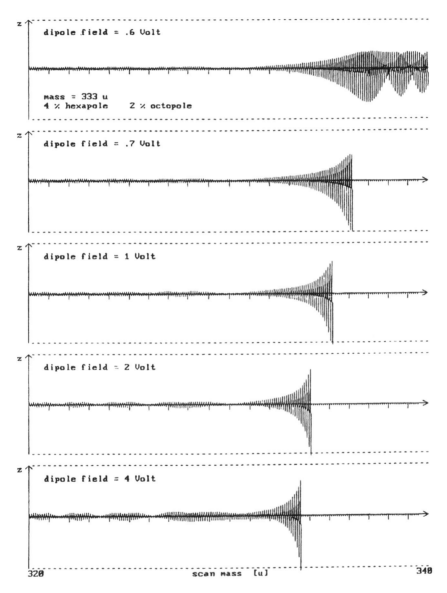

FIGURE 52

Ion ejection with combined superimposed hexapole and octopole fields. The ejection is unipolar for all dipolar excitation voltages, and the necessary voltages are extremely low. The voltage of 0.6 V does not eject the ion (top), but 0.7 V already is successful (second from top). The values of 2 and 4 V (below) result in sharp ejections, causing only very low beating oscillations before the point of ejection is reached (compare ejection voltages and beat structure with the octopole case of Figure 48).

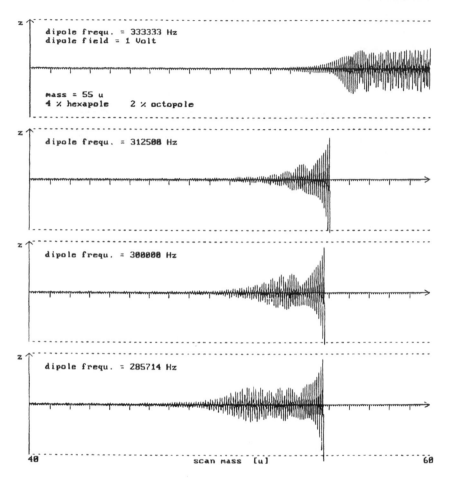

FIGURE 53

For ion ejection, dipolar excitation must be performed before the ion frequency matches the nonlinear resonance. In the case of matching dipole excitation and nonlinear resonance frequencies (top), the 1 V dipolar voltage is not sufficient to eject the ion. With smaller dipolar frequencies (lower scans), ejection takes place.

5. Influence of the Beat on the Mass Scan

The starting point of the beat and the beat frequency have a dramatic influence on the ejection process. Figure 55 shows how the scan start (start of the beat) influences the point of ejection on the mass scale. When a scan is started at the same mass of the mass scale, subsequent ion masses are not ejected in the sequence of their masses, but highly irregularly, caused by formation of the beat structure.

If, however, the scan start is shifted by one mass for each subsequent ion mass, the ejection becomes highly regular, as is seen in Figure 56.

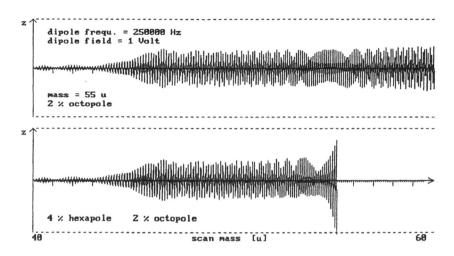

FIGURE 54

The dipolar excitation can be performed relatively long before the nonlinear resonance ejects the ion. The dipole frequency matches the nonlinear resonance of the octopole at one fourth the drive frequency without ejecting the ion (top). When a hexapole field of sufficient strength is added, the ejection takes place when the hexapole resonance is reached (bottom). This is a wonderful example for the power of the hexapole resonance.

These simulation experiments do not really reflect the true situation in real ion traps. In practical cases, the ions undergo collisions with the damping gas, smearing out the structure of the tiny beats formed long before the ion is ejected. Furthermore, there are Coulomb interactions among the many ions in a cloud. Ions of the same mass in the cloud may even somewhat synchronize their movement. Ions of different masses stir the residual cloud. A kind of threshold may exist for the beat formation.

6. Influence of Negative Octopole Fields on the Nonlinear Hexapole Resonance Ejection Scan

As already known from the mass-selective instability scan, negative octopoles have a deteriorating influence on the resolution of resonance scan methods. This situation is true also for the hexapole nonlinear resonance ejection scan. Figure 57 presents an investigation of the influence of negative and positive octopole field superpositions on the hexapole nonlinear resonance ejection scan.

Strong negative octopole field superposition causes a slow increase of the oscillation amplitudes by the resonance, and some hunting oscillations of the amplitude. As soon as resonance is reached by the scan, the amplitude of the secular oscillation begins to increase. The octopole field weakens the basic quadrupole field in the outer regions of the ion trap; therefore, the secular oscillation becomes slower and falls off resonance. The superimposed beat structure causes a decrease of the amplitude. The

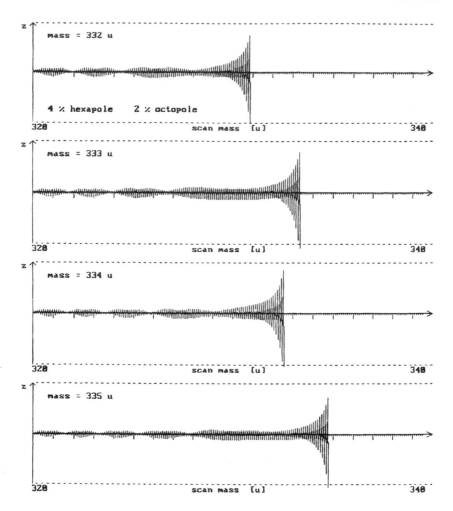

FIGURE 55
The beat structure causes a dependence of the ejection time on the scan start. Subsequent masses from 332 to 335 u are ejected in an irregular sequence, caused by different time distances from the common scan start at mass 320 u.

progression of the scan brings the secular frequency again closer to resonance, and the amplitude begins to increase once more. The amplitude growth, however, again shifts the secular frequency away from resonance. This circle creates the hunting oscillation of the amplitude.

With weak octopole superposition, the ejection of ions may even be sharpened because the ions can be held in the resonance condition for a longer time. Thus, the dipolar excitation voltage amplitude may be lowered. Strong positive octopole field superpositions suppress the ejection, as has been shown and as is illustrated in Figure 51.

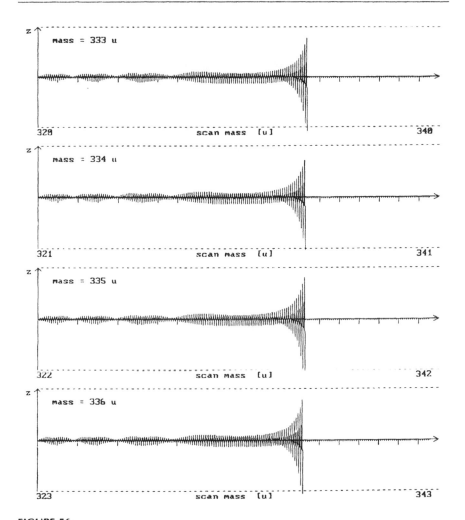

FIGURE 56

Dependence of ejection time on scan start and beat structure. When the scan start for subsequent ion masses is shifted by one mass each, the ejection for the ion masses becomes regular. In practical ion trap studies, this experiment cannot be performed. However, in practical ion trap work, other effects such as friction from gas collisions and friction by Coulomb repulsion between ions will diminish this dependence on the scan start conditions.

The dipolar excitation voltage used for the hexapole nonlinear resonance ejection scan produces a beat of the secular frequency before resonance is attained, as shown in Figure 58. The amplitude of this beat depends on this voltage and affects the width of the hunting oscillation, as is known from the mass-selective instability scan. In favored cases, the hunting oscillation can disappear completely.

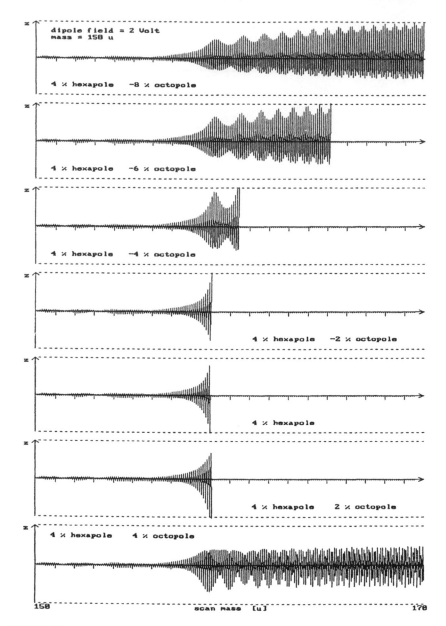

FIGURE 57

Influence of the octopole field superposition on the nonlinear hexapole resonance ejection scan. Negative octopole fields have the same deteriorating effect on the resolution already known from the mass-selective instability scan. Strong negative octopole field superpositions produce a hunting oscillation on the slowly increasing amplitude, because the growing amplitude shifts the secular frequency back from the resonance, whereas the scan shifts the secular frequency back to the point of resonance. Strong positive octopole fields suppress the ejection, as is known from Figure 48.

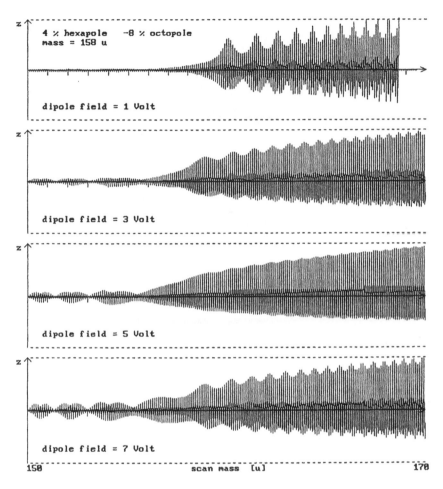

FIGURE 58
Influence of the dipolar excitation voltage on the width of the hunting oscillation caused by the negative octopole field during hexapole ejection scans. The voltage produces a beat structure before the resonance takes place, and the amplitude of the beat influences the hunting oscillations.

Not only does the dipolar voltage influence the hunting oscillation, the starting width of the secular amplitude must also have the same effect. The upper three tracks in Figure 59 confirm this, but the fourth track shows some strange behavior. The rather large starting amplitude forms a nice beat structure which quite naturally slows down as the secular frequency approaches resonance. The decreasing amplitude after the last beat before resonance (which is arrived at later because of the higher amplitude) suddenly increases the secular frequency in such a way that the frequency shifts behind the secular frequency. Contact with resonance is

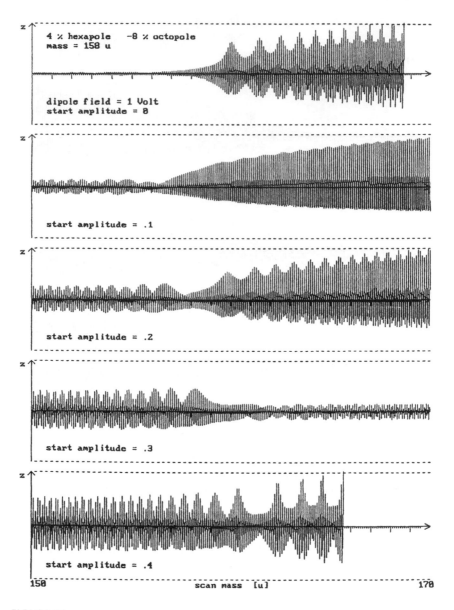

FIGURE 59

The influence of the start amplitudes on the hunting oscillation effect is somewhat confusing. The upper three tracks show the expected behavior. The fourth track, however, shows a surprising effect. The decreasing amplitude after a beat maximum shifts the secular frequency to a higher frequency value, thereby jumping over the resonance condition. The amplitude remains small after this event.

lost, the amplitude cannot recover to larger values, the beat structure becomes faster with the progressing scan, and the secular oscillations remain small. These findings confirm the model of the effect of negative octopole fields on ejection scans. Negative octopole superpositions must be avoided.

7. Full Space Simulation

The z-axis simulation is blind to all effects concerning the coupling of r- and z-oscillations, the sum or difference type resonances, and the r-type resonances. To study these effects, full space simulations must be performed. Simulations of this kind have been performed extensively for pure quadrupole fields. Simulations with nonlinear ion trap fields reveal the utility of the octopole sum resonance for ion ejection, and the exchange of energy between the r- and z-direction oscillations.[18] The simulations were performed with superposition of pure octopole fields and with the potential distribution of the modified angle ion trap. The relative strengths of the nonlinear resonances can be estimated in such simulations using appropriate methods.

VI. EXPERIMENTAL RESULTS WITH A SPECIAL NONLINEAR ION TRAP

The modified hyperbolic angle ion trap,[15] operated at an RF drive frequency of 1 MHz, was used to demonstrate the fast mass-selective ejection of ions with the hexapole resonance $\beta_z = 2/3$ which appears at 333 kHz ($\beta_z = 0.6$) in the "RF-only" mode of operation.

A. Fast Ejection Scan by Nonlinear Resonances

A pre-excitation of the ions in the z-direction by a dipolar frequency of 333 kHz ($\beta_z = 2/3$, equal to one third of the RF drive frequency of 1 MHz) was used to resonate the ions out of the center before they see the nonlinear resonance condition. This dipolar frequency was fixed in phase to the 1 MHz RF drive frequency, the phase being adjusted for optimum ion ejection.

The cloud of ions of a distinct mass encountering the nonlinear resonance increases its secular oscillation amplitude in a hyperbolic way, and after a few subsequent oscillatory swings, leaves the trap through small holes in one of the end-caps. The ions exiting thus form short pulses of roughly 100 ns in duration, which are about 3 µs apart, corresponding to the secular oscillation frequency of the ion cloud. The data acquisition and measurement electronics were equipped with a fast preamplifier connected to the last dynode of the secondary electron multiplier, a sample-and-hold device, and an analog-to-digital converter operating in exactly this period of 3 μs. The phase of the measurement periods was locked to the dipolar frequency.

The ion trap was operated with normal air as the damping gas, at a pressure of about 20 mPa. The ionization periods could be varied between 5 μs and 20 ms; the subsequent damping periods amounted to 5 ms in most experiments. The scan was performed by ramping the RF drive voltage amplitude. By a variation of the scan rate between 7 and 30 secular oscillations per mass unit, the optimum speed for optimum mass resolution was found to be about 12 oscillations per mass unit. (The fastest scan rate for a so-called unit mass resolution with 10% valley turned out to be 8 oscillations per mass unit, i.e., 24 μs per mass unit, or about 40,000 u s⁻¹.)

Figure 60 shows a selected section of a full spectrum resulting from a single scan, taken with the selected scan speed of 12 oscillations per mass. The section of the spectrum with its 45 mass units was measured in exactly 1.62 ms. It can be seen that about 70% of the ions are ejected in three

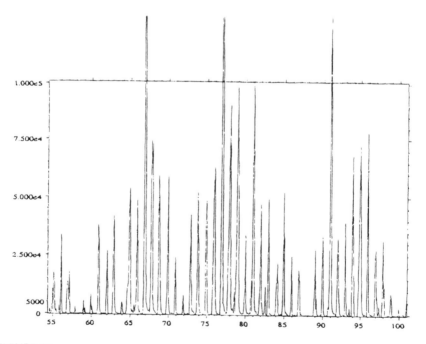

FIGURE 60

Section of a single full spectrum acquisition. Spectrum measurements are taken every 3 μs, synchronized with the ejection of ion pulses from the trap. The scan rate is adjusted to 12 measurements per atomic mass unit, thus the shown segment with 45 mass units is measured in exactly 1.62 ms. A full spectrum from 50 to 500 u is taken in 16.2 ms.

secular oscillations only, 95% are ejected in five oscillations, and roughly 100% leave in seven oscillations. The five 3-μs intervals until the next mass remain free when the ion mass is not severely overcrowded. Thus, the mass resolution $R = m/\Delta m$ equals about 5 m. The mass resolution proved to be constant throughout the entire mass range up to mass 500 u.

Stray single ions, however, leave the trap between the ion mass peaks. These background ions are more frequent the greater the intensity of the surrounding mass peaks. The stray ions seem to stem from those intense mass peaks brought out of synchronous oscillation with the other ions by collisions with the damping gas molecules. When several spectra are summed, the stray ions begin to fill the valleys between the peaks, as can be seen in Figure 61.

Figure 61 shows the addition of 50 scans of the mass spectrum. Compared with a single spectrum measurement, the mass resolution has suffered slightly by the fact that either the electronic scan with its 16-bit precision for control of the RF drive voltage was not exactly reproducible, or that instead the ion ejection events were not exactly reproducible. However, the effect is scarcely noticeable and the overall resolution is still very satisfying. The line widths have increased by about 20%. The stray ions between the mass peaks have summed to produce smooth webs, with the result that the baseline between the mass peaks is only visible in areas of small peaks.

Figure 62 presents a section of a mass spectrum from a higher mass range spectrum. In fact, a continuation of the spectrum of Figure 61 is shown here. This section of the 50 added mass spectra shows the pentamer group from the bleeding in of dimethylsilicone, which was used as an inlet membrane. The second peak in each of the three groups consists of about ten ions each. These ten ions are ejected statistically during the measurement of the 50 subsequent spectra, and they sum correctly to construct a mass peak. This figure confirms that single ions are ejected at the correct places on the mass scale, at least they do when they appear after peaks of higher intensity.

We observed, however, that the first mass peak within a group of peaks in a spectrum appeared to be somewhat distorted. The cooled ion cloud in the center may have condensed into some near-distance order in nonejecting time periods of the spectrum, and seemed to require some energy-consuming "stirring" by the ions of the leading peak in a group before correct ejection of subsequent mass peaks could take place.

B. Ion Isolation Support by Nonlinear Resonances

We have seen that nonlinear resonances can be used to sharply eject ions of a single mass from the ion cloud inside the trap. In the isolation procedure for ions of a selected mass, it is often difficult to eliminate fully

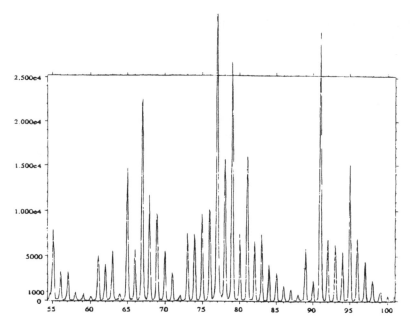

FIGURE 61

Section of a full spectrum with 50 added subsequent spectrum acquisitions. The mass resolution is only weakly deteriorated. The full spectrum with 50 acquisitions took 2.2 s, including 20 ms ionization time each.

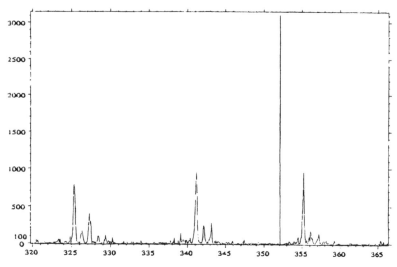

FIGURE 62

Section of the same spectrum as in Figure 61, but in the higher mass range. The smaller peaks (for example, 342 and 343 u) consist of roughly 10 ions only, measured in 50 spectrum acquisitions, showing that single ions can be ejected at correct times so that they add up to spectrum peaks.

the ions of the neighboring mass at the high mass side. Here, the sharp ejection by nonlinear resonances can be particularly useful. The RF drive voltage is adjusted in such a way that the neighboring ions match the nonlinear resonance at $\beta_r + \beta_z = 1$. When the high mass range is cleaned out by a frequency sweep of the dipole excitation voltage, the neighboring mass peak disappears "magically" by remote resonance, causing a small oscillation beat of the ion sufficient to start a nonlinear resonance. The lower mass range may be cleaned out as usual by a normal scan.

Figure 63 shows the original mass spectrum at the left-hand side. At the right-hand side is shown the result of the ion isolation process for the small ion peak of mass 78 u, which is enclosed between the much bigger peaks of masses 77 and 79 u. About 40% of the original ions of mass 78 u remain stored in the trap by this isolation process. The isolation procedure also works when the ion trap is overloaded with about 20 times the amount of ions. Thus, isolated ions of one mass can be stored in the ion trap with number densities corresponding to 5- to 10-fold those that yielded the original spectrum.

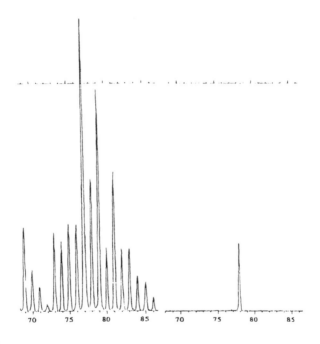

FIGURE 63

Spectrum segments before and after the isolation of ion species. At the right-hand side, we see a spectrum of background air in the laboratory with strong ion peaks for masses 77 and 79 u, at the left-hand side, a spectrum after isolation of mass 78 u ions. About 40% of the ions could be saved during the isolation process. A nonlinear resonance was used to clean the trap from ions of mass 79 u.

Figure 64 gives the result of the same isolation procedure for an ion of higher mass. The selected ion of mass 192 u was situated between the peaks of much higher intensity of masses 191 and 193 u. Again, a yield of about 40% was achieved.

C.   Storage of Ions Over Long Periods of Time

The modified hyperbolic angle ion trap can store ions without measurable losses over long periods of time. Long storage times are possible even in the presence of medium-molecular weight damping gases such as air. The superimposed octopole field reduces all losses by unwanted disturbances caused by noise or hum on the RF drive voltage, or by nonlinear resonances caused by field faults. Furthermore, the modified-angle ion trap has apertures between the electrodes which are much closer than with the stretched separation traps.

In Figure 65, the results of several ion storage experiments have been assembled. A steady flow of air with a low concentration of toluene was fed continuously to the mass spectrometer over more than a day from a Tedlar bag. The ions of m/z 91 were produced by EI within the ion trap, then isolated and stored, in preprogrammed experiments over increasing numbers of hours. The ions were not perfectly isolated; a small amount of m/z 92 ions were kept in order to measure the mass resolution.

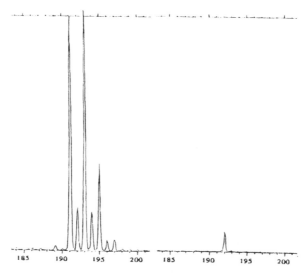

FIGURE 64
The isolation has the same yield at higher masses.

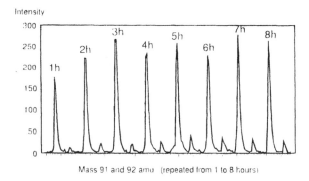

Mass 91 and 92 amu (repeated from 1 to 8 hours)

FIGURE 65
In the modified angle ion trap, ions can be stored over long period of times, even with air
as damping gas. Repeated experiments to store isolated ions were performed automatically,
with increasing storage times from 1 to 8 h. The subsequent single acquisition measurements show no trend of ion losses. The fluctuations agree with those of single spectrum acquisitions.

The results of successive experiments with storage periods from 1 to
8 h were assembled in the same figure (using spreadsheet graphics). The
mass peaks shown here are single-acquisition spectra, which show some
fluctuations in intensity. Within the precision given by these fluctuations,
no measurable decrease of the number of stored ions was detected.

The reproducibility of the ion ejection time along the mass scale in
these experiments is shown in Figure 66, in which the profiles of the ion
signals are projected on top of one another. For the projection, the count
of the measurements from the scan start was used for relative adjustment.
It is seen that all the ions leave the trap with a time precision of ± 3 μs,
measured from the beginning of the scan. The figure also gives an im-
pression of the mass resolution achievable with a scan rate of 28,000 u/s.

D. Improvement of Fragmentation by Octopole Superposition

It should be mentioned here that fragmentation of ions within the
ion trap is plausible with octopole superposition. Only the octopole-in-
duced effect of a relatively high threshold voltage for the ion ejection by
dipolar resonant excitation permits an adequate acquisition rate of ion
kinetic energy to cause fragmentation subsequently in collisions with molecules without touching the end-caps.

The modified angle ion trap device is especially well suited for such
fragmentation because it can be operated with air and other damping
gases of medium molecular weight. The higher mass of the damping gas
molecules, compared with the usual helium damping gas, increases the
probability of ion fragmentation processes subsequent to ion/molecule
collisions.

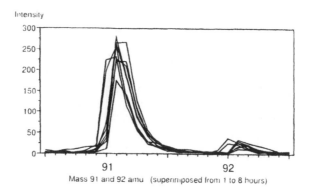

Mass 91 and 92 amu (superimposed from 1 to 8 hours)

FIGURE 66
The resulting peaks of the long-time storage experiment are superimposed to show the re-producibility of the mass scale, even after long periods of time.

All daughter ions generated by collisional fragmentation can be stored inside the ion trap, as can be seen from Figure 67, provided that the daughter ions do not see a black canyon effect, as discussed above.

VII. POTENTIAL DISTRIBUTION IN NONLINEAR TRAPS

When investigating the properties of a nonlinear ion trap with a given electrode structure, it is very important to know the potential generated by the these electrodes. From the potential, valuable information on the characteristics of the motion of the ions can be deduced, for example, the pseudopotential well depths, the secular frequencies, the various multipole contributions, etc. In principle, the calculation of the potential $\Phi(x, y, z)$ is straightforward.

A. Field Expansion in Multipole Terms

The potential is a solution of the Laplace equation

$$\Delta\Phi = 0 \qquad (26)$$

with the appropriate boundary conditions. These boundary conditions are given by the potentials applied to the electrode structure, and by the geometrical shape of the electrode structure itself. However, the electrodes of an ion trap, in general, do not completely enclose the trapping volume. Therefore, the region outside the trap also influences the potential inside the trap. It is useful to expand the desired potential $\Phi(x, y, z)$ in a series of orthogonal functions, each of which is a solution of Equation 26. The set of orthogonal functions is selected most conveniently so that the

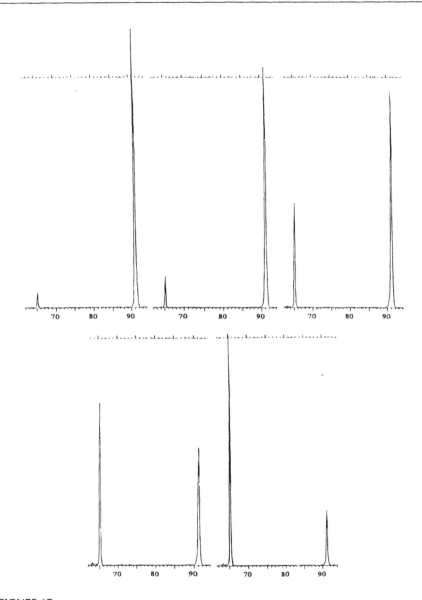

FIGURE 67
The isolated ions of mass 91 u with some spontaneous decay to mass 65 u (left-hand spectrum) are partially fragmented to show the high yield of daughter ions (right-hand side). The yield of about 100% can be obtained only in the absence of black canyon effects.

symmetry of these functions resembles the symmetry of the problem. For problems with cylindrical symmetry, Bessel functions are well suited, whereas for problems with both spherical and rotational symmetries, Legendre polynomials are preferred.[24]

If P_n is the Legendre polynomial of order n, then the potential expansion is given by

$$\Phi(\rho, \theta, \varphi) = \Phi_0 \sum_{n=0}^{\infty} A_n \frac{\rho^n}{r_0^n} P_n(\cos \theta) \tag{27}$$

where r_0 is some scaling length; ρ, θ, and φ are the spherical coordinates; A_n are weight factors for the different terms, and Φ_0 is a position-independent quantity which, in RF traps, depends on time. The various terms (index n) of the sum in Equation 27 are called the multipole components of the potential. Due to the assumed rotational symmetry, the potential $\Phi(\rho, \theta, \varphi)$ does not depend on the angle φ.

The $n = 0$ term in Equation 27 gives a position-independent contribution, but because only potential differences have any effect in reality, we may discard this term. In general, the $n = 1$ term (dipole) is not present in the storage field of ion traps, so we also neglect it. The $n = 2$ term is the quadrupole term, and the higher terms correspond to hexapoles ($n = 3$), octopoles ($n = 4$), decapoles ($n = 5$), dodecapoles ($n = 6$), etc. Due to the orthogonality of the Legendre polynomials $P_n(\cos \theta)$, the weight coefficients A_n can be determined unequivocably from the boundary conditions. They are real properties of the field. It is normal to write the multipole components of Equation 27 in cylindrical coordinates ($r = (x^2 + y^2)^{1/2}$, z) with the abbreviation $\rho^2 = z^2 + r^2$ as:

$$P_2(\cos \theta) = \frac{(2z^2 - r^2)}{2\rho^2} \tag{28}$$

$$P_3(\cos \theta) = \frac{(2z^3 - 3zr^2)}{2\rho^3} \tag{29}$$

$$P_4(\cos \theta) = \frac{(8z^4 - 24z^2 r^2 + 3r^4)}{8\rho^4} \tag{30}$$

$$P_5(\cos \theta) = \frac{(8z^5 - 40z^3 r^2 + 15zr^4)}{8\rho^5} \tag{31}$$

$$P_6(\cos \theta) = \frac{(16z^6 - 120z^4 r^2 + 90r^4 z^2 - 5r^6)}{16\rho^6} \tag{32}$$

B. Multipole Components of Different Ion Traps

In this section the weight coefficients A_n of the multipole components are calculated for three types of ion traps, namely:

1. The truncated Paul ion trap (Figure 10) with finite electrode size
2. The modified hyperbolic angle ion trap (Figure 9b)
3. The stretched end-cap separation ion trap (Figure 9a)

It is assumed that an RF potential V is applied to the ring electrode, and $-V$ to both end-cap electrodes. The calculation is performed by approximating the electrode surfaces of the three types of traps with the equipotential surfaces of the potential given by Equation 27. The expansion is taken to $n = 80$. All three types of traps have both rotational and planar symmetry, so the coefficients A_n for uneven n vanish, and the results are presented for the quadrupole ($n = 2$), octopole ($n = 4$), and dodecapole ($n = 6$) coefficients.

1. Ideal Paul Trap and Truncated Paul Trap

The electrodes of a Paul trap have the form of the equipotential surfaces of a pure quadrupole potential (Equation 28). These surfaces are hyperboloids of revolution which the surface of the ring electrode is given by

$$r = \pm\sqrt{2z^2 + r_0^2} \tag{33}$$

and the surfaces of the end-caps are given by

$$z = \pm\sqrt{r^2/2 + z_0^2} \tag{34}$$

Here, r_0 and z_0 are the inscribed radius of the ring electrode and half the separation of the end-cap electrodes, respectively, in the points nearest to the trap center. These surfaces have a common asymptotic cone with an angle of $\tan \alpha = 1/\sqrt{2}$ ($\alpha = 35.26°$) to the radial plane.

A Paul quadrupole ion trap can be built for any values r_0 and z_0, but all Paul traps have the same angle α between the asymptotic cone and the radial plane[13] (see also Figure 5). However, in a real Paul trap there are unavoidable deviations from the equations (Equations 33 and 34) that describe the surfaces. They are due, for example, to holes in the electrodes which are used for ion or electron introduction, ion ejection, etc.[25] Other field faults result from the finite size of the electrodes. The corresponding multipole components have been calculated[26] for a truncated Paul ion trap whose electrodes are cut off in a distance ρ_0 from the trap center (see Figure 10). The results of the calculation are given in Table 3.

The coefficients A_n for the uneven multipoles (hexapole, decapole, etc.) are zero due to the assumed plane symmetry of the electrode structure, and the other coefficients decrease with increasing ρ_0. For the quite common value of $\rho_0 = 2r_0$, the relative coefficients A_4/A_2 and A_6/A_2 are of the order of some tenths of a percent.

2.   Modified Hyperbolic Angle Ion Trap

The surfaces of the modified angle ion trap are hyperboloids of revolution. Compared to the Paul quadrupole trap, the angle α of the asymptotic cone has been increased. The equations of the surfaces are

$$r = \pm\sqrt{\Theta z^2 + r_0^2} \qquad (35)$$

for the ring and

$$z = \pm\sqrt{r^2/\Theta + z_0^2} \qquad (36)$$

for the end-caps. The angle α is related to the quantity θ by $\tan \alpha = 1/\sqrt{\theta}$. A favorable value for θ is 1.9 such that the corresponding angle is $\alpha = 35.96°$. It is possible to build a modified hyperbolic angle ion trap with any values of r_0 and z_0, but a particularly favored relation is $(r_0/z_0)^2 = \theta$. The weights of the multipole components for an ion trap with a cutoff length $\rho_0 = 2$ and shape parameter $\theta = 1.9$ are given in Table 2.

The multipole components of this trap are shown in Figure 68 as a function of the cutoff distance ρ_0. Compared to the truncated Paul trap with the same ρ_0, the octopole is much larger and has opposite sign. The sign of the octopole is equal to the sign of the quadrupole field; the octopole field is added. The dodecapole has about the same magnitude as in the truncated Paul trap, but with reversed sign (again, the dodecapole field now has the same sign as the quadrupole field). This arrangement shows that the octopole component of the potential of the modified angle trap is significantly higher than that of the truncated Paul traps due to the change in angle, and not to the finite size of its electrodes.

Figure 69 displays the dependence of the multipole components on the value θ (defined by $\tan \alpha = 1/\sqrt{\theta}$). By changing the angle, even multipole superpositions of any sign and of higher magnitudes than in use so far can be generated.

3.   Stretched End-Cap Separation Ion Trap

Another type of nonlinear ion trap can be produced by stretching the end-cap separation in the z-direction by a certain amount, without changing the shape of the electrodes.[7] Commercially available ion traps are stretched by $\Delta z = 0.108 * z_0$. The surfaces of the stretched separation ion trap are given by

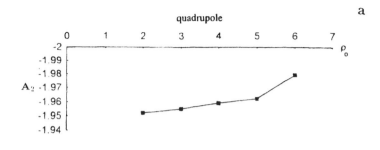

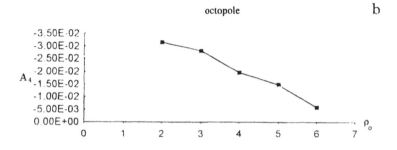

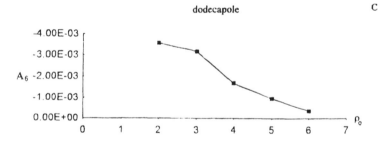

FIGURE 68

Multipole field strengths as a function of the truncation parameter ρ_0 for the modified angle ion trap with shape parameter $\Theta = 1.9$.

$$r = \pm\sqrt{2z^2 + r_0^2} \tag{37}$$

for the ring and

$$z = \pm\left[\sqrt{r^2/2 + z_0^2} + \Delta z\right] \tag{38}$$

for the end-caps. The surfaces are hyperboloids of revolution, but the ring and end-cap electrodes no longer have common asymptotes. By shifting the end-caps away from the ring, the slit between the ring and end-cap

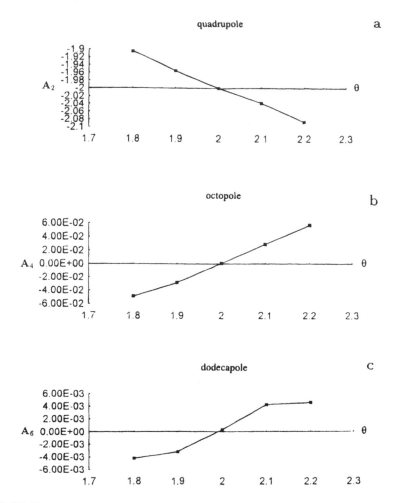

FIGURE 69

Multipole field strengths as function of the shape parameter Θ, for the modified angle ion trap with truncation $\rho_0 = 3$.

electrodes is larger than for a truncated Paul or modified angle trap of the same r_0, z_0, and ρ_0. The multipole components of the stretched-separation trap are shown in Figure 70 as a function of Δz.

Table 2 presents some weights of multipole components for the case $\rho_0 = 2$ and $\Delta z = 0.108 * z_0$. If we compare the modified angle ion trap with $\theta = 1.9$ and the stretched separation ion trap with $\Delta z = 0.11 * z_0$, we find that the signs of the octopole and dodecapoles are the same. The magnitude of the octopole is also about the same, whereas the magnitude of the dodecapole is larger for the stretched separation trap than for the modified angle trap. The quadrupole contribution is larger for the modified angle trap than for the stretched-separation trap.

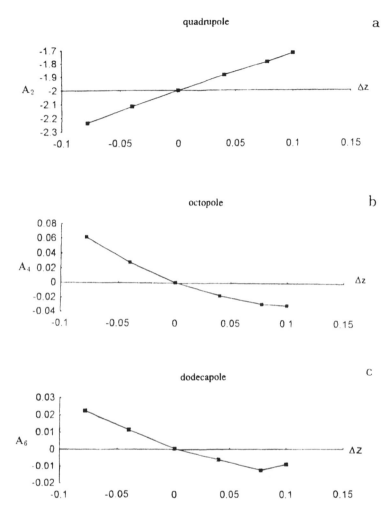

FIGURE 70
Multipole field strengths for the stretched end-cap ion trap as a function of the stretching distance Δz. The truncation parameter is $\rho_0 = 3$.

C. Variational Calculation of the Modified Angle Ion Trap Potential

In the previous section, the potential of some ion traps was described in terms of coefficients of the multipole expansion (Equation 27). This section reviews another method to determine the potential; this method yields an analytical expression for the potential of the modified angle trap.

The Laplace differential equation with the curved boundary conditions (Equations 35 and 36) for the modified angle ion trap are usually difficult to solve. The variational calculus method transforms the problem connected with the boundary into the problem of searching for an extremum of the variation about a functional, and hence generates an approximated finite analytical expression for the solution of the differential equation. It can be proved that the function

$$J[u(x, y, z)] = \iiint_\Omega \left\{ \left(\frac{\partial u}{\partial x} \right)^2 + \left(\frac{\partial u}{\partial y} \right)^2 + \left(\frac{\partial u}{\partial z} \right)^2 \right\} dx \, dy \, dz \qquad (39)$$

corresponds to an Euler equation which equals the Laplace equation (26). Here, Ω denotes the volume in which the potential $u(x,y,z)$ is defined, and on whose boundary the boundary conditions are given. Ritz[27] has suggested the following solution. Let $\chi_1(x,y,z), \ldots, \chi_n(x,y,z)$ be a complete set of functions which satisfies the given boundary condition. Its linear combination u_n with the coefficients $a_1, \ldots, a_n$,

$$u_n = \sum_{k=1}^{n} a_k \chi_k (x, y, z) \qquad (40)$$

can be substituted into Equation 39 and integrated. The function J is now a function of the variables $a_1, \ldots, a_n$, which can be obtained by finding the extremum of J. As shown by Friedrichs,[28] u_n from Equation 40 converges to an exact solution $u(x,y,z)$ of the Laplace equation with the boundary conditions given for $n \to \infty$. In general, the Ritz method is very efficient. For an appropriate choice of χ_k, only a small number n ($n = 1,2$) is necessary to reach a sufficient precision for the calculation.

Using the Ritz method, an analytical expression for the potential of the modified angle trap has been derived:[29]

$$u(r, z) = \Phi_0 \left(\frac{r^2 - \Theta z^2}{r_0^2} \right)$$

$$+ c(\Theta) * (2 - \Theta) * \Phi_0 * \frac{r_0^4 - (r^2 - \Theta z^2)^2}{r_0^4} \qquad (41)$$

$$* \exp\left(-\frac{a(\Theta) r^2 + b(\Theta) \Theta z^2}{r_0^2} \right)$$

Here, θ is related to the angle of the asymptotic cone α by $\theta = 1/\tan^2\alpha$. The parameters a, b, and c depend on θ; their dependence is shown in Figure 71. For comparison, the potential distribution for a pure quadrupole ion trap with $\theta = 2$ and a nonlinear ion trap with $\theta = 1.8$ are shown in Figure 72.

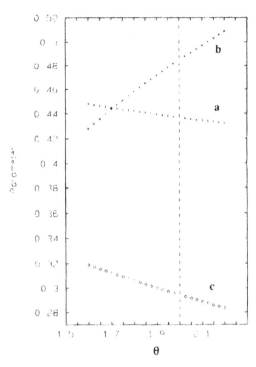

FIGURE 71
Modified angle ion trap:
parameters a, b, and c (see text)
as a function of the angle
parameter Θ.

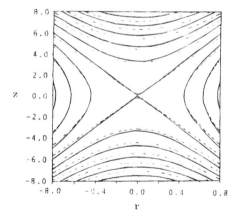

FIGURE 72
Comparison of the potential
distributions between ideal quadrupole
trap (dashed lines) and the modified
angle ion trap with Θ = 1.8 (solid lines).

VIII. THEORY OF NONLINEAR RESONANCE

A. Introduction

In Section VII, the potential of a nonlinear ion trap was discussed, and the multipole components were given for some selected ion trap designs. In this chapter, we address the following question: Can we understand theoretically the influence which these multipole contributions have on the motion of the ions?

The calculations in the foregoing section have shown that the multipole components are weak compared to the quadrupole component. Therefore, one might be tempted to believe that weak multipole components lead to only slight changes in the motion of the trapped ions. This belief seems to be well founded in general, because ion traps with a far different design from those discussed in the previous chapter have been used successfully to store ions. However, the principle that small changes have small consequences is not always true. The famous problem of vanishing denominators shows that for some special conditions, small, nonlinear components of the field induce large changes in the motion of the ions. This phenomenon is called nonlinear resonance. At first, such resonances were discovered in problems with time-independent fields.[30] Nonlinear resonances in time-independent fields have gained in interest during the study of the stability of trajectories in particle storage rings.[31,32] In RF quadrupole fields, nonlinear resonances were first observed by von Busch and Paul.[1] In quadrupole ion traps, Whetten and Dawson[2,3] performed measurements and numerical calculations on the mass peak splitting which arises from nonlinear resonances.

The authors present here the main steps of a recent theoretical treatment of these resonances.[17] We begin with the potential of a nonlinear ion trap. The ion motion is treated with the Hamilton formalism, and a series of canonical transformations is applied in order to find the characteristics of the nonlinear resonances. The first canonical transformation eliminates the explicit time dependence for the quadrupole component; it results in a harmonic oscillation at the trap's secular frequencies perturbed by the higher-order multipole components of the trapping potential. The second canonical transformation is nonlinear; its purpose is to eliminate the nonlinear contributions in the Hamilton function. A perturbation expansion is made for this nonlinear transformation, and the expansion is solved order by order. It turns out that this procedure does not always work, i.e., in some parameter constellations there is no perturbation expansion for the transformation. These parameter constellations are simply the nonlinear resonances we are looking for cases in which small changes do not have small consequences.

It turns out that a necessary condition for a nonlinear resonance to occur is a certain relationship between the secular frequencies ω_r and ω_z of the ions and the RF drive frequency Ω of the trap

$$n_r\omega_r + n_z\omega_z = \nu\Omega \tag{42}$$

where ν, n_r, and n_z are integer numbers. This condition gives rise to certain resonance lines within the stability diagram; however, the condition in Equation 42 is necessary but not sufficient for the appearance of resonances. Some of the expected resonance conditions disappear due to special symmetries of the multipole components. The occurrence of resonance lines is discussed in detail for the hexapole and octopole components.

B. Trapping Potential and Hamilton's Equations

It is assumed that the trapping potential ϕ of the nonlinear ion trap is rotationally symmetric. In spherical coordinates (ρ, θ, φ), the dependence on φ vanishes, and from Equation 27 one obtains

$$\Phi(\rho, \theta) = (U - V \cos(\Omega t)) \sum_{n=2}^{\infty} A_n (\rho/r_0)^n P_n(\cos \theta) \tag{43}$$

where $P_n (\cos \theta)$ is the nth order Legendre polynomial (see Equations 28 to 32 for the Legendre polynomials of orders 2 through 6) and r_0 is some scaling length. The monopole and the dipole term, which usually are not present in the storage field, have been omitted from Equation 43. The term $n = 2$ is the quadrupole term, $n = 3$ is the hexapole, $n = 4$ is the octopole, etc. It is assumed that the quadrupole term is much larger than the sum of all the other terms, i.e., the quadrupole ion trap is studied with a small, nonlinear perturbation. Thus, the potential $\Phi(\rho, \theta)$ can be separated into two parts: the quadrupole part $\Phi^2 (\rho, \theta)$ and the higher-order multipole part $\Phi'(\rho, \theta)$

$$\Phi'(\rho, \theta) = (U - V \cos(\Omega t)) \sum_{n=3}^{\infty} A_n (\rho/r_0)^n P_n(\cos \theta)$$
$$= (U - V \cos(\Omega t)) \, \phi'(\rho, \theta) \tag{44}$$

$$\Phi^{(2)}(\rho, \theta) = (U - V \cos(\Omega t)) \, A_2 (\rho/r_0)^2 P_2(\cos \theta)$$
$$= (U - V \cos(\Omega t)) \, \phi^2(\rho, \theta) \tag{45}$$

The term $(U - V \cos(\Omega t))$ describes the strength and the time dependence, whereas $\Phi'(\rho, \theta)$ and $\Phi^2 (\rho, \theta)$ represent the spatial distributions of the higher multipoles, and the basic quadrupole field, respectively.

We apply the abbreviations

$$a_z = \frac{8eU}{mr_0^2\Omega^2} = 2a_r \tag{46}$$

$$q_z = -\frac{4eV}{mr_0^2\Omega^2} = -2q_r \tag{47}$$

$$\xi = \Omega t/2 \tag{48}$$

and the Hamilton function h, given in Cartesian coordinates (x,y,z), can be split into the quadrupole term and the higher-order multipole part by

$$h = h^{(2)} + h' \tag{49}$$

where

$$\begin{aligned}
h^{(2)} = \tfrac{1}{2}(p_x^2 + p_y^2 + p_z^2 + \left(a_r - 2q_r \cos(2\xi)\right) \\
* \left(x^2 + y^2\right) + \left(a_z - 2q_z \cos(2\xi)\right) * z^2)
\end{aligned} \tag{50}$$

represents the quadrupole part, and the higher-order multipoles are given by

$$h' = \left(r_0^2/2\right)\left(a_z - 2q_z \cos(2\xi)\right) \phi'(r, z) = \sum_{k=3}^{\infty} h^{(k)} \tag{51}$$

In Equation 50, p_x, p_y, and p_z are the canonical momenta given by

$$p_x = \frac{\partial x}{\partial \xi} \qquad p_y = \frac{\partial y}{\partial \xi} \qquad p_z = \frac{\partial z}{\partial \xi} \tag{52}$$

The Hamilton function h consists of a main part h^2, which includes the kinetic energy and the potential energy due to the quadrupole part of the trapping potential Φ^2, and a perturbing part h', which describes the potential energy due to the higher-order multipole components Φ'.

From the Hamilton equations, we arrive at the equations of motion, which are written in Cartesian coordinates (x, y, z):

$$\frac{\partial^2 x}{\partial \xi^2} + \left(a_r - 2q_r \cos(2\xi)\right) x = r_0^2\left(a_r - 2q_r \cos(2\xi)\right)\frac{\partial \phi'}{\partial x} \tag{53}$$

$$\frac{\partial^2 y}{\partial \xi^2} + \left(a_r - 2q_r \cos(2\xi)\right) y = r_0^2\left(a_r - 2q_r \cos(2\xi)\right)\frac{\partial \phi'}{\partial y} \tag{54}$$

$$\frac{\partial^2 z}{\partial \xi^2} + \left(a_z - 2q_z \cos(2\xi)\right) z = -\frac{r_0^2}{2}\left(a_z - 2q_z \cos(2\xi)\right)\frac{\partial \phi'}{\partial y} \tag{55}$$

When $A_n = 0$ for $n > 2$, i.e., when we study a pure quadrupole trap, the right-hand side of these equations of motion is zero, and the equations have the form of Mathieu equations. The equations decouple from each other, and the movements in the three Cartesian directions are independent of each other. In addition, the equations are linear in the coordinates x, y, z.

The higher-order multipole components appearing on the right-hand side introduce a coupling of the motion in the three spatial directions, and they introduce a nonlinearity.

C. Linear Canonical Transformation

In this section, we discuss the solutions of the equations of motion (Equation 53 to 55) without the nonlinear perturbation h' (ϕ'), i.e., with the right-hand side of these equations set to zero. The solutions of the resulting Mathieu equation are well known.[33] We restrict the discussion to the first area of stability: $0 < \beta_r, \beta_z < 1$. We write the coordinates x, p_x, y, p_y, z, p_z in a vector

$$
\mathbf{x} = \begin{pmatrix} x \\ p_x \\ y \\ p_y \\ z \\ p_z \end{pmatrix}
\tag{56}
$$

The equations of motion are then given by

$$
\frac{d\mathbf{X}}{d\xi} = \begin{pmatrix}
0 & 1 & 0 & 0 & 0 & 0 \\
-(a_r - 2q_r\cos(2\xi)) & 0 & 0 & 0 & 0 & 0 \\
0 & 0 & 0 & 1 & 0 & 0 \\
0 & 0 & -(a_r - 2q_r\cos(2\xi)) & 0 & 0 & 0 \\
0 & 0 & 0 & 0 & 0 & 1 \\
0 & 0 & 0 & 0 & -(a_z - 2q_z\cos(2\xi)) & 0
\end{pmatrix} \mathbf{X}
\tag{57}
$$

We apply a linear transformation L to the coordinates X, the new coordinates being η:

$$
\mathbf{X} = L\eta
\tag{58}
$$

with

$$
L = \begin{pmatrix} L_x & 0 & 0 \\ 0 & L_y & 0 \\ 0 & 0 & L_z \end{pmatrix}
\tag{59}
$$

and

$$L_z = \begin{pmatrix} \sum_{n=-\infty}^{\infty} C_{2n,z} \cos(2n\xi) & \sum_{n=-\infty}^{\infty} C_{2n,z} \sin(2n\xi) \\ \sum_{n=-\infty}^{\infty} (2n + \beta_z) C_{2n,z} \sin(2n\xi) & \sum_{n=-\infty}^{\infty} (2n + \beta_z) C_{2n,z} \cos(2n\xi) \end{pmatrix} \tag{60}$$

$$L_x = L_y = \begin{pmatrix} \sum_{n=-\infty}^{\infty} C_{2n,r} \cos(2n\xi) & \sum_{n=-\infty}^{\infty} C_{2n,r} \sin(2n\xi) \\ \sum_{n=-\infty}^{\infty} (2n + \beta_r) C_{2n,r} \sin(2n\xi) & \sum_{n=-\infty}^{\infty} (2n + \beta_r) C_{2n,r} \cos(2n\xi) \end{pmatrix} \tag{61}$$

The coefficients $C_{2n,r}$ and $C_{2n,z}$ are given in Chapter 2. The equation of motion in the coordinates η reads:

$$X = \begin{pmatrix} 0 & \beta_r & 0 & 0 & 0 & 0 \\ -\beta_r & 0 & 0 & 0 & 0 & 0 \\ 0 & 0 & 0 & \beta_r & 0 & 0 \\ 0 & 0 & -\beta_r & 0 & 0 & 0 \\ 0 & 0 & 0 & 0 & 0 & \beta_z \\ 0 & 0 & 0 & 0 & -\beta_z & 0 \end{pmatrix} \eta \tag{62}$$

This is a simple harmonic oscillation at the secular frequencies of the ion trap. Note that this is an exact result; no pseudo-potential approximation is involved. The unperturbed Hamilton function for the coordinates η is

$$h^{(2)} = \tfrac{1}{2}\beta_r(\eta_1^2 + \eta_2^2) + \tfrac{1}{2}\beta_r(\eta_3^2 + \eta_4^2) + \tfrac{1}{2}\beta_z(\eta_5^2 + \eta_6^2) \tag{63}$$

D. Nonlinear Canonical Transformation

This section deals with the full Hamiltonian h, including both the quadrupole and the higher-order multipole parts of the potential. We start with the Hamiltonian $K(\eta, \xi)$ expressed in the variables η:

$$K(\eta, \xi) = h(X, \xi) + \frac{\partial F}{\partial \xi}(\eta_2, x, \eta_4, y, \eta_6, z, \xi) \tag{64}$$

where $F(\ldots)$ is the generating function. Because L is a linear transformation, F is a quadratic form in its variables with the coefficients given by the elements of the matrix L. These coefficients are periodic in ξ with period π. From Equation 64 we obtain

$$K(\eta, \xi) = \tfrac{1}{2}\beta_r(\eta_1^2 + \eta_2^2) + \tfrac{1}{2}\beta_r(\eta_3^2 + \eta_4^2) + \tfrac{1}{2}\beta_z(\eta_5^2 + \eta_6^2) + h'(L\eta, \xi) \tag{65}$$

We seek a nonlinear canonical transformation from the variables η to the variables y with the generating function $S(\eta_1, y_2, \eta_3, y_4, \eta_5, y_6)$. The transformation is selected to keep the quadratic part of the Hamiltonian K unchanged while simplifying the cubic, fourth order, and higher terms. We make the perturbation expansion

$$S(\eta_1, y_2, \eta_3, y_4, \eta_5, y_6, \xi)$$

$$= \eta_1 y_2 + \eta_3 y_4 + \eta_5, y_6 + S^{(3)} + S^{(4)} + \ldots + S^{(n)} + \ldots \quad (66)$$

It follows that

$$y_1 = \frac{\partial S}{\partial y_2} = S_{y_2} = \eta_1 + S^{(3)}_{y_2} + S^{(4)}_{y_2} + \ldots \quad (67)$$

$$\eta_2 = \frac{\partial S}{\partial \eta_1} = S_{\eta_1} = y_2 + S^{(3)}_{\eta_1} + S^{(4)}_{\eta_1} + \ldots \quad (68)$$

Similar equations hold for the other variables y_3, η_4, y_5, η_6. The transformed Hamilton function is

$$g(\mathbf{y}, \xi) = K(\mathbf{y}, \xi) + \frac{\partial}{\partial \xi} S(\eta_1, y_2, \eta_3, y_4, \eta_5, y_6, \xi)$$

$$= \frac{1}{2}\beta_r \left(y_1^2 + y_2^2 + y_3^2 + y_4^2 \right) + \frac{1}{2}\beta_z \left(y_5^2 + y_6^2 \right) + \sum_{k=3}^{\infty} g^{(k)}(\mathbf{y}, \xi) \quad (69)$$

Substitution of K, S, and Equations 67 and 68 into Equation 69 and equating terms with the same order leads to

$$\frac{\partial S^{(n)}}{\partial \xi} + \beta_r \left(y_2 S^{(n)}_{\eta_1} - \eta_1 S^{(n)}_{y_2} + y_4 S^{(n)}_{\eta_3} - \eta_3 S^{(n)}_{y_4} \right) + \beta_z \left(y_6 S^{(n)}_{\eta_5} - \eta_5 S^{(n)}_{y_6} \right)$$

$$= g^{(n)}(\eta_1, y_2, \eta_3, y_4, \eta_5, y_6, \xi)$$

$$- h^{(n)}(\eta_1, y_2, \eta_3, y_4, \eta_5, y_6, \xi)$$

$$+ p^{(n)}(\eta_1, y_2, \eta_3, y_4, \eta_5, y_6, \xi) \quad (70)$$

In this equation, the $h^{(n)}$ are known functions (see Equation 51). $p^{(n)}$ include all terms with power n in the products between η_1, y_2, η_3, y_4, η_5, y_6, $S^{(1)}, \ldots, S^{(n-1)}, p^{(1)}, \ldots, p^{(n-1)}$, and $h^{(1)}, \ldots, h^{(n-1)}$. $p^{(n)}$ are also known functions because the calculation is performed from low n to high n iteratively.

The next step is a change of variables from y to z with:

$$z_1 = \eta_1 + i y_2 \quad \bar{z}_1 = \eta_1 - i y_2$$

$$z_2 = \eta_3 + i y_4 \quad \bar{z}_2 = \eta_3 - i y_4$$

$$z_3 = \eta_5 + i y_6 \quad \bar{z}_3 = \eta_5 - i y_6 \quad (71)$$

The functions $S^{(n)}$, $g^{(n)}$, $p^{(n)}$, and $h^{(n)}$ are expanded into a Taylor series:

$$s^{(n)} = \sum_{(n)} s^{(n)}_{jklmop}(\xi)\, z_1^j \bar{z}_1^k z_2^l \bar{z}_2^m z_3^o \bar{z}_3^p \tag{72}$$

$$g^{(n)} = \sum_{(n)} g^{(n)}_{jklmop}(\xi)\, z_1^j \bar{z}_1^k z_2^l \bar{z}_2^m z_3^o \bar{z}_3^p \tag{73}$$

$$h^{(n)} = \sum_{(n)} h^{(n)}_{jklmop}(\xi)\, z_1^j \bar{z}_1^k z_2^l \bar{z}_2^m z_3^o \bar{z}_3^p \tag{74}$$

$$p^{(n)} = \sum_{(n)} p^{(n)}_{jklmop}(\xi)\, z_1^j \bar{z}_1^k z_2^l \bar{z}_2^m z_3^o \bar{z}_3^p \tag{75}$$

In Equations 72 to 75, the following relations are valid for the exponents:

$$j + k + l + m + o + p = n; \qquad n > 2; \qquad j, k, l, m, o, p \geq 0 \tag{76}$$

Substitution of Equation 71 into Equation 70 leads to

$$\frac{\partial}{\partial \xi} s^{(n)}_{jklmop}(\xi) - i\left[\beta_r(j + l - k - m) + \beta_z(o - p)\right] s^{(n)}_{jklmop}$$

$$= \frac{i}{2} g^{(n)}_{jklmop}(\xi) - \frac{i}{2} h^{(n)}_{jklmop}(\xi) + p^{(n)}_{jklmop}(\xi) \tag{77}$$

All the Taylor expansion coefficients of $s^{(n)}$, $g^{(n)}$, $p^{(n)}$, and $h^{(n)}$ are time periodic with period π. Therefore, they are written as a Fourier series

$$s^{(n)}_{jklmop} = \sum_{v=-\infty}^{\infty} s^{(n)}_{v,\,jklmop} e^{2iv\xi} \tag{78}$$

$$g^{(n)}_{jklmop} = \sum_{v=-\infty}^{\infty} g^{(n)}_{v,\,jklmop} e^{2iv\xi} \tag{79}$$

$$h^{(n)}_{jklmop} = \sum_{v=-\infty}^{\infty} h^{(n)}_{v,\,jklmop} e^{2iv\xi} \tag{80}$$

$$p^{(n)}_{jklmop} = \sum_{v=-\infty}^{\infty} p^{(n)}_{v,\,jklmop} e^{2iv\xi} \tag{81}$$

Insertion of Equations 78 to 81 into Equation 77 gives the final result:

$$s^{(n)}_{jklmop}(\xi) = -i \sum_{v=-\infty}^{\infty} \frac{p^{(n)}_{v,\,jklmop} + \dfrac{i}{2} g^{(n)}_{v,\,jklmop} - \dfrac{i}{2} h^{(n)}_{v,\,jklmop}}{2v - \left[\beta_r(j + l - k - m) + \beta_z(o - p)\right]} e^{2iv\xi} \tag{82}$$

with $j + k + l + m + o + p = n$ and $j, k, l, m, o, p \geq 0$.

The expression (Equation 82) for the canonical transformation s diverges if the denominator vanishes. Thus, if the condition

$$\beta_r(j + l - k - m) + \beta_z(o - p) = 2v \tag{83}$$

is fulfilled, the perturbation expansion Equation 66 breaks down and a nonlinear resonance occurs.

E. Discussion of Nonlinear Resonances

We set $n_r = j + l - k - m$ and $n_z = o - p$, and Equation 83 gives

$$n_r\beta_r + n_z\beta_z = 2v \tag{84}$$

where n_r, n_z, and v are whole numbers. By multiplication with the frequency Ω of the driving field, the fundamental relation for nonlinear resonances is obtained

$$n_r\omega_r + n_z\omega_z = v\Omega \tag{85}$$

$\omega_r = \beta_r\Omega/2$ and $\omega_z = \beta_z\Omega/2$ are the secular frequencies in the r- and z-directions of the trapped particle. The relation was already introduced above as Equation 42.

This relation tells that a nonlinear resonance may occur, when the sum of a whole number of secular frequencies matches the driving frequency. As shown above, this is a resonance between overtones of the secular frequencies and sideband frequencies. These overtones result from the presence of nonharmonic terms in the trapping potential.

From Equation 85 it seems to follow that the presence of any nonlinear component in the trapping potential gives rise to all resonance lines. In fact, this is not true. Equation 85 is rather a necessary condition for the occurrence of a nonlinear resonance. It is possible that the numerator in Equation 82 also vanishes at a possible nonlinear resonance condition. In this case, no resonance occurs; such cases are discussed in the next section.

We recall some designations. We call

$$N = |n_r| + |n_z| \tag{86}$$

the order N of the resonance. It must be distinguished from the order n of the multipole. When n_r and n_z have the same sign and $n_r, n_z \neq 0$, we speak of a sum resonance; when n_r, n_z have different signs and $n_r, n_z \neq 0$, we call it a difference resonance. When additionally $v = 0$, we have a pure coupling resonance. When $n_r \neq 0$ and $n_z = 0$, we encounter an r-direction resonance; when $n_r = 0$ and $n_z \neq 0$, it is a z-direction resonance. From Equations 77 and 83 to 85, the following properties can be deduced:

1. If in the original Hamilton function (or in the higher multipole potential contribution, Φ' respectively) terms of order $n \geq 3$ are present, all can in principle excite resonance lines of any order. However, a term of order n will have the strongest effect on the resonance conditions of order $N = n, n - 2, n - 4$, and smaller effects on the other conditions.
2. When n_r and n_z have different signs, and $v = 0$ (pure coupling resonance), energy is only exchanged between the r- and z-directions.[32] When $v \neq 0$ (difference resonance), the energy exchange between the r- and z-directions is the main effect, but a small amount of energy may be taken up from the RF drive field.

F.   Nonlinear Resonances of the Hexapole and Octopole Superposition

This section is a discussion of the nonlinear resonance lines for the hexapole and the octopole superpositions within the first stability area ($0 \leq \beta_r, \beta_z \leq 1$).

In Section I, the multipole fields were classified as even and odd. Usually the hexapole and the octopole contributions are the largest and, therefore, the most essential in a given ion trap. The hexapole field is usually the strongest odd and the octopole the strongest even multipole field contribution. This particular situation pertains to the truncated Paul ion trap, and the stretched end-cap and modified angle ion traps. Because of their symmetry with respect to the plane $z = 0$, the octopole field should be dominant here, and the hexapole field should be absent. In practice, however, the asymmetry of the end-cap perforations and of external fields in front of the holes, and the asymmetry induced by assembly tolerances, cause some hexapole field contributions which usually cannot be neglected.

Applying the general resonance condition (Equation 84), and a maximum order $N \leq n$ of the resonance, as given by the rule stated in paragraph (1) above, the nonlinear resonance lines for the hexapole ($n = 3$) are

$$3\beta_r = 2 \quad (n_r = 3, n_z = 0)$$
$$3\beta_z = 2 \quad (n_r = 0, n_z = 3)$$
$$\beta_r + 2\beta_z = 2 \quad (n_r = 1, n_z = 2)$$
$$2\beta_r + \beta_z = 2 \quad (n_r = 2, n_z = 1) \tag{87}$$

For the octopole they are

$$4\beta_r = 2 \quad (n_r = 4, n_z = 0)$$
$$4\beta_z = 2 \quad (n_r = 0, n_z = 4)$$
$$2\beta_r + 2\beta_z = 2 \quad (n_r = 2, n_z = 2)$$
$$\beta_r + 3\beta_z = 2 \quad (n_r = 1, n_z = 3)$$
$$3\beta_r + \beta_z = 2 \quad (n_r = 3, n_z = 1) \tag{88}$$

Note that the condition Equation 84 is necessary, but not sufficient. Due to the rotational symmetry of the trapping potential, the numerator in Equation 82 vanishes for some resonance lines. For example, the hexapole potential (Equation 29) contains only even powers of r. Due to linearity of the transformation L, the Hamiltonian K contains only terms with $(\eta_1)^j(\eta_2)^k$ and $(\eta_2)^l(\eta_1)^m$ with $(j + k)$ and $(l + m)$ even. Furthermore, from Equation 71, $n^{(n)}$, in Equation 74 has only terms $(z_1)^j(_1)^k$ and $(z_2)^l(_2)^k$ with $(j + k)$ and $(l + m)$ even. Therefore, if $(j + k)$ or $(l + m)$ is odd, $h^{(3)}_{jklmop} = 0$ and $g^{(3)}_{jklmop}$, $s^{(3)}_{jklmop}$, and $p^{(3)}_{jklmop}$ in Equation 77 also can be selected as zero. Consequently, resonance lines will not exist when $nr = j - k + l - m$ is odd. The remaining lines are for the hexapole:

$$3\beta_z = 2$$
$$2\beta_r + \beta_z = 2 \tag{89}$$

and for the octopole:

$$4\beta_r = 2$$
$$4\beta_z = 2$$
$$2\beta_r + 2\beta_z = 2 \tag{90}$$

In particular, no hexapole resonance exists for $\beta_r = 2/3$ with rotationally symmetrical potentials. The nonlinear hexapole resonance $\beta_r = 2/3$ has been wrongly stated in previous literature.[3] The hexapole resonance lines are plotted in Figure 11 within the first area of stability.

G. Strength and Width of Nonlinear Resonances

The conditions for the occurrence of nonlinear resonances have been established. The general condition is Equation 84; due to special properties of the nonlinearity ϕ' some of these resonance lines may disappear. The lines for pure hexapole and pure octopole superpositions are summarized in Equations 89 and 90, respectively. In this section, we investigate other properties of the nonlinear resonances, namely their effect on the motion of the ions and their width.

The effect of a nonlinear resonance on the motion of trapped ions is an increase of the mean energy of these ions, and may be shown in the following way. We define the intensity I of a particular resonance as the mean rate of energy increase in which the average is taken over a period π of the variable ξ.

$$I = \left\langle \frac{dh}{d\xi} \right\rangle = \left\langle \frac{\partial h}{\partial \xi} \right\rangle \qquad (91)$$

With the help of Equation 64 we obtain

$$I = \left\langle \frac{\partial K}{\partial \xi} \right\rangle - \left\langle \frac{\partial^2 F}{\partial \xi^2} \right\rangle \qquad (92)$$

F is a quadratic form in its variables, so $\partial^2 F / \partial \xi^2$ does not produce any resonance lines within the first area of stability. However, it contributes to the boundaries at β_r, $\beta_z = 0$, or 1. Thus, we may consider the time average over the period π of $\partial^2 F / \xi^2$ to be zero. Therefore,

$$I = \left\langle \frac{\partial K}{\partial \xi} \right\rangle = \left\langle \frac{\partial h' \left(A\boldsymbol{\eta}, \xi \right)}{\partial \xi} \right\rangle \qquad (93)$$

This equation tells us that any increase of the mean energy of a trapped ion is due to nonlinear components in the trapping field. In particular, for a pure quadrupole field, the energy of a trapped ion averaged over the period π of the variable ξ (equal to one period of the RF drive frequency Ω) is zero, and there is no heating. The RF heating effect does not occur in this treatment because it is due to phase-changing collisions of trapped ions with the background gas.[34] Such collisions are not included in the present model of ion motion.

We evaluate Equation 93 for the example of h' containing only the hexapole term. The solution of the nonlinear equation system Equation 62 is

$$\begin{aligned}
\eta_1 &= a \cos(\beta_r \xi + b) & \eta_2 &= -\beta_r a \sin(\beta_r \xi + b) \\
\eta_3 &= c \cos(\beta_r \xi + d) & \eta_4 &= -\beta_r c \sin(\beta_r \xi + d) \\
\eta_5 &= e \cos(\beta_z \xi + f) & \eta_6 &= -\beta_z e \sin(\beta_z \xi + f)
\end{aligned} \qquad (94)$$

where a, b, c, d, e, f describe the initial conditions. From the matrix L, we take only the terms with C_0, C_2, C_{-2} because the higher coefficients C_{2n} decrease very rapidly with n. The resulting equations are a first-order solution for the nonlinear equations of motion (Equation 53 to 55); they are inserted into Equation 93 and it is found that $I = 0$ unless one of the res-

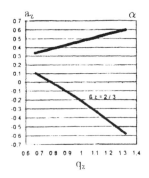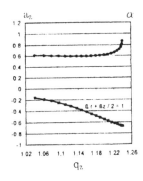

FIGURE 73

Relative strength (intensity) of the two hexapole resonance conditions $3\beta_z = 2$ and $\beta_z + 2\beta_r$ $= 2$. The lower curves describe the resonance condition in a_z/q_z parameters, the upper curves give the relative strength of the resonances along the q_z-axis.

onance conditions, Equation 89, is fulfilled. The result

$$I = f_1(a, b, c, d, e, f) * f_2(a_z, q_z, \Omega) \tag{95}$$

is a product of a term depending only on the initial conditions and a term depending on the trapping parameters. The expressions for f_1 and f_2 are given in the literature.[17]

In Figure 73 the function f_2 is plotted for the two hexapole resonance lines. f_2 increases toward the boundaries of the first area of stability, but shows no large variation.

Another important property of a resonance is its width. A finite width can result from damping the ion's motion by, for example, a buffer gas. In addition, a finite width can arise when ions remain in resonance for a finite time only. Such is generally the case because the energy increase gives rise to an amplitude increase which, due to the nonlinear components of the trapping field, shifts the secular frequency of the ions. Therefore, the action of the nonlinear resonance shifts the ions out of resonance. This effect is more pronounced for even multipoles than for the uneven.

The resonance width reported in Figure 3.2 of Reference 22 amounts to 1.5 kHz at a secular frequency in the z-direction of 330 kHz; therefore, the line Q is about 200, which shows that reasonable Q-factors can be achieved.

REFERENCES

1. von Busch, F.; Paul, W., Z. Phys. 1961, 164, 588.
2. Whetten, N.R.; Dawson, P.H., J. Vac. Sci. Technol. 1969, 6, 100.
3. Dawson, P.H.; Whetten, N.R., Int. J. Mass Spectrom. Ion Phys. 1969, 2, 45.
4. Franzen, J., Int. J. Mass Spectrom. Ion Processes. 1991, 106, 63.
5. Franzen, J.; Gabling, R.-H.; Heinen, G.; Weiß, G., U.S. Patent 5,028,777 (German Patent filed December 1987).
6. Franzen, J., Int. J. Mass Spectrom. Ion Processes. 199, 106, 63.
7. Louris, J.; Schwartz, J.; Stafford, G.; Syka, J.; Taylor, D., Proc. 40th ASMS Conf. Mass Spectrom. Allied Topics. Washington, D.C., 1992; p. 1003.
8. Stafford, G.C.; Kelley, P.E.; Stephens, D.R., U.S. Patent 4,540,884 (filed December 1982).
9. Landau, L.D.; Lifshitz, E.M., Lehrbuch der theoretischen Physik. Vol. 1, Mechanik, Akademie-Verlag: Berlin, 1990.
10. Wineland, D.J., in Proc. of the Cooling, Condensation and Storage of Hydrogen Cluster Ions Workshop. Bahns, J. T., Ed. SRI International: Menlo Park, CA, 1987; p. 181ff.
11. Walz, J.; Siemers, I.; Schubert, M.; Neuhauser, W.; Blatt, R., Europhys. Lett. 1993, 21, 183.
12. Paul, W.; Steinwedel, H., U.S. Patent 2,939,952 (German Patent filed December 1953).
13. Knight, R.D., Int. J. Mass Spectrom. Ion Processes. 1983, 51, 127.
14. Finnigan MAT ion trap series (ITD™, ITS40™, and ITMS™).
15. The modified hyperbolic angle ion trap was developed by Bruker-Franzen Analytik GmbH, under contract DAAA15-87-C0008 awarded to Teledyne CME, for the U.S. Army ERDEC (Aberdeen Proving Ground) as "Chemical-Biological Mass Spectrometer" CBMS.
16. Wang, Y., Ph.D. thesis, University of Bremen, 1992; Wang, Y., European Patent 90903006.6 (1992); Wang, Y.; Wanczek, K.-P., J. Chem. Phys. 1993, 98, 2647; Arnold, J., private communication.
17. Wang, Y.; Franzen, J.; Wanczek, K.-P., Int. J. Mass Spectrom. Ion Processes. 1993, 124, 125.
18. Wang, Y., Int. J. Mass Spectrom. Ion Processes. 1991, 106, 72.
19. March, R.E.; Hughes, R.J., Quadrupole Storage Mass Spectrometry. Chemical Analysis Series, Vol. 102; Wiley Interscience: New York, 1989.
20. Guidugli, F.; Traldi, P., Rapid Commun. Mass Spectrom. 1991, 5, 343.
21. Morand, K.L.; Lammert, S.A.; Cooks, R.G., Rapid Commun. Mass Spectrom. 1991, 5, 491.
22. Guidugli, F.; Traldi, P.; Franklin, A.M.; Langford, M.L.; Murrell, J.; Todd, J.F.J., Rapid Commun. Mass Spectrom. 1992, 6, 229.
23. Yost, R.A., 9th Asilomar Conference on Mass Spectrometry. September 27 to October 1, 1992.
24. Abramovitz, M.; Stegun, I.A., Handbook of Mathematical Functions. U.S. National Bureau of Standards: Washington, D.C., 1964, Dover: New York, 1965, and later editions.
25. Dawson, P.H., Quadrupole Mass Spectrometry and its Applications. Elsevier: Amsterdam, 1976.
26. Wang, Y.; Franzen, J., Int. J. Mass Spectrom. Ion Processes. 1994, 132, 155.
27. Ritz, W., J. Reine Angew. Math. (Crelle). 1908, 135.
28. Friedrichs, K., Am. J. Math. 1946, 68, 4.
29. Wang, Y.; Franzen, J., Int. J. Mass Spectrom. Ion Processes. 1992, 112, 167.
30. For an introduction, see Schuster, H.G., Deterministic Chaos. Physik Verlag: Weinheim, 1984.

31. Moser, J., in *CERN Symp.* 1956, *290*, 2.
32. Hagedorn, R., *CERN Rep.* 57-1, 1957.
33. McLachlan, N.W., *Theory and Applications of Mathieu Functions.* Clarendon Press: Oxford, 1947.
34. Blatt, R.; Zoller, P.; Holzmüller, G.; Siemers, I., *Z. Phys.* 1986, *D4*, 121.

Chapter 4

COMMERCIALIZATION OF THE QUADRUPOLE ION TRAP

John E. P. Syka

CONTENTS

I. INTRODUCTION

This chapter* is an historical account of the problems that were encountered and overcome by Finnigan Corporation in the realization of a novel mass spectrometric detector in the early 1980s. The account is essentially a story of the people involved and their individual and collective struggles to understand, at least in part, the behavior of gaseous ions stored in the quadrupole ion trap, and to produce a working mass spectrometer with the device. In an effort to portray a truth that is absent from more formal scientific articles—that technological development is accomplished by real people who have names and whose actions are often

* From a talk "The Geometry of the Finnegan Ion Trap: History and Theory", presented by author at 9th Asilomar Conference on Mass Spectrometry, Sept. 1992.

0-8493-4452-2/95/$0.00+$.50

shaped their individual natures—the characters in this story are referred to by name.

As is well known by now, the Finnigan ion trap is not of an ideal geometry, in that the electrode structure is not proportioned to produce homogeneous quadrupole fields. A cross-section of the electrode structure for an ideal ion trap is shown in Figure 1 together with a cross-section of the Finnigan "stretched" ion trap showing the enhanced separation of the end-cap electrodes. Figure 2 shows once again the cross-sections of the ideal ion trap with the theoretical spacing of the end-cap electrodes and of the Finnigan ion trap with the stretched end-cap electrode spacing; the dashed curved lines in the latter represent the theoretical shapes for the end-cap electrodes appropriate to the end-cap electrode separation, while the solid curved lines (which intersect with the z-axis) represent the actual end-cap electrodes. The actual end-cap profiles are seen to be too steep in comparison with the dashed lines representing the ideal theoretical shapes.

This account traces the evolution of the Finnigan ion trap geometry. Included is a complete discussion of the "mass shift" problem which was solved by introducing the previously described deviation to the ideal Paul ion trap geometry. An explanation is offered both as to why mass shift was observed with an ideally proportioned ion trap and why the nonideal ion trap geometry solved this problem. It is also intended that this account accurately portray the context in which the decision was made in 1984 to maintain the Finnigan ion trap geometry as a trade secret.

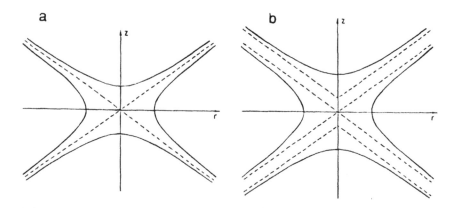

FIGURE 1
Cross-sectional view of the quadrupole ion trap: (a) theoretical end-cap electrode spacing; (b) Finnigan "stretched" end-cap electrode spacing.

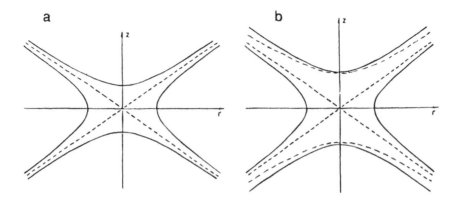

FIGURE 2
Cross-sectional view of the quadrupole ion trap: (a) theoretical end-cap electrode spacing; (b) the curved, dashed lines in this cross-section of the Finnigan "stretched" ion trap show the theoretical shapes of the end-cap electrodes appropriate to the end-cap electrode separation.

II. DEVELOPMENT OF THE FIRST FINNIGAN ION TRAP

The origin of Finnigan's involvement with ion traps dates from June–July 1979, after George Stafford attended a lecture given by Ray March at the Seattle meeting of the American Society for Mass Spectrometry. He became interested in ion traps and, after some reflection had an idea that represented an entirely new strategy for the use of the quadrupole ion trap for mass analysis. George's new idea, of course, was what is now known as the mass-selective instability scan. As his work incorporating his negative ion conversion dynode into existing Finnigan quadrupole mass filter instruments was winding down, George was able to convince Michael Story, who was at that time in charge of the Finnigan Research Department, to clear the way with the upper management of Finnigan to fund a research project on ion traps. As George was one of four scientists in the department, this represented no small commitment on the part of Finnigan. It was decided to try to build a mass-selective instability-based ion trap mass spectrometer. John Syka, an engineering student employed during the summer as a technician in the research department, was assigned as George's helper. Both George and John immersed themselves in the papers by John Todd et al. and Ray March et al. and Peter Dawson's book *Quadrupole Mass Spectrometry* in an effort to comprehend the field in which they were inserting themselves. George had expressed his interest in ion traps to Ray March, and Ray briefly loaned one of his cylindrical trap assemblies to George for inspection. However, the collective experience within the Finnigan research department regarding the importance of mechanical accuracy for quadrupole mass filters strongly influenced George and John to construct, for their first device, the best hyperbolic trap structure their modest budget would allow—the best hyperbolic ion trap electrode structure the in-house machinist, Denis Collier, could machine manually. John's primary task that summer was the mechanical design of the ion trap analyzer assembly. George went off to the IMSC meeting in Oslo (the 8th International Mass Spectrometry Conference, 1979) while John stayed behind and made the initial mechanical drawings. Before departing, George specified that John should design a trap that would achieve a mass range of just over 500 u when operated with a 2.281-MHz radiofrequency (RF) voltage supply capable of delivering 6000 V peak to the ring electrode (this requirement corresponded to the RF output capability of the Finnigan 4000 mass spectrometer electronics).

John developed a design for a conventionally proportioned trap, $r_0^2 = 2z_0^2$, which had an r_0 on the order of 5 mm. When George returned from his trip, he reviewed a scale layout drawing of the trap with John. Both concluded, relying primarily on intuition, that this r_0 seemed too small. It was agreed that an r_0 slightly under 10 mm was more appropriate. The desired mass range would be achieved by modifying the electronics. As it was time for John to return to college for the fall term, George simply altered the dimensions on the existing drawings, and reran a program that John had written which calculated a series of coordinates for Denis Collier to use in machining the ring and end-cap electrodes. These coordinates took into account the approximate cutting tool radius, thus yielding more accurate hyperbolic electrode surfaces. Denis's considerable craftsmanship yielded trap electrodes with profiles that were accurate to better than 0.05 mm. The overall tolerance of the trap assembly was probably ±0.2 mm. That original trap in fact worked better than most traps made as part of the eventual product development effort. It served as a sort of standard to which traps fabricated with numerically controlled machines were compared and often found wanting. The many hundreds of Finnigan-designed traps sold over the intervening 14 years, excepting one important modification, have retained essentially the same dimensions as the first.

By the spring of 1980, George and Bill Fies, Finnigan's quadrupole electronics expert, had assembled a simple system with the analyzer driven by a Finnigan 4000 quadrupole electronics module modified to produce single-ended rather than differential RF. In those days the systems were not inherently computerized so that the initial scan function was accomplished with analog circuitry. The mass filter scan generation circuitry was modified to provide a fixed RF voltage during ionization prior to the mass analysis RF ramp. A knob was provided which allowed adjustment of the initial storage RF voltage during ionization in addition to the usual first mass, last mass, and scan rate control knobs. Initially, the ion trap was driven with an RF frequency of 2.281 MHz, thus yielding a mass range of only 100 u or so. The first mass spectra were obtained with nominal mass resolution of ca. 125. The most noteworthy observation (other than the fact that George's scan scheme worked at all) was that the system could be scanned quite rapidly. Near-unit resolution was obtained at scan rates of ca. 90 u/μs. The bandwidth limitations of the ion current amplifier prevented observation of undistorted mass peaks at higher scan rates. Also, considering the lack of precision of the trap structure relative to that required for mass filters, the resolution attained was actually quite promising.

Experimentation with the ion trap continued throughout the summer of 1980. John returned for another summer of work as George's assistant. The only means of data acquisition was a Tektronix oscilloscope camera, so they filled a notebook with notes of early mass spectra and oscilloscope camera pictures on which the instrument settings had been scratched. John became particularly adept at using multiple exposures to capture the results of a series of experiments on one picture; this technique allowed for more ready comparison of data and conserved film.

Early in the summer, Bill Fies modified the system RF electronics to run at 1 MHz, which brought the mass range up to in excess of m/z 500. While m/z 502 of FC-43 calibrant could be observed, it was not very well resolved, having a typical peak width of ca. 5 u. Resolution at the low end of the mass range was only marginally less than that observed at the original operating frequency. Then a brief study of the effects of scan rate on spectral resolution was made. The results indicated that modest improvements in resolution were available with reduced scan rates at the cost of reduced signal-to-noise ratio. The total number of ions detected in a mass peak (area) was observed to remain constant.

From the outset it had been observed that detected ion signal strengths were rather weak in comparison to what would have been expected from mass filter instruments under similar conditions. In successive attempts to improve instrument sensitivity, Denis Collier was asked to add ever more perforations in the mechanically fragile exit end-cap diaphragm. The primary results of this effort were a machinist pushed to the limits of his patience and an exit end-cap that resembled a miniature colander. Ion signal strengths were increased, at most, only marginally.

A key observation made that summer was the strong degree of control that the amplitude of the RF voltage during ionization exerted over line shape and resolution, the relative abundance between ions of low and high masses, and the overall sensitivity of the instrument. It became routine to "tune" the ion trap for optimal resolution and sensitivity for a particular mass peak or mass range by adjusting the ionization RF amplitude. Neither George nor John fully understood what they were observing and they knew that they did not understand it. Both George and

John were well aware of the relevant passages in Dawson's book and elsewhere describing how space charge could alter stability boundary positions as well as those describing how the trapping conditions (a_z, q_z, a_r, and q_r) could affect ion accumulation efficiency (acceptance) during ionization. They had observed mass peak broadening when sample pressures were elevated which they directly attributed to space charge. They had also observed, at very high sample pressures, what George immediately recognized as ion/molecule reactions (self-chemical ionization, i.e., self-CI), a phenomenon later often confused with space charge-induced line broadening. While the observation that higher mass ions consistently tuned at generally higher ionization RF voltages was consistent with what was expected from ion acceptance theory, the actual ionization RF voltages necessary for optimal tuning varied daily. An experiment designed to measure the relationship between ionization exposure (electron gun filament emission) and peak intensity added to their confusion, as it yielded a nonlinear relationship, even though the line shape of the mass peak of interest gave little indication of any influence of space charge. Complete comprehension of what they were observing would elude them for 4 more years.

In recent years, George and John have more than once looked back at their longstanding confusion regarding these effects and have marveled at how they could have been so thickheaded. However, thickheadedness was not the problem; George and John, as well as others that later joined in the development project, were dealing with the mental reorientation that everyone must undergo who moves from working on beam-type mass spectrometers to ion traps. With expectations shaped by experience with quadrupole filter instruments, which have a number of voltages that must be adjusted to tune the instrument, it seemed perfectly natural that the ion trap would have at least one parameter requiring adjustment to achieve an acceptable tune. The fact that there seemed to be only one such parameter was noteworthy.

George and John also studied combined RF-DC (direct current) mass-selective instability scans. No advantage in terms of resolution or sensitivity was observed. Because the addition of a DC component to the applied ring voltage added complication and reduced the mass range that could be scanned in one experiment, the use of DC was quickly abandoned.

The problem that consumed the majority of that first summer's effort was that of reducing noise caused by the gate electrode of the electron gun failing to inhibit completely the introduction of electrons into the trap during the mass analysis part of the scan sequence. Electrons in the ion trap during this period would create spurious ions which were ejected, then detected. Eventually, after much time and effort was expended, an electron gun was built that gated properly. While not significant in any fundamental sense, this episode was typical of the kinds of problems one runs into in actually trying to build an instrument. It is one thing to conceive of a gating electron gun; it is another thing to build one, particularly the first time.

In the autumn of 1980 John returned to college to finish his degree. As George was also involved in the design of a new ion source for one of the quadrupole mass filter products, it was recognized that George would need help if there were to be progress made on the ion trap project. Eventually, Walt Reynolds, who had developed considerable experience with mass spectrometer electronics as well as with computerization of mass spectrometers while working in Carl Djerrasi's laboratory, at Stanford University, joined the company as a contractor and became George's full-time assistant. In May 1981, upon completing his degree work, John Syka accepted a permanent position with Finnigan with the expectation of continuing his work with George on the ion trap; however, as George already had Walt working with him, John was assigned to work on mass filter systems. Although John would not be officially assigned to work on ion traps until 1985, he retained a close association with George and the ion trap project.

Up to this time, few people within Finnigan were aware of the ion trap project. Furthermore, for competitive reasons, nobody outside the company had been made aware of Finnigan's ion trap development project. However, George and Mike Story decided that it might be helpful if George could have access to the wisdom of somebody with experience in the field. Because John Todd, at the University of Kent (England), had the oldest active ion trap research program related to mass spectrometry, Finnigan approached him with the offer of a consulting relationship. John made an initial visit to the Finnigan facility in the spring of 1981, and a relationship was established between the Kent and Finnigan groups.

John Todd made a second, longer visit in September of the same year and was able to participate in an experiment that Walt and George had been preparing for months. Walt had constructed a set of RF drive electronics that allowed one to perform a mass-selective instability scan by scanning the RF frequency rather than the RF voltage amplitude. The apparatus was less than ideal, as the RF amplitude changed somewhat as a function of the RF frequency during the frequency scan. A mass spectrum was generated with a resulting composite RF frequency-amplitude scan. Mass peaks from a fluorinated calibrant corresponding to m/z ratios in the vicinity of 1000 u were observed. The observed resolution was poor even in comparison with spectra generated with the ion trap operated in the RF voltage amplitude scanning mode. These less than promising results, in combination with the high RF power consumption and complexity involved in maintaining the required RF voltage amplitude while scanning the RF frequency, caused George and Walt to abandon this scanning technique.

George went back to using his original RF voltage amplitude scanning apparatus. During the ensuing year or so, George was compelled to divide his attention between his ion trap work and various other projects. He was able to get assistance from the engineering department in the design and fabrication of sets of March-type cylindrical trap electrodes as well as a set of circularly profiled trap electrodes. Brief experimental evaluation of these structures indicated that they produced inferior resolution. George returned to using his original hyperbolic profiled trap.

It was during this same period that Walt turned his attention to putting the ion trap instrument under computer control. Walt chose for the hardware platform a Z80-based S100 bus microcomputer which was the state of the art at that time. Mike Knowles, a programmer from Finnigan's software development group, was enlisted to do the programming for this experimental instrument control and data system. John Syka's recollection is that Mike chose to work in the Forth language as an exercise to assess the suitability of that language for use in mass spectrometer control and data processing. The fact that all commercial ion trap instruments sold to this day run on Forth-based data systems can be traced directly to this decision.

The advent of computerization was a major step in the development of modern ion trap technology. Not only did it allow George to acquire mass spectral data, it also allowed the implementation of ever-more complex scan functions, eventually to include data dependent scan functions. It is no accident that the rapid progress in ion trap technology throughout the 1980s correlates with the extraordinary advances in microprocessor technology made during the same period.

However, during the year or so it took Walt and Mike to put the ion trap instrument under computer control, technological progress on the mass-selective instability ion trap appeared nearly to have halted. The most obvious deficiency in the instrument's performance was its limited resolution (100 to 200 FWHM), which had not seen any improvement since the initial experiments in 1980. By the summer of 1982, even a true believer in ion traps such as John Syka was beginning to have doubts whether an ion trap operated in the mass-selective instability mode would be a commercially viable instrument. It appeared to John that in order to attain resolutions with George's ion trap, which would be comparable with those available routinely with quadrupole mass filter-based instruments, it would be necessary to build three-dimensional ion traps to the same precision as said quadrupole mass filters. Thus, the three-dimensional device did not appear to offer significant advantages by way of cost or ease of manufacture. In the absence of any real cost advantage, it was not clear to John what advantage the ion trap technology would offer over the well-established quadrupole mass filter technology. John had been greatly impressed by the early ion trap work of Paul, Fisher, and Rettinghaus using mass-selective detection methodology. John entertained the notion of applying Fourier transform (FT) techniques to the mass-selective detection mode of operation of the ion trap. After encouraging discussions in September of 1981 with Bob McIver, a leading practitioner of Fourier transform ion cyclotron resonance (FTICR) mass spectrometry from the University of California at Irvine, John came to believe increasingly that a FT ion trap was the real future in ion trap technology. The potential for high resolution that a FT ion trap would offer would be a technological advantage that would more than offset the costs associated with a precision electrode structure. This vision would lead John to become involved with a joint program between Finnigan, Bob McIver, and Don Hunt of the University of Virginia to build a tandem quadrupole FTICR instrument. John expected that the experience gained from this project would be a necessary prerequisite to his long-term goal of building a FT quadrupole ion trap.

However, prevailing economic conditions were indirectly going to give the ion trap effort a needed boost. The U.S. economy was in deep recession. Compounding the situation were the extraordinarily high inflation rates that in turn resulted in high interest rates. The combination of recession and high capital costs caused demand for capital equipment to drop precipitously. As mass spectrometers are capital equipment, sales of Finnigan instruments, as well as corporate profitability, suffered. By April of 1982 the situation was so serious that Finnigan was compelled to reduce its U.S. work force by 5%.

During such lean times, the best response a company can make is to introduce new products that offer increased value in terms of capabilities, productivity, and, most importantly, price. Finnigan had recently introduced new products offering new or improved capabilities, including the first commercial tandem quadrupole instrument and a single quadrupole instrument boasting an extended mass range (1800 u). However, Finnigan had no new product offerings that were truly low in price and none were under development.

Naturally, Finnigan management was looking for a new low cost product. Upper management started to view George's ion trap project as having been a worthwhile effort that just had not panned out. Mike Story argued against this view, and persuaded T. Z. Chu, the company president and CEO, that perhaps the ion trap could be the basis for a very low cost product. George was asked to evaluate the suitability of the ion trap as the mass spectrometer in a gas chromatograph/mass spectrometer (GC/MS) system. The goal would be an ion trap GC/MS instrument having limited mass resolution and limited mass range that could be offered at perhaps one third of the price of the least expensive mass filter based systems. Presumably the low mechanical tolerances and the mechanical simplicity offered by having the mass analyzer also serving as the ion source would yield cost savings that would allow such aggressive pricing.

In the summer of 1982, Paul Kelley was asked to join George in the ion trap GC/MS evaluation effort. Paul had worked originally as a demonstration chemist, but had moved into engineering and had recently been engaged in adapting one of the small quadrupole systems to have chemical ionization capability. George and Paul commenced their task by examining the effects of helium in the ion trap in preparation for coupling the ion trap to a gas chromatograph. Paul modified the ion trap instrument to accommodate the admission of helium to the analyzer vacuum chamber. Within Finnigan, everyone associated with instrument development assumed that the effects of adding gas to the ion trap, be it helium or any other gas, would be deleterious—at least that was what most of the published literature seemed to indicate. Furthermore, considerable collective direct experience with quadrupole mass filters operating at elevated pressures only reinforced this view. The experiment was intended to determine the extent of these deleterious effects and to establish a maximum operating pressure for the instrument. Instead, as has been well-described elsewhere, when the partial pressure of helium within the analyzer was increased to ca. 1 mtorr, Paul and George observed a seemingly miraculous improvement in both mass resolution and sensitivity. Near-unit mass resolution was attained over the instrument's entire mass range, and the intensity of the mass peaks increased by an order of magnitude or more. In the wake of this result, a GC was coupled quickly to the system. As Mike Knowles's ion trap data system had evolved by then to a state of sufficient functionality to acquire GC/MS data, Paul was able to establish quickly, using "clean" standards, that this ion trap GC/MS system had quite respectable sensitivity and linearity.

III. PROJECT STARLET

In September 1982, less than 1 month after George and Paul made their remarkable discovery, a crash engineering project was started to build an ion trap-based low-cost GC detector. The goal was to build an instrument that could be sold for under $20,000 (in 1983 dollars). The product was to be announced in March 1983 at the Pittsburgh Analytical Instrument Conference and instrument shipments were expected to begin by the middle of 1983. The code name for this project, STARLET, reflected management's high expectations for this first ion trap product. This product, by virtue of its expected low cost, simplicity, and ease of use, was anticipated to dominate the low-end GC/MS instrument market and return Finnigan to profitability. Some of the initial projections of annual unit sales for this product were nearly as large as the entire lifetime production of previous mass filter-based products. These high expectations were in part driven by Finnigan's continuing grim financial situation. In November 1982, 2 months into the STARLET project, Finnigan was compelled to implement a second round of layoffs, reducing its U.S. work force by a further 5%. Those employees who retained their jobs were put on furlough 1 day out of every 2 weeks. Because of the critical nature of the STARLET project, participants were exempt from this reduced work schedule.

George Stafford, lacking project management experience, was not put in charge of the engineering project. Instead, Don Bradford, who had managed two successful quadrupole instrument-related projects, was given this job and a project team was assembled. The design of the instrument's digital electronics and firmware was assigned to Walt Reynolds and Tek Yong, a young digital electronic engineer. The design of the instrument's analog electronics would be the responsibility of Ed Marquette, a relatively young analog engineer, and Bill Fies. Bill Fies, as he had for all previous Finnigan mass spectrometers, would be the primary architect of the RF voltage generation and control electronics. Design of the instrument's analyzer, vacuum system, and GC interface would be the responsibility of Dave Stephens, a mechanical designer with much experience designing ion sources and quadrupole mass filter analyzer assemblies. The industrial design and packaging of the system was the responsibility of John Spurr, a contractor with considerable experience in instrumentation product design. The task of writing the software for the instrument's data system initially was solely the responsibility of Mike Knowles. Later Don Hoekman and Stephen Bradshaw would supersede Mike Knowles in this responsibility and complete the software for the project. As it was to be several months before any prototype instruments would be built and available for use, George Stafford and Paul Kelley were expected to continue to work with the original ion trap instrument and address various outstanding technical issues related to ion trap operation.

Even though it had been over 3 years since George conceived of the mass-selective instability scanning method, no patent application had been filed. A tutorial describing the theory of operation of the ion trap, along with data showing the remarkable enhancement in performance obtained by using 1 mtorr of helium, was provided to the corporation's patent council. Nearly 1 year later this same material would be reformulated as the first publication describing the mass-selective storage technique. Included in the material provided to the patent attorney were drawings of Dave Stephens' initial mechanical design concept of the ion trap electrode structure for the ion trap product. In an effort to reduce the parts count, Dave had produced an ingenious design in which the ion trap electrode structure was also the vacuum manifold. Elastomeric seals were located between the trap electrodes and the ceramic rings used as insulating electrode spacers, rendering the structure vacuum-tight. The Stafford, Kelley, and Stephens patent application (patent no. 4540884) was filed at the end of 1982, 3 months prior to the scheduled product announcement.

One major decision made very early in the project was to use the IBM PC as the hardware platform for the new ion trap instrument's data system. At the time, this was not an obvious choice as there were other microprocessor-based computers which were more widely accepted, such as the Apple computer and the Z80-based systems running under the CPM operating system. There were those within Finnigan who argued that the new instrument, which was intended to be marketed as a GC detector rather than as a mass spectrometer, needed only the most rudimentary data handling capabilities. It was even suggested that an analog output for connection to a strip chart recorder would be satisfactory. Walt Reynolds successfully argued that the combination of IBMs corporate stature and the open architecture of the IBM PC would make this machine the standard of the future and therefore the best choice for the new product. The ensuing rapid advancement of the IBM PC technology and application software was to generate rapidly increasing expectations for the ion trap data system, causing the scope of the software part of the STARLET project to expand continually as the project progressed. While this proved to be a burden on those who were responsible for writing the instrument's data system, it did result in a software and instrument control architecture that would eventually allow for relatively rapid incorporation of advanced functionality such as CI scan functions and automatic gain control (AGC) to the basic instrument.

In an effort to reduce the power consumption of the ring RF generation circuitry, and thereby reduce both the size and cost of a major instrument subsystem, it was decided to make the first prototype analyzer smaller and drive it at a lower RF frequency than the original experimental ion trap structure. The first prototype ion trap electrode structure had an r_0 of ca. 7 mm and was conventionally proportioned. It was operated at an RF frequency that was somewhat less than 1 MHz. While the original experimental instrument had many perforations in the ion exit end-cap electrode, this was deemed unnecessary and a more readily manufacturable seven-hole arrangement was adopted. This exit end-cap hole pattern located one hole on the tip of the electrode, with the other six holes symmetrically positioned in a hexagonal pattern about this central hole. This ion exit end-cap hole arrangement is still used in Finnigan's commercial ion trap instruments. The diameter of these holes as well as the single hole on the electron entrance end-cap electrode was ca. 1.2 mm. The spectra obtained with this first prototype analyzer exhibited very disappointing resolution. Mass peaks generated with this first prototype had peak widths that were $1\frac{1}{2}$ to 2 times as wide as those obtained with the original research system. There were many factors that could have contributed to the observed poor resolution other than the reduced trap electrode dimensions, such as instabilities in the electronics and firmware controlling the instrument. After a brief examination of these possibilities, it was concluded that the surest way to build an instrument that replicated the performance of the original research ion trap was to build a trap structure that was similar in size to the electrode structure of the original ion trap. This forced a redesign of both the ion trap electrode assembly and the RF electronics. The second version of the prototype ion trap electrode structure was conventionally proportioned and had an r_0 of 10 mm, which was slightly larger than the r_0 of the original ion trap electrode structure. To further ensure that this second prototype delivered the desired resolution, the frequency of the ring electrode RF voltage was increased to 1.1 MHz, which was 10% higher than the frequency used for the original ion trap. The ion exit end-cap for the second version of the trap retained the seven-hole pattern.

Initially, the introduction of helium to the trap produced one negative side effect: a substantial increase in the detection of spurious ion signals during the mass analysis scan. The problem was eventually traced to internally excited helium atoms produced in the vicinity of the filament. As neutrals, these excited particles could traverse the analyzer unaffected by the electron gun gate and ion trap electric fields to the vicinity of the ion detector where they would strike surfaces or other neutral molecules with low ionization potentials and create ions. These charged secondary species produced the spurious ion signals. Bill Fies solved this problem by reducing the filament bias potential from −70 to −12 V, thus preventing electrons outside the ion trapping field from acquiring sufficient energy to create excited helium species. Because no electrons were introduced into the ion trap during the mass analysis scan, the helium-related noise was completely eliminated. However, the low RF voltage applied to the ring electrode during ionization caused electrons to attain energies of only 12 to 45 eV inside the ion trap. Mass spectra obtained on the ion trap showed substantially less fragmentation than those obtained using quadrupole instruments with 70 eV electron impact ion sources.

Another problem that was to garner substantial attention during the early months of the STARLET project was the issue of tunable mass range. Although the practice of tuning the instrument by adjusting the amplitude of the ring RF voltage during ionization allowed one to achieve unit resolution at any particular mass, it did not allow unit resolution across the whole mass range while maintaining acceptable sensitivity for high mass ions. At ionization RF voltage amplitudes high enough to provide good sensitivity and resolution for mass peaks in the uppermost third of the instrument's mass range, the far more abundant low mass ions were so efficiently accumulated that the ion trap was heavily space charged. This space charge problem remained until the low mass ions were scanned out of the trap, causing the low mass peaks to have poor resolution. At ionization RF voltage amplitudes low enough to reduce the efficiency of low mass ion accumulation so that space charge effects would be absent, relatively high mass ions were so inefficiently trapped that their corresponding mass peaks would be very weak or totally absent.

As mentioned before, the origin of these problems is now well understood, but there was considerable confusion at the time. One idea was that the tuning effect was perhaps related to some residual influence of the initial phase of the RF field during ion creation. In an attempt to test this idea, a fast switching electron gate circuit was constructed that allowed electrons to enter the ion trap only during a ca. 100 ns portion of an RF voltage cycle. The RF phase during which ionization was allowed to occur could be freely adjusted. Initial results indicated that the RF phase during ionization did have an effect on the tunable mass range and for a moment it appeared that the problem was near solution. However, further analysis revealed that the observed improvement was attributed to the reduction in space charge in the trap produced by the reduced ionization time associated with ionizing during only a fraction of each RF cycle.

Eventually, a compromise solution to the tunable mass range problem was adopted in which the mass range of the instrument was scanned in segments. The RF voltage amplitude during the ionization period preceding each scan segment was appropriately adjusted to yield the desired tune. The data from each segment were concatenated to produce a full mass spectrum. No one was particularly satisfied with this procedure, but it seemed, at least initially, to produce satisfactory results.

Segmentation of the mass scan is still used in current ion trap instruments although the reason for this practice has changed. Segmentation of the mass range now serves to reduce the length of time ions remain stored within the ion trap, thereby increasing the sample partial pressure at which ion molecule reaction (self-CI) product ions begin to be observed. Adjustment of the relative ionization times for each mass range segment is also commonly used as a means of adjusting relative abundances of fragment ions in the spectra of calibration compounds to conform to specified relative abundances dictated by analytical protocols originally established for quadrupole mass filter instruments.

The STARLET instrument was introduced at the Pittsburgh Conference in March 1983 as the ITD™ 700 (Ion Trap Detector™). The introduction consisted of a few oral presentations which discussed very simplified descriptions of the novel mass-selective instability technique and its suitability for application in GC/MS. A prototype ITD™ 700 instrument was displayed but not operated, and Finnigan immediately began to accept orders. Between March 1983 and December 1984, when the first instrument was shipped, Finnigan accepted in excess of 70 orders for ITD™ 700 instruments. Most of these orders were taken within the first year after introduction. While the acceptance of the product was gratifying, the existence of deliverable orders, in many cases prepaid, served to intensify the pressure on those responsible for the development of the product.

During the spring of 1983 the mechanical, electrical, and software subsystems had developed to a sufficient level of maturity to allow extensive testing of ITD™ 700 prototypes for their intended application, GC/MS. The performance of the prototype instrument as a GC/MS was problematic because optimal ionization RF voltage tune settings as well as sensitivity were unstable. Sometimes strange peaks appeared in the mass spectra of well-known compounds. Also, the partial pressure of water vapor within the trap was much higher than it should have been. The high level of water vapor resulted in a highly space-charged trapping environment during ionization which affected the accumulation efficiency of sample ions. Because such low ionization RF voltages were used, small changes in trap space charge during ionization could cause strong changes in the accumulation efficiency of high mass sample ions. After much effort, it was determined that elastomeric seals used in the analyzer became permeable to water at the elevated operating temperature (150 to 200°C) of the trap. Because condensation of samples eluting from the gas chromatograph would occur within the analyzer if its operating temperature was reduced, the elastomeric seals had to be replaced with metal seals. However, the use of metal seals proved incompatible with the original design concept which had the analyzer serving as the vacuum manifold. This forced Dave Stephens to completely redesign the analyzer and vacuum system during the summer of 1983. Dave's second design was more conventional, with a small discrete vacuum manifold that accepted the ion trap analyzer mounted on a flange. This second vacuum system and analyzer design proved enduring as basically the same design is used in dedicated ion trap based GC/MS instruments currently manufactured. It also proved to be less expensive to manufacture than the first design. The use of metal seals dramatically reduced the partial pressure of water within the ion trap, thus improving tuning and sensitivity. The ion trap was improved, but was not yet satisfactory.

By the end of the first year the ion trap detector project was seriously behind schedule. The ITD™ 700 was not yet ready for delivery and it was not clear when it would be. Clearly, the size of the task had been greatly underestimated. Not only were there the usual myriad problems that always arise when a new instrument is created from scratch, but there were also major, unresolved issues regarding the fundamental operation of the ion trap. The demands and frustrations of the project began to affect the participants. Serious differences of opinion arose between the principal figures in the project regarding the best way to address and resolve the problems with the trap's behavior. In September 1983, George Stafford began a 9-month sabbatical leave and returned to the University of Virginia, where he had done his graduate work, to teach. Paul Kelley increasingly devoted his time to organizing the construction of a pair of ion trap instruments designed specifically for experimental purposes. These ion trap instruments were driven with a mixture of ITD™ and Finnigan 4000 electronics and had spacious cradle-type vacuum systems that made them ideal for experimental work. Walt Reynolds and Pedro Ceja, a most able technician, assisted Paul in putting these systems together. Upon completion, one of these instruments was shipped to George Stafford at the University of Virginia, and the second instrument was for Paul's use. The original ion trap instrument was now obsolete and was cannibalized.

During the autumn of 1983 the people primarily responsible for characterization and solution of the problems with the ITD were Don Bradford and a new addition to the group, Dennis Taylor, a chemist with years of experience in finding and resolving problems with prototype instruments. As Don Bradford was also managing the project, Dennis did most of the experimental work. Although completely new to ion traps and working in isolation, Dennis proceeded to characterize the various ion trap pathologies. Dennis confirmed that the occurrence of spurious mass peaks that had sometimes been observed with the elastomerically sealed trap structure persisted with the redesigned analyzer and vacuum system. Dennis also identified another particularly vexatious problem. The ratios of the isotopically different molecular ion peaks were sometimes observed to change substantially across a GC peak. In some cases at the top of the GC peak the molecular ion peak containing a single ^{13}C appeared to grow more intense than the purely ^{12}C isotopic peak. Paul Kelley had observed one instance of this phenomenon the previous spring, but concluded it was some sort of space charge effect. Dennis showed that space charge was not the likely culprit as the resolution of the isotopic peaks was maintained even though the ratios changed. These serious problems were present in addition to a residual variability in the instrument's tuning and sensitivity caused by daily variations in background gas pressure and column bleed.

In January 1984, the problems with the performance of the prototype ITD™ 700 instrument appeared so intractable that the engineering department manager and the ITD™ project manager decided to propose that the engineering project be suspended until the research department could offer solutions to these problems. The problem of the dynamically changing isotope ratios and an alleged inadequacy in sensitivity were the primary justifications for this step. A summit meeting to consider this extreme step was held on Super Bowl Sunday in January 1984. The principals in the ion trap development effort (George Stafford, Paul Kelley, and

Don Bradford) and the senior management of Finnigan attended this meeting. Also, Finnigan's most trusted consultants from universities were invited to the meeting. John Todd traveled from the U.K. for this 1-day meeting. The stakes, of course, were very high as the suspension of the ITD™ project would adversely affect both the financial prospects and the reputation of the company.

When the data were presented during the meeting, it was immediately recognized by a number of the university consultants that the problem with dynamically changing isotope ratios was caused by the generation of protonated molecular ions due to ion/molecule reactions that occurred at relatively high sample partial pressures. Because sample partial pressures will vary by orders of magnitude during elution from a GC, it was not at all surprising that the relative abundance of protonated molecular ions often changed across a GC peak. It was asserted that the observation of these protonated molecular ions was a unique and advantageous feature of the ion trap. Furthermore, the data also indicated that the ITD™ 700 had more than adequate sensitivity for a commercial product. Based on these conclusions, Finnigan management decided to continue with the ITD™ 700 project.

The day after the summit meeting, both the engineering department manager and the ITD™ project manager remained unconvinced as to the inherent sensitivity of the ion trap. George Stafford, who remained in San Jose after the meeting, put in an all-night effort in the laboratory and obtained data that demonstrated unambiguously that the ITD™ 700 had much better sensitivity than was called for in the instrument's specifications. This high sensitivity was obtained by using a sufficiently high ionization RF voltage amplitude.

IV. MASS SHIFT

Sometime during the summer of 1984, somebody, perhaps an application chemist, noticed that something was wrong with the mass spectra of nitrobenzene produced with the ion trap. The molecular ion mass peak was consistently appearing at m/z 122 rather than at m/z 123. Dennis Taylor, as the project's system chemist, began to investigate this problem. Dennis quickly reproduced the problem on his own instrument. It should be noted that the ion trap data system did not allow acquisition of profile mass spectral data. All mass spectra were presented as "sticks" and each mass assignment was integral. No provision was made to represent fractional m/z ratios. Detected ion current signal was allocated to the nearest nominal mass. Upon showing his data to the other members of the development group, he was confronted with considerable skepticism. Because of the combined effects of ion/molecule reactions and line spreading due to space charge, incorrect mass assignments were not particularly unusual at high sample concentrations. The general view was that what Dennis observed was probably related to one of these pathologies. There was also the possibility that some error had been introduced into the display and acquisition software.

Dennis, trusting in his data, persisted in his investigation of this problem. Dennis acquired ion trap mass spectra of codeine, which has a fragment ion at m/z 124, and therefore could be compared directly with the nitrobenzene molecular ion. However, m/z 124 of codeine was assigned correctly. This observation indicated strongly that there were no peculiarities in the scan generation or spectral acquisition and display software.

Dennis also set about obtaining a profile mass spectrum of the molecular ion region of nitrobenzene. The only means for obtaining a profile spectrum with the instrument software was to make a screen capture of the profile mass peaks displayed when the instrument was run in its tuning mode. Dennis resorted to keeping his instrument in the tuning mode during the GC run. If he was alert, he was able to capture one or two snapshots of the molecular ion region of nitrobenzene during the few seconds it could be observed as it eluted from the gas chromatograph. It was in this way that Dennis obtained the first profile ion trap mass spectrum of nitrobenzene. His spectrum was very similar to the one shown in Figure 3. In this figure the nitrobenzene molecular ion is clearly centered near the threshold between the intervals corresponding to m/z 122 and m/z 123. The FC-43 fragment ion at m/z 131 present in the same spectrum appears centered within the assignment interval corresponding to its true m/z ratio. It was now abundantly clear why nitrobenzene molecular ion peak was frequently assigned to m/z 122. As it was effectively

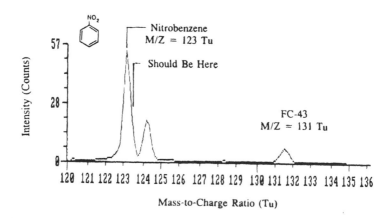

FIGURE 3
Mass spectrum of nitrobenzene obtained with a quadrupole ion trap having the theoretically correct electrode geometry.

shifted −0.4 u from its correct position, any small change in instrument calibration could easily cause most of the peak to be attributed to m/z 122.

The resident experts convened, and a list was made of the possible sources of the problem. The fractional nature of nitrobenzene's mass shift completely precluded ion/molecule reactions as the cause of this disturbing effect. The absence of any loss in resolution associated with high space charge conditions within the trap made it seem very unlikely that space charge was the cause. However, a more definitive experiment would need to be done before space charge could be eliminated completely as a cause. Again, some suggested that systematic errors in the RF amplitude control electronics were the cause. Programming errors in the software that both controlled the instrument and performed data acquisition and display were also suggested as possible culprits.

However, more substantive ideas were also put forth. It was suggested that perhaps the observed mass shift was caused by differences in the rate of relaxation of the ion displacements to the center of the trap brought about by differences in the collision cross-sections between differing ion species. The idea was that more weakly damped ion species, if provided insufficient time to relax prior to being ejected from the trap, might remain at larger displacements than other more strongly damped ions and, therefore, would leave the trap prematurely.

It was also suggested that mass shift was due to field errors. Perhaps field perturbation caused by the electron multiplier potentials penetrating through the holes in the end-cap electrodes was the cause. Another possibility was field perturbation caused by surface charging of the ceramic spacers. Early in the project such charging had caused random drifts in resolution and mass assignment. However, this problem had been resolved entirely by designing the ion trap electrode assembly so that the ion trapping region was effectively shielded from any dielectric material. It was also suggested that this "mass shift" was perhaps caused by systematic errors in the fabrication of the electrode structure.

During September and October of 1984, Dennis Taylor, Paul Kelley, and others systematically eliminated all but the last of these possibilities. John Syka recalls one rather grim autumn afternoon when all interested parties gathered around Paul Kelley's ion trap research instrument to observe an experiment. An oscilloscope had been connected to the instrument to allow simultaneous observation of both the RF voltage applied to the ring electrode and the raw ion current signal at the output of the instrument's electrometer. By comparison between the positions of the m/z 123 of nitrobenzene relative to nearby FC-43 fragment ion peaks, the shifted position of the nitrobenzene ion was established. No irregularities in the RF amplitude ramp were observed. The leak valve used to control the partial pressure of nitrobenzene in the trap vacuum chamber was adjusted so that the intensity of the nitrobenzene ion, which was initially quite intense, was reduced by about two orders of magnitude. Neither the resolution nor the position of the m/z 123 peak changed. This particular experiment established definitively that mass shift was not a space charge-related phenomenon. Furthermore, direct observation of the ion current signal and the applied RF voltage eliminated the possibility that mass shift was an artifact generated by electronics or software.

Other experiments showed that neither the multiplier potential field nor surface charging of the ceramic spacers were the origin of the problem. Experiments involving variation of the delay between ionization and ion detection seemed to eliminate the possibility that insufficient ion relaxation had any association with the problem.

During this same period, Dennis Taylor carried out the monumental task of examining a multitude of other compounds to see what other substances exhibited mass shift in their mass spectra. While the majority of the substances that he examined exhibited no irregularities in their mass spectra, he did find other instances of mass shift. He identified the molecular ion of anthracene as one of these cases. The anthracene molecular ion exhibited a positive mass shift and was observed at m/z 178.5 rather than at m/z 178, as is shown in Figure 4. The most extreme case of mass shift that Dennis identified was the molecular ion of pyrene which sometimes exhibited a mass shift of almost 1 u so that this ion was observed at m/z 203, rather than at m/z 202, as is shown in Figure 5. The hexachlorobenzene molecular ion, m/z 282, exhibited a positive mass shift of 0.5 u. The hexachlorobutadiene molecular ion exhibited no mass shift, but a fragment ion from this compound at m/z 223 exhibited a positive mass shift of 0.5 u. Upon inspection, the calibration compound, FC-43

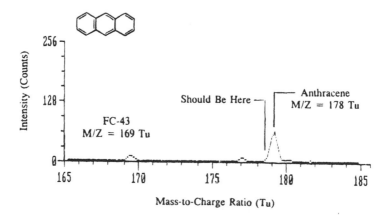

FIGURE 4
Mass spectrum of anthracene obtained with a quadrupole ion trap having the theoretically correct electrode geometry.

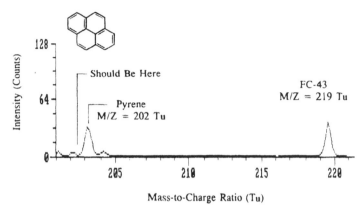

FIGURE 5
Mass spectrum of pyrene obtained with a quadrupole ion trap having the theoretically correct electrode geometry.

(perfluorotributylamine) exhibited mass shift as its fragment ion peak at m/z 119 appeared to have a slight negative mass shift of perhaps 0.2 u. Prior to the identification of mass shift as a phenomenon, it had always been noticed that this particular peak did not quite calibrate properly whenever the calibration procedure was run. However, because the deviation was on the order of one ADC sample interval, this observation had been no cause for concern. Table 1 is a summary of the mass shifts identified by Dennis Taylor while using an ion trap having the theoretical geometry.

By the end of October 1984, the mass shift problem had become a major crisis. The viability of the ITD™ 700 and the entire ion trap program was threatened. In the absence of a solution to the mass shift problem, the ion trap was not truly a mass spectrometer.

As all other possibilities seemed to have been eliminated, everyone involved began to concentrate on the idea that the mass shift problem was caused by some unknown field imperfection. Limited consideration was given to the physical properties associated with ion structure, such as dipole moments or physical cross-section that in combination with field inhomogeneity would somehow cause the observed ion structure-dependent dispersion. The task of identifying the field fault that was supposedly responsible for causing the mass shift problem was partitioned between three people.

Dennis Taylor was to tackle the problem of verifying that the elements of the ion trap structure were actually built to the dimensions given in the mechanical drawings of these parts. Dennis was also to determine experimentally whether the slight variations in the relative positioning of analyzer parts allowed by assembly clearances could affect the magnitude of the observed mass shifts.

TABLE 1

Summary of Mass Shifts for Ions from a Variety of Compounds
as Observed with an Ion Trap Having the Theoretical Geometry

Parent compound	m/z (/Th)	Mass shift
Nitrobenzene	123	−0.5
Codeine	124	0.0
Pyrene	202	+0.7
Anthracene	178	+0.5
Hexachlorobenzene	282	+0.5
Hexachlorobutadiene	258	0.0
	223	+0.5
Perfluorotributylamine (FC-43)	69	0.0
	100	0.0
	119	−0.2
	131	0.0
	219	0.0
	264	0.0

John Syka, who was working on a project unrelated to ion traps, was drafted to check the engineering drawings to ensure that the design of the ion trap was going to produce an ideal quadrupole field. It was intended that John's mechanical engineering background in combination with his theoretical understanding of ion trap operational theory would alert him to anything in the mechanical design that might be problematic. As the person with the strongest mathematics and physics background, John was asked to develop some theory as to what sort of field errors might lead to the observed mass shifts. In John's opinion, this latter task could be interpreted as a measure of just how desperate the situation had become.

Paul Kelley, who was by then the engineering manager of the project, took on the task of trying to effect some change in the observed mass shift by experimenting with the ion trap geometry. While minor differences in the degree of the mass shift had been observed with different ion traps, as yet nobody had been able to induce any change in the observed mass shifts. Even if Paul could alter the ion trap so as to produce bigger mass shifts, this would be a start. Paul's ion trap research instrument with its cradle vacuum system was ideally suited for this line of experimentation.

Immediately John set about developing a physical model that would explain mass shifts. An interesting aspect of the mass-shifted peaks, and one which had guided John in his thinking, had been apparent from the beginning: the mass resolution of shifted and nonshifted peaks was the same. It really was like two different ions having the same m/z ratio, but different structures were experiencing different values of q_z although, of course, at the same RF voltage. They behaved appropriately in the sense that when the value for q_z increased to 0.908, each ion species was ejected with the expected mass resolution. However, because ejection occurred at slightly different RF voltages, each ion species produced peaks that corresponded to different inferred m/z values. The question was: how could ion species experience different values of q_z as a result of their structures?

John commenced with an examination of the equations of ion motion. In the most idealized case, ion motion within a perfect three-dimensional ion trap is independent in each dimension. In any dimension in the three-dimensional ion trap, the motion is governed by the Mathieu equation

$$\frac{d^2u}{d^2\xi} + \left(a_u - 2q_u\cos(2\xi)\right)u = 0 \qquad (1)$$

However, this model is valid only if there are no field perturbations due to space charge or electrode structure, no collisions, and the magnitudes of the applied fields remain constant. Clearly, the Finnigan ion trap violated all of these conditions at least part of the time. The greatest violation of these conditions was that the trap was being operated with a bath gas pressure of ca. 1 mtorr. At this pressure, numerous collisions between trapped ions and the helium bath gas atoms were a certainty. While a completely rigorous representation of the effects of collisions would introduce coupling between all three dimensions of motion, a simplified model, in which the ion motion in each dimension was still treated as being independent would modify the Mathieu-type equations of motion by adding two terms

$$\frac{d^2u}{d\xi^2} + K(\xi)\frac{du}{d\xi} + \left(a_u - 2q_u\cos(2\xi)\right)u = H(\xi) \qquad (2)$$

The first new term, $K(\xi)\,du/d\xi$, is a statistical damping term reflecting that, for an ion in motion, collisions produce an opposing force on the ion in proportion to its velocity, $du/d\xi$. Generally, this is the only term that collisions are considered to add to the equation of motion. Alone, this term would cause stable ion trajectories to damp to the center of the trap. However, John believed that even in the absence of space-charge effects, ion motion could not collapse into an arbitrarily small area at the center of the ion trap. John convinced himself that even a single ion in the presence of damping gas would not relax to a full stop at the absolute center of the ion trap, because the random bombardment of the ion by neutral particles would continually kinetically excite the ion. For this reason John added the second term, $H(\xi)$, to represent the random excitation of an ion due to random bombardment by bath gas atoms and other neutral species. From a qualitative point of view, Equation 2 expresses the existence of a balance between collisional focusing and collisional excitation. Inspection of this equation indicates that the term $K(\xi)\, du/d\xi$ causes initially large ion trajectories to collapse toward the center of the field until an equilibrium is reached between the dissipation and excitation terms, and a terminal ion trajectory having some average amplitude is established.

Because the terms $K(\xi)$ and $H(\xi)$ depended in part on the collision cross-section of an ion, a physical attribute clearly associated with the structure of an ion thus introduced ion structure as a relevant parameter that might affect ion motion. While it seemed clear to John that ion species with different collision cross-sections would relax to trajectories with different average amplitudes, he did not see how this translated into the observed mass shift. John believed that there would be an insufficient number of collisions occurring during the short period of rapid growth in the trajectory of an ion prior to ejection to alter the ejection voltage of an ion. Furthermore, variations in the average z-dimension oscillation amplitude, associated with differing collision cross-sections, would most likely result in different peak widths for different ion species, something that had not been observed. However, if there were a significant field error, motion in the x, y, and z dimensions would be coupled, and perhaps ion distribution in the radial dimension would result in different perceived values of q_z. This possibility led John to consider the nature of ion distributions in the radial dimension.

John was well aware from the literature that in order to represent ion motion in the radial dimension accurately using a polar coordinate system, it is necessary to account for angular momentum. In the ideal case, the equations of motion for the x- and y-dimensions which have the form of Equation 1 combine to create a single equation describing motion in r which is given as

$$\frac{d^2r}{d\xi^2} + \left(a_r - 2q_r \cos(2\xi)\right) r - \frac{k^2}{r^3} = 0 \qquad (3)$$

where

$$k = \left[r^2 \frac{d\theta}{d\xi} \right]_{\xi=0} \qquad (4)$$

The additional term associated with angular momentum behaves as if it were a strong repulsive force acting on the ion when it approaches the central axis, and which always prevents the ion from reaching the central axis. The only time that an ion would not be subject to this "force" would be the rare case when an ion was initially directed either toward or away from the central axis of the ion trap, giving the ion no angular momentum. While this discussion of the effect of angular momentum has neglected the effects of collisions, the basic conclusions should still apply. John reasoned that as collisions cause ions to relax to equilibrium trajectories, collisions would also establish an equilibrium average angular momentum which would "force" ions to be distributed around rather than on the central axis.

John also appreciated that even if one ignores particulars of ion motion in the x, y plane, any ion distribution, when represented solely as a function of r, will by definition yield a nonzero average radial displacement. To demonstrate this to himself, John arbitrarily supposed that at equilibrium all the trapped ions of a common type were uniformly distributed within a maximum radius, r_{max}. Such a distribution is shown in Figure 6a. In this figure the number ions per unit area as a function of r is constant up to a r_{max} of $0.2r/r_0$. However, when this distribution was plotted in terms of the number of ions per unit of radial displacement, as in Figure 6b, the distribution became skewed toward the higher displacements. As more area existed at the larger radial displacements more ions were found at larger radial displacements.

Thus, when John considered the effects of angular momentum in combination with the notion that different ion species having the same m/z ratio but different collision cross-sections would dampen to different av-

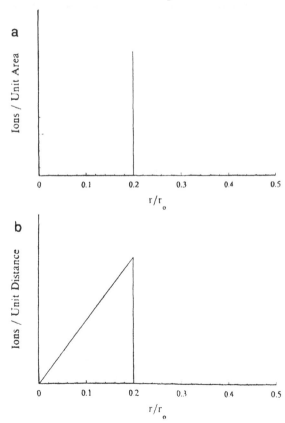

FIGURE 6
Radial ion distribution at equilibrium: (a) uniform distribution, in terms of ions/unit area, up to $r = 0.2r_0$; (b) linear dependence of ions/unit distance up to $r = 0.2r_0$.

erage oscillation amplitudes, it seemed plausible that ion species having the same m/z ratio but different structure could have distinct average radial displacements. John envisioned the terminal radial distributions of two populations of ions having the same m/z ratio but differing collision cross-section to be as shown in Figure 7. If this were so (and John realized this was a big if), then a field error which would cause ions to experience different values of q_z as a function of radial displacement could cause the observed mass shifts.

John began his search for a type of field error that would cause q_z to be a function of r by considering the perturbation of the trapping field caused by displacing one end-cap electrode axially as shown in Figure 8. Perceiving that this change to the structure would induce perturbing fields of primarily odd order, he decided to treat the more purely odd order case in which both end-caps were displaced in the same direction along the z-axis. Treating the potential field as a power series expansion in r and z, John eliminated all terms not allowed by the radial symmetry of the device and the uneven symmetry of the perturbation in the structure's geometry. After differentiating with respect to z to obtain a power series expansion of the z component of the electric field, he then elimi-

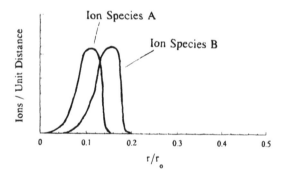

FIGURE 7
Illustration showing different radial distributions for ion species, A and B, having the same mass-to-charge ratio, yet with different structures.

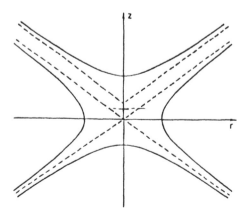

FIGURE 8
Cross-sectional view of an ion trap in which a single end-cap electrode has been displaced axially in order to introduce odd-order potential field terms.

nated all terms except for constants and those terms that varied linearly with z. The justification for this step was that he was looking for a dependence of q_z on r and that q_z only had meaning as a coefficient of a term that varies linearly with z. He then used this abbreviated expansion to derive an approximate z-dimension equation of motion for ions in the perturbed ion trap. After applying a coordinate substitution

$$z' = z - \Delta z \tag{5}$$

that effectively shifts the z-axis origin by the average amount that the end-caps were displaced, Δz. The resulting equation of motion, shown as Equation 6, was obtained:

$$\frac{d^2 z'}{d\xi^2} - 2q_z \cos(2\xi)\, z' = 0 \tag{6}$$

However, because Equation 6 was simply another Mathieu equation, it was clear that such odd order field perturbations failed to yield any dependence of q_z on r.

John then began to consider the effects of the perturbation of the trapping field that would be caused by symmetric axial displacement of the end-cap electrodes from the center of the device as illustrated in Figure 9. Such an error, because of its even symmetry, would introduce purely even order terms to the potential field. Using the same procedure he had previously applied in the odd symmetry case, he developed Equation 7 as the approximate equation of motion for ions in the z-dimension

$$\frac{d^2 z}{d\xi^2} - 2q_z\left(1 + \epsilon_2\, r^2 + \epsilon_4\, r^4 + \cdots\right)\cos(2\xi)\, z = 0 \tag{7}$$

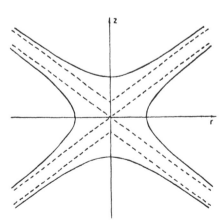

FIGURE 9
Cross-sectional view of an ion trap in which both end-cap electrodes have been displaced axially in order to introduce even-order potential field terms.

Inspection of Equation 7 indicated that this time the desired result had been obtained, as q_z was clearly represented as a function of r. Evaluation of the coefficients, ε_n, of the power series expansion of q_z, would necessitate obtaining an approximate polynomial expansion for the trapping field produced by a particular end-cap displacement. However, comparison of the ideal end-cap profile corresponding to the new z_0 established by the altered spacing of the end-caps with the actual profile of the end-caps did allow a qualitative description of the dependence q_z on radial displacement, as is depicted graphically in Figure 10. It should be noted that Figure 10 is drawn by hand. The purpose of this figure is to portray the idea that if the end-cap electrodes are displaced inward, larger average ion radial displacements would cause ions to experience higher relative q_z values, whereas when the end-caps are displaced outward, larger average ion radial displacements would result in ions experiencing lower relative q_z values. These displacements indicated that relative mass shifts could be reversed if the end-caps were spaced out rather than spaced in. Certainly, one should expect to see changes in observed mass shifts if the end-cap spacing was altered.

However, what was truly important was that this collection of ideas appeared to identify a class of electrode geometry errors which could cause the observed mass shifts. When John came to this notion late one night, he became highly excited as he was absolutely convinced that either he or Dennis was going to find a geometric error in the trap structure that would introduce even order field terms. Such errors could be

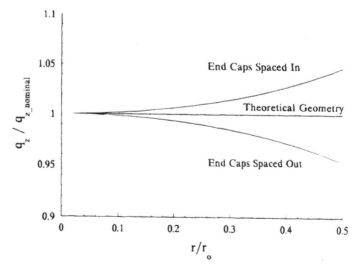

FIGURE 10
Variation of q_z with the radial position of an ion as a function of end-cap electrode separation.

introduced readily in either the design or fabrication of the parts. Incorrect dimensioning of the ceramic insulators or the electrode parts could also easily result in such errors. A systematic error in electrode profiling or dimensions introduced by errors in the programming or operation of the numerically controlled lathe that created these parts was another possibility. John was convinced that a solution to the mass shift problem was at hand. When he attempted to explain his ideas to Dennis, Paul, and George they were skeptical.

Within a few days, Dennis returned with the results from measurements of an ion trap analyzer assembly. These measurements, obtained with a coordinate measuring machine, indicated that the dimensions of individual parts and of the total assembly were within the tolerances given by the mechanical drawings. By then, John had completed his examination of the entire set of analyzer mechanical drawings, and he had found them to be error free. John was forced to conclude that the ion trap electrode structure, as designed, should have been producing a good theoretical three-dimensional quadrupole field. The results from Dennis's experimental work was equally disappointing. Dennis had examined a number of different ion trap structural error conformations by shifting the relative positions of the ion trap electrodes as much as was allowed by the assembly clearances. He had been unable to induce any changes in the observed mass shifts in this manner. The inability to identify any systematic errors in the ion trap electrode assembly meant that no solution to the mass shift problem was at hand. John's ideas, which seemed to offer a physical explanation for the problem, were beginning to appear completely invalid. No one believed that the known dimensional errors of the parts and of the ion trap structure were large enough to cause the mass shift problem. In any case, at this point in the project, building a more accurate trap was simply not an option.

John can recall reviewing the situation late one afternoon with Dennis and Paul. Both John and Dennis were somewhat grim; Paul was musing about the experiments that he had been doing in the laboratory which involved tilting one of the end-caps. He had moved out an end-cap electrode and had tilted it so that it involved even stranger-order field terms but, basically, it would introduce odd-order field terms. He achieved this tilting by placing a shim beneath one of the ceramic insulating spacers between the ring and end-cap electrodes; the magnitude of the tilt was 0.25 mm or so, as shown in Figure 11. As he had observed no change in behavior of the ion trap, the next thing that he wanted to try was to space out both end-caps. John immediately brightened up at this prospect and encouraged Paul with this approach (not that encouragement was necessary, as Paul was going to do it anyway). John still harbored the hope that his beautiful ideas were not wrong. Paul's results using an ion trap with a tilted end-cap were, in fact, consistent with John's explanation as

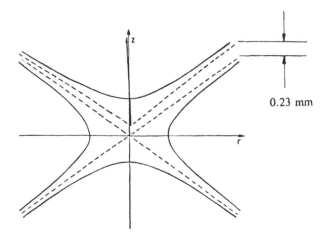

0.23 mm

FIGURE 11
Cross-sectional view of an ion trap in which a single end-cap electrode has been tilted in an attempt to introduce potential field terms which may influence mass shift.

the symmetry of the field error introduced by a single tilted end-cap was odd and therefore should not cause mass shifts. However, Paul's proposed experiment would introduce the exact sort of even symmetry field error that according to John's explanation, should produce mass shifts. If there were any validity to John's ideas, Paul's proposed experiment would produce a change in observed mass shifts.

Paul returned to the laboratory and shifted both end-caps out by an extraordinary amount, some 0.75 mm as shown in Figure 12, which was a little more than 10% of the value of z_0 of the ion trap. The ion trap was of field symmetric geometry with $r_0^2 = 2z_0^2$. Paul's shims, which he placed between the end-caps and the ring electrode, were relatively thick, and when he looked at the mass spectrum of nitrobenzene, it was clear that he had effected a change in ion trap performance. The molecular ion of nitrobenzene, as shown in Figure 13, was exactly where it ought to be, right on $m/z = 123$!

This result produced a profound sense of relief in all who were involved with ion trap development. Paul immediately went on vacation. John and Dennis continued with the investigation of Paul's "spaced out" ion trap to ensure that the effect applied to other compounds, and proceeded to confirm that the molecular ion of anthracene was at the correct mass assignment (Figure 14), as was pyrene (Figure 15); pyrene had exhibited the largest mass shift, so the elimination of the mass shift for pyrene in the shimmed ion trap was sufficient to convince everyone that the problem had been solved.

Experience over the intervening 8 years indicates that the problem was indeed solved. During this period Finnigan has delivered over 2000

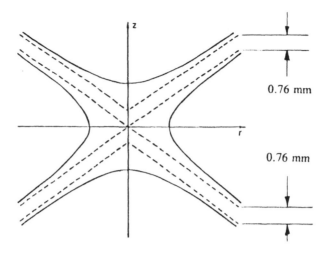

FIGURE 12

Cross-sectional view of an ion trap in which both end-cap electrodes have been symmetrically displaced by 0.76 mm in an attempt to introduce potential field terms which may influence mass shift.

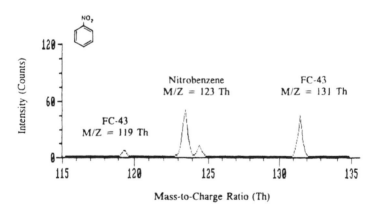

FIGURE 13

Mass spectrum of nitrobenzene obtained with an ion trap having the electrode geometry shown in Figure 12.

ion trap-based instruments incorporating both a "spaced out" electrode geometry and the use of helium bath gas. There has never been an instance in which a user complaint related to mass assignment with one of these instruments was the result of a true mass shift.

Dennis and John then examined a couple of other symmetric displacements of the end-cap electrodes from the ring electrode; they found that when the displacement was only some 0.5 mm (as compared to some

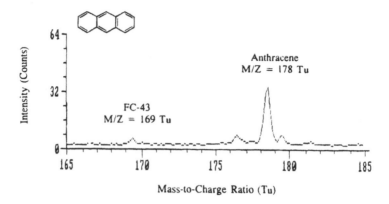

FIGURE 14

Mass spectrum of anthracene obtained with an ion trap having the electrode geometry shown in Figure 12.

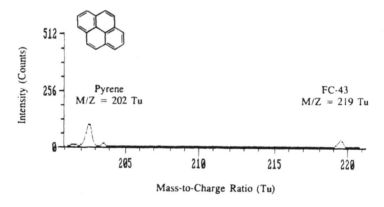

FIGURE 15

Mass spectrum of pyrene obtained with an ion trap having the electrode geometry shown in Figure 12.

0.75 mm) the mass shifts could be improved, i.e., reduced, but the mass shifts for nitrobenzene, pyrene, and anthracene were still evident, as shown in Table 2. When the symmetric displacement of the end-cap electrodes from the ring electrode was increased beyond 0.75 mm, no mass shift was observed; they were half-expecting to see some form of reversal in the direction of the mass shift with a grossly extended ion trap, but they did not. Such a result would have been consistent with the ideas represented in Figure 10.

At that time, they were certainly familiar with descriptions in the literature regarding the nonlinear resonance effects that introducing nonideality in the electrode geometry was likely to induce. Prior to the observation of the mass shift effect, it was thought that the damping provided by the presence of helium bath gas quenched these effects. While no ill effects due to the use of end-cap shims had been observed, no one had absolute confidence that damping would entirely quench effects such as nonlinear resonances associated with such a significant deviation from the ideal geometry. There was no time to engage in a research project. A decision needed to be made and caution dictated that they should make the minimum change in trap geometry needed to resolve the mass shift problem. Thus, the dimensions arrived at, and which are given in Table 3, were $r_0 = 10$ mm and $z_0 = 7.83$ mm. Of course, the effective value of r_0 is 10.55 mm, as obtained from Knight's equation. The ring electrode hyperboloid profile corresponds to a value of r_0 of 10 mm exactly, and the end-cap electrode hyperboloid profiles correspond to a value of z_0 of 7.07 mm exactly.

Why did the shimmed ion trap work? What had happened? Why did spacing out the end-cap electrodes improve mass assignment rather than induce a mass shift? Upon reflection, John recalled that there were holes in the end-cap electrodes that they had neglected. He began to think about the holes located at the tips of the end-caps. He deliberately neglected the other six holes located on the exit end-cap, reasoning that they would introduce field perturbations having the wrong symmetry. The axially positioned holes would produce an equation of motion such as

$$\frac{d^2z}{d\xi^2} - 2q_z\left(1 + \epsilon_2' \, r^2 + \epsilon_4' \, r^4 + \cdots\right)\cos(2\xi)\, z = 0 \qquad (8)$$

This equation of motion is identical in form to the equation of motion obtained for the case in which the end-cap electrodes are symmetrically displaced (Equation 7). The two equations of motion would differ only by the magnitude of the coefficients in the series expansion of q_z in terms of r. John expected that the holes in the end-cap electrodes would have the

TABLE 2

Summary of Mass Shifts for Ions from a Variety of Compounds as Observed with Ion Traps Having Various End-Cap Electrode Displacements

Parent compound	m/z (/Th)	Observed mass shift (/Th); end-cap electrode spacing			
		0 mm	0.51 mm	0.76 mm	1.01 mm
Nitrobenzene	123	−0.5	−0.2	0.0	0.0
Pyrene	202	+0.7	+0.3	0.0	0.0
Anthracene	178	+0.5	+0.2	0.0	0.0
Hexachlorobenzene	282	+0.5	0.0	0.0	—
Hexachlorobutadiene	223	+0.5	0.0	0.0	—

TABLE 3

Summary of the Finnigan Ion Trap Dimensions

r_0 = 10.00 mm
z_0 = 7.83 mm
$r_{0,equivalent}$ = 10.55 mm
Ring electrode hyperbolic profile corresponds to r_0 = 10.00 mm
End-cap electrode hyperbolic profiles correspond to z_0 = 7.07 mm

effect of flattening the field isopotential surfaces in the vicinity of the central axis of the ion trap. This flattening effect would translate to a dip in the effective value of q_z near the central axis as is depicted in Figure 16. Ions near the central axis would perceive a somewhat lower value of q_z than those having larger radial displacements. Figure 17 depicts what John concluded would be the combined effects of the holes in the end-caps and successively larger end-cap electrode spacing on the variation of q_z with radial displacement. If this were true, then displacing the end-caps outward from their ideal position would effect a flattening out of the dip in the dependence of q_z on radial displacement induced by holes at the tips of the end-cap electrodes. It seemed reasonable to John that with an appropriately chosen end-cap spacing, one could compensate for the effects of the holes located on the tips of the end-cap electrodes in the region near the center of the trap. John convinced himself that displacing the end-caps effectively improved the homogeneity in the quadrupole field near the center of the ion trap at the expense of the field at higher radial displacements. However, this situation seemed highly advantageous as damping, caused by the helium buffer gas, restricted the location of ions during the mass analysis RF scan to a region close to the center or at least the central axis of the device. This explanation, for lack of any alternative, became the best explanation within Finnigan of why mass shifts were observed and why adopting a spaced out trap geometry solved the problem.

V. EPILOGUE

The mass shift problem was the last major issue preventing shipments of the ITD™ 700 instruments. The resolution of the mass shift problem in November 1984 allowed Finnigan to begin shipments of the ITD™ 700 by the end of 1984.

Prior to commencing shipments, which under U.S. and E.P.C. patent law would constitute public disclosure, the filing of a patent application was considered in order to protect the knowledge relating to the solution of the mass shift problem. Although the modification of the ion trap geom-

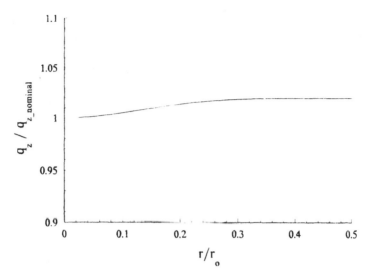

FIGURE 16

Variation of q_z with the radial position of an ion as induced by the presence of holes in the in the end-cap electrodes.

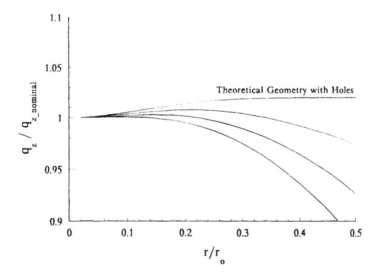

FIGURE 17

Variation of q_z with the radial position of an ion as induced by the presence of holes in the tips of the end-cap electrodes combined with increasing end-cap electrode separation. The lowest curve corresponds to the largest end-cap electrode separation.

etry that solved the problem was trivial, it was knowledge Finnigan gained at considerable pain and expense. It was information that was of little or no use to users of the instrument, but of enormous value to potential competitors. No one involved with the mass shift problem believed that the spaced out ion trap electrode geometry was unique in its ability to suppress mass shift effects. It seemed highly likely that there were other electrode geometries, based on different electrode profiles, that would have similar effect. This posed the problem of how to go about writing patent claims that would cover all such geometries. Finnigan's corporate patent counsel advised that retaining this knowledge as a trade secret was the best policy, and this advice was accepted. All Finnigan employees who had knowledge of the modification to the electrode geometry signed a document effectively pledging them to maintain that knowledge as a company secret.

This secret was maintained until 1992, when it was decided that Finnigan users needed to be informed of the true geometry of their ion traps, as various users of Finnigan ion trap instruments had begun to observe effects such as "black canyons" that were potentially related to the modified geometry of the Finnigan ion trap. Also, the ion trap development group at Bruker-Franzen had by then described extensively various virtues of nonideal trap geometries. The primary secret was that the Finnigan ion trap was nonideal. Thus, it fell to John Louris to make a public announcement concerning the trap geometry at the 40th Conference of the American Society for Mass Spectrometry and Allied Topics, in Washington, D.C.

The delivery of the ITD™ 700 instruments precipitated a year of rapid advances in ion trap technology at Finnigan. Very early in 1985 George Stafford demonstrated that if the ring electrode RF voltage during ionization was fixed at an amplitude such that m/z 20 was at the threshold of instability, the sensitivity of the instrument was increased by more than an order of magnitude and "tuning" the ion trap was no longer necessary. It was this experiment that finally led to the collective understanding of how the effects of ion acceptance, mass-to-charge ratio, trapping field intensity, and space charge had combined to produce the tuning effect. This understanding allowed Dennis Taylor and Stephen Bradshaw, beginning that following summer, to develop a data-dependent scan function which dynamically regulated the accumulation of ions during ionization. This AGC scan function increased by more than two orders of magnitude the range of sample concentrations that could be introduced into the trap without deterioration in resolution due to space charge effects. This increase greatly expanded the range of applications for which ion trap instrumentation could be used.

Many ideas conceived during the ITD™ project had not been pursued because of more immediate concerns. Not only was there now time to explore these ideas, but the project had provided the hardware and software tools with which to pursue them. The experimental ion trap instrument built for Paul Kelley's use during the project became the vehicle for most of this work. In the early spring of 1985 the first experimentation with MS/MS scan functions began. During preparation for MS/MS experiments in which resonance excitation was employed to induce ion fragmentation, experiments incorporating resonance excitation into the mass selective instability scan were performed. At that time it was understood that this technique, now often referred to as axial modulation, could be used to extend the mass range of the instrument. However, the discovery by Michael Weber-Grabau of the ability of this technique to provide enhanced resolution while analyzing approximately 30-fold larger stored ion populations was 3 years in the future. In August 1985, John Syka's vision of a FT ion trap was realized on the same instrument. Bill Fies and John Syka demonstrated the feasibility of such a technique by producing mass spectra from transient ion image currents.

In the spring of 1985, John Louris, a graduate student of Graham Cooks, came to Finnigan to participate in the continuing MS/MS work and assist in the construction of a third research ion trap instrument which he would take back to Purdue University upon his departure. Not only did John contribute extensively to the MS/MS work, but he was the primary architect of the first CI scan functions. The instrument that returned with John to Purdue in September 1985 was the direct precursor of the ITMS™, the instrument designed specifically for ion trap research which Finnigan offered commercially soon thereafter.

The rapid development of ion trap technology brought about by the introduction of the first commercial ion trap instrument was restricted to the Finnigan research laboratory only briefly; research into and application of ion trap technology was no longer restricted to those few who had the time, resources, inclination, and fortitude to build an instrument before proceeding. For those who were to use commercial instruments, as well as those who chose to build their own, the concepts realized in the ITD™ and its sibling, the ITMS™, would be catalysts for further development and wide-ranging application of the ion trap. The greater story of this continuing expansion of endeavor involving ion traps is evidenced by the enormous amount of research described in this monograph.

ACKNOWLEDGMENT

The preceding story would never have reached print if it were not for the insistence and patience (backed up with considerable effort both in writing and in editing) of R. E. March. Thanks are also due to those in the Trent group who helped in preparing the manuscript and to K. Cox, S. Quarmby, and D. Taylor for editorial and moral support. The author bears full responsibility for the veracity and accuracy of the preceding text. Any errors or omissions are his responsibility alone.

PART 2

Ion Activation and
Ion/Molecule Reactions

Chapter 5

EFFECTS OF COLLISIONAL COOLING ON ION DETECTION

Jennifer S. Brodbelt

CONTENTS

I. INTRODUCTION

Since the introduction of the mass-selective stability mode of operation,[1] it has been recognized that the storage and detection of ions in a quadrupole ion trap are optimized by operation of the trap with a high background pressure (≥ 1 mtorr) of an inert buffer gas, typically helium. Multiple collisions between analyte ions and helium neutrals dampen the kinetic energies of ions and effectively collapse the ion cloud to the

0-8493-4452-2/95/$0.00+$.50

center of the trap, resulting ultimately in more efficient detection and resolution of the ion signal. Ion extraction is related directly to the position of ions in the ion trap prior to ejection, and it has been estimated that the most effective external detection occurs for ions located within a 2-mm cylinder in the center of the trap.[2]

The theory of ion motion within the quadrupole ion trap has been presented in detail in Chapter 2. In addition, some effects of collisions on the motion, distribution, and energy of ions in quadrupole ion traps[3] have been reported. In general, the presence of a buffer gas increases the lifetime of ions and reduces their average kinetic energies. The importance of the buffer gas pressure on the optimization of ion storage was recently evaluated experimentally.[4] In this chapter, four parameters that affect the storage of ions are discussed: the pressure of the buffer gas, the q_z value which establishes the ion trapping environment, the cooling time during which ions may collapse to the center of the trap, and the supplementary radiofrequency (RF) voltage used to translationally excite selected ions. For evaluation of collisional cooling and detection, benzene was the chosen analyte because benzene molecular ions do not undergo facile dissociation in the quadrupole ion trap, and thus competitive dissociation reactions are minimized. For examination of the effect of buffer gas pressure on collisional activation, the dissociation of n-butylbenzene ions was studied because the two predominant fragmentation pathways of n-butylbenzene molecular ions are well characterized.[5]

The conceptual model used to describe the effects of collisional cooling on ion containment is based on the three-dimensional ion cloud representation reported previously.[6] In the cloud model, the distributions of ions stored in the presence of a buffer gas are nearly Gaussian, and collisions between ions and a light buffer gas cause viscous damping of ion motion. The damping collisions serve to minimize excursions of ions from the center of the trap. Alternate models[7] to describe the motion and energies of ions in quadrupole ion traps have been reported (such as the pseudopotential well model discussed in Chapter 2), but these models typically are derived for "ideal" conditions in which the behavior of an isolated ion is examined in the absence of space charge or collisional effects.

The effects of the four parameters (helium pressure, cooling time, q_z, and kinetic excitation) on collisional cooling and detection of ions are described in the following sections.

II. EFFECT OF BUFFER GAS PRESSURE AND STORAGE TIME

The pressure of buffer gas used during the storage and detection of ions affects not only the number of collisions that ions will experience during their residence in the trap but also the rate of collisional cooling (in effect, the time between collisions). The second variable is the cooling time period between ion formation and detection, which also determines the number of collisions that ions undergo prior to detection. All collision frequencies (used to estimate the number of collisions required for optimum ion detection) were calculated from the kinetic collision rate theory for gases, assuming a cross-section of 0.8 nm^2 and a temperature of 500 K.[8] For example, in an environment of 1 mtorr helium, an ion undergoes about 30 collisions per millisecond. The ion current of benzene molecular ions was recorded as a function of cooling time at five different helium buffer gas pressures. Benzene ions were formed, apex isolated, and then cooled at $q_z = 0.4$. These results are shown in Figure 1. For cases in which buffer gas pressures of ≥ 0.5 mtorr are used, no appreciable increase occurs in the ion signal as the cooling period is extended beyond a few milliseconds. This result is consistent with the suggestion that under standard conditions (1 mtorr helium in the trap), the ion cloud cools within 1 to 2 ms to the center of the ion trap which maximizes detection of ions during the analytical scan. In contrast, when a buffer gas pressure of only 0.2 mtorr is used, there is a substantial increase in the ion signal as the cooling period is increased. In fact, after 10 ms of cooling, the ion current has increased by 250% from the ion current detected after zero delay. This result is especially significant because it indicates that under low pressure conditions (such as 0.1 mtorr) without cooling intervals, the quadrupole ion trap is filled with many more ions than are ultimately detected.

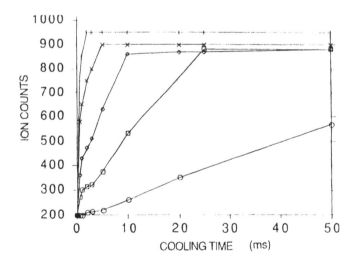

FIGURE 1

Effect of time-dependent collisional cooling on detection of benzene molecular ions at various helium pressures: O, 0.09 mtorr; ☐, 0.20 mtorr; ◊, 0.35 mtorr; x, 0.50 mtorr; and +, 0.70 mtorr. $q_z = 0.4$ at all pressures. (From Wu, U.-F. and Brodbelt, J.S., *Int. J. Mass Spectrom. Ion Processes*, 115, 67–83, 1992. With permission.)

However, with the addition of an appropriate cooling interval, the trap can operate very effectively even at low helium pressures. From an analytical perspective, collisional cooling can lead to significantly larger signals than otherwise thought possible for low pressure experiments. As shown later, this effect is particularly relevant for collisional activation studies at lower helium pressures (<0.5 mtorr).

First-order rate constants for collisional cooling at various pressures were calculated from a secondary study of time-dependent cooling (i.e., the experiment described for the results in Figure 1 was repeated for a larger number of pressures), and the rate constants are plotted as a function of pressure in Figure 2. The cooling rate constants were calculated from the initial rate of increase in ion current as a function of time, assuming that the initial rate is directly proportional to the detected ion current. The general approximation $k = \ln(I_f/I_0)/t$, where I_f is the ion current measured at the end of the cooling interval of duration, t, and I_0 is the ion current measured without a cooling delay, was applied to derive the rate constants. There is a near-linear increase in the collisional cooling rate with pressure up to 0.7 mtorr, at which point a leveling in the increase occurs. This leveling is attributed to the attainment of cooling rates faster than the minimum time scale of the experiment (i.e., there is a finite time for multiplier turn-on and RF scanning in which collisional cooling continues to occur despite the desire for a zero cooling time experiment). The absolute number of collisions required to maximize the ion signal apparently remains relatively constant for the different buffer gas pressures. For example, with 1 mtorr of helium in the trap, the benzene ions experience about 50 to 100 collisions in the 3 ms interval prior to detection. At 0.2 mtorr, a storage time of about 20 ms is required to attain a similar number of collisions. This cooling rate relationship is reflected in the curve shown in Figure 2.

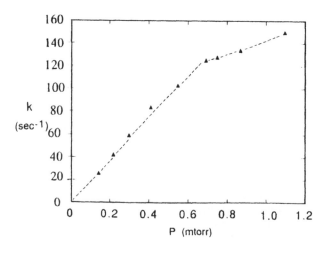

FIGURE 2
Relative rate constants for collisional cooling of benzene molecular ions as a function of helium pressure; $q_z = 0.4$ at all pressures. (From Wu, U.-F. and Brodbelt, J.S., *Int. J. Mass Spectrom. Ion Processes, 115*, 67–83, 1992. With permission.)

The rationalization for the collisional cooling phenomena was attributed to several factors. First, at lower helium pressures (<0.5 mtorr) and at low storage times (<10 ms), ions that passed the outskirts of the diffuse ion cloud during the analytical scan (the RF ramp for ejection) may more likely be accelerated into an end-cap electrode than through one of the end-cap holes. The end-cap holes are small perforations concentrated in the center of the electrode, and one can imagine that only ions positioned nearly on-axis will pass through a hole and strike the external detector. Alternatively, it has been suggested from a study of the kinetic energies[9] of ions ejected from a quadrupole ion trap that the ions at the fringes of the ion cloud may not receive the necessary kinetic energy from the RF field to facilitate efficient and coherent extraction during the analytical scan. These two factors are both consistent with the observation of increased detectable ion current at longer cooling times. For both of these factors, as the storage interval is increased, the diffuse ion cloud has sufficient time to contract due to collisional reduction of ion kinetic energies and damping of motion, and thus more ions are aligned for successful ejection and detection.[10]

III. RADIOFREQUENCY VOLTAGE EFFECTS: q_z

The q_z value as determined by the radiofrequency voltage $V_{(0-p)}$ applied to the ring electrode is the third parameter. The equation that defines q_z as introduced in Chapter 2 is

$$q_z = \frac{8eV}{m(r_0^2 + 2z_0^2)\,\Omega^2} \tag{1}$$

where m is the molecular weight of the ion, e is the charge, r_0 is the radius of the trap (1 cm), and Ω is the drive frequency (1.1 MHz). The q_z parameter establishes the ion trapping environment, influences the frequency of ion motion, and also determines the number of ions that can be stored. Additionally, the relative average kinetic energy of a trapped ion prior to any collisional cooling is directly related to the q_z value, and thus at higher q_z values, collisions are more energetic. For these experiments, helium was admitted to only 0.1 mtorr to maximize the appearance of collisional cooling effects. The ionization time was adjusted to 10 ms (at $q_z = 0.2$) to establish an ion signal of sufficient intensity in the absence of cooling. The benzene ions were stored at a range of RF levels corresponding to $q_z = 0.2$ (approx. 200 $V_{(0-p)}$) to $q_z = 0.8$ (790 $V_{(0-p)}$). The ion currents measured before and after a 100-ms collisional cooling period are shown in Table 1. There is an overall increase in detected ion current as q_z increases, until a significant drop occurs at $q_z = 0.8$. The in-

TABLE 1

Effect of q_z on Detection of Benzene Ions

	Ion current (arbitrary units)	
q_z	No cooling period	Cooling period (100 ms)
0.2	65	170
0.3	65	220
0.4	65	245
0.5	65	255
0.6	65	265
0.7	65	265
0.8	65	200

Source: From Wu, U.-F., and Brodbelt, J.S., *Int. J. Mass Spectrom. Ion Processes, 115,* 67–83, 1992. With permission.

creasing portion of the trend is rationalized by considering two factors related to q_z. First, the initial benzene kinetic energies are greatest for higher q_z values, and thus the ion cloud requires the most cooling to ensure effective damping of ions to the center of the trap. Upon cooling, the ion cloud collapses to the center and results in a more concentrated ion population. Such ion cloud focusing effects are much less important for the lower q_z values (lower frequency ion motion), in which the ions have lower initial kinetic energies prior to any collisional cooling, and thus the addition of a 100-ms cooling time will have little overall effect on the detected ion current at low q_z values. Second, the number of ions that can be trapped is greatest at the largest q_z values, and collisional cooling should maximize the storage capabilities at higher q_z values. The drop-off in the ion current at $q_z = 0.8$ is likely related to a transition in the trapping environment past some high ion kinetic energy threshold beyond which ions cannot be effectively cooled on a time scale, competitive with ion loss processes (i.e., neutralization by impingement on an electrode). At standard helium pressures (1 mtorr), the enhancement of ion detection at variable q_z due to collisional cooling is not observed because the ion cloud has already sufficiently collapsed in the first few milliseconds. These results indicate that the value of q_z plays a particularly important role in determining the trapping environment at low pressures, and that higher q_z values are more effective for retaining larger ion populations.

IV. SUPPLEMENTARY TRANSLATIONAL EXCITATION

The fourth experimental variable of interest is the supplementary RF voltage applied at the resonant frequency of a selected ion for a specified period of time. This voltage is used to increase the kinetic energy of ions during their residence in the trap and is most often used for collisional activation experiments. Because collision-activated dissociation (CAD) is such a useful tool for characterization of ion structures, it is important to understand how the kinetic excitation of ions used to induce dissociation ultimately affects ion detection. For examination of the collisional cooling of translationally energized ions, benzene ions were again chosen as an initial model system. Three helium pressures were used, and the cooling period prior to detection was varied. Because of the different helium pressures used for activation, the molecular ions experience different numbers of collisions during the activation interval. At 0.1 mtorr, the benzene ions undergo approximately 15 collisions in a 5-ms interval. At 1 mtorr, benzene ions experience about 150 collisions in a 5-ms interval. Despite this disparity in collision numbers during activation, a qualitative examination of the effects of collisional cooling on kinetically activated ions is still relevant. In fact, in most CAD experiments performed in a quadrupole ion trap, many of the excited precursor ions dissociate long before they experience the full amount of activating collisions. Thus, the absolute collision numbers during the activation period are less important than the amplitude of the RF voltage that defines the extent of acceleration of precursor ions. Overall, the magnitude of the activation time is a much more important parameter for determining the conversion efficiency of precursor ions to fragment ions than for establishing an ultimate limit on energy deposition.

In this set of experiments, the benzene ions were formed, apex isolated, then cooled for a variable cooling time at $q_z = 0.4$ to allow the ion cloud to collapse to the center of the trap. An RF voltage of 0.1 to 1.0 $V_{(p-p)}$ was applied at the axial frequency of motion of the ion of interest (i.e., at $q_z = 0.4$, the frequency for benzene ions is 155 kHz). The amplitude of the RF voltage was sufficiently great to excite and/or eject the benzene ions without causing dissociation to detectable fragment ions. The results of this time- and pressure-dependent study are shown in Figure 3, normalized to account for the variation in trapping efficiencies with pressure.

As with the trends shown in Figure 1, all the curves show an increase in ion intensity with time. Once again, the slowest increase is for the lowest helium pressure case (0.1 mtorr). This is expected because at lower helium pressures the time between collisions is greatest, and thus the benzene ions are accelerated to proportionally greater velocities during the activation period, resulting in an outward diffusion of the ion cloud. Moreover, the subsequent collisional deactivation rate is slowest, as shown in Section II. At the higher helium pressures (>0.5 mtorr), the greater collision frequency prevents the activated ions from accelerating appreciably between collisions; therefore, the ions do not undergo significant excursions from the center of the trap. This pressure dependence means that collisional dampening of the ion cloud during the cooling period plays an especially dramatic role in recovering kinetically hot ions at lower helium pressure (<0.5 mtorr) conditions. The fundamental

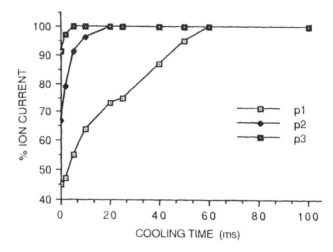

FIGURE 3

Effect of time-dependent collisional cooling of kinetically activated benzene molecular ions at various helium pressures: p1 = 0.1 mtorr; p2 = 0.5 mtorr; p3 = 1 mtorr. q_z = 0.4, 200 $mV_{(p-p)}$) activation, 5 ms activation at all pressures. (From Wu, U.-F., and Brodbelt, J.S. *Int. J. Mass Spectrom. Ion Processes*, 115, 67–83, 1992. With permission.)

importance of these results lies in the observation that kinetically activated ions can be retained in the trap and subsequently detected after cooling, even if their initial kinetic energies and/or trajectories make them undetectable at short times. Previously, any CAD experiment that resulted in detection of "no fragment ions" was interpreted as an indication of an empty trap (i.e., no fragment ions formed and stored), or an otherwise unsuccessful experiment. However, the results shown herein suggest that perhaps appropriate conditions were not chosen to sample effectively the diffuse fragment ion population, and thus fragment ions were formed but not detected.

The q_z parameter during ion activation and storage was also varied. For this set of experiments, the helium pressure was set to 0.1 mtorr, benzene ions were formed at q_z = 0.2, apex isolated (q_z = 0.78), cooled (q_z = 0.3), and activated at variable q_z. The resulting ion current was then detected prior to and after a 100-ms cooling interval. At higher q_z values, collisions are more energetic because of the corresponding increase in the frequency of ion motion. The effect of variation of q_z on the detection of activated benzene ions is shown in Table 2. The most significant increase in the ion current detected after the cooling interval occurs for q_z = 0.3 to 0.6, whereas this increment is modest when q_z = 0.2 or 0.7. The magnitude of the increase in ion current is rationalized by considering the depth of the pseudopotential well and its capability to constrain translationally excited ions. A low q_z value corresponds to a shallow trapping well, one which is unsatisfactory for confinement of fast-moving activated ions.

TABLE 2

Effect of q_z on Detection of Activated Benzene Ions

	Ion current (arbitrary units)	
q_z	Without cooling period	With 25 ms cooling period
0.2	100	400
0.3	100	700
0.4	100	700
0.5	100	700
0.6	100	650
0.7	100	550

Source: (From Wu, U.-F., and Brodbelt, J.S. *Int. J. Mass Spectrom. Ion Processes*, *115*, 67–83, 1992. With permission.)

Thus, even a long cooling period cannot effectively collapse the ion cloud because the translationally hottest ions pass out of the shallow potential well and likely are neutralized via impingement upon an electrode. At higher q_z values, the deeper potential wells can trap effectively more of the translationally energized ions until they undergo sufficient collisional cooling and collapse to the center of the RF field. Possibly at the highest value of q_z, $q_z = 0.7$, the benzene ions undergo very energetic collisions during activation and then are too kinetically hot to be effectively cooled, even with a long delay period, thus accounting for the drop in ion counts at higher q_z.

V. DETECTION OF FRAGMENT IONS GENERATED BY COLLISION-ACTIVATED DISSOCIATION

To extend the study of translationally excited benzene ions, the effects of collisional cooling on the formation, storage, and detection of fragment ions generated by CAD were also evaluated in some preliminary experiments. Collision-activated dissociation is typically performed by the application of a low amplitude (0.5 $V_{(p-p)}$) voltage across the endcap electrodes at the resonant frequency of the ion of interest. Previous studies have shown that low energy dissociation pathways, in particular rearrangement processes, are favored in the quadrupole ion trap. Presumably this enhancement of low energy pathways occurs because the high collision frequency with helium prevents the ions from accelerating substantially and obtaining large kinetic energies between collisions. Thus, an upper bound is set on the amount of kinetic energy available for conversion to internal energy during the activation process. Additionally, the time between collisions (typically 0.1 ms at 1 mtorr helium) is sufficiently long that ions with enough energy to dissociate via the low energy pathways do so long before another activation step occurs. Higher kinetic energies (and thus higher internal energies) can be attained by reducing the pressure in the trap; however, without additional experimental modifications, operation at low pressures causes loss of ion current due to the decreased buffering of ion motion.

In order to evaluate the effects of collisional cooling on the storage and detection of fragment ions, collisional activation was used to induce dissociation of *n*-butylbenzene molecular ions. Historically, *n*-butylbenzene often has been the molecule of choice for examination of energy deposition of various ionization and CAD techniques.[11] The activation energies for the two major fragmentation pathways of these compounds are well characterized.[5] A direct cleavage leading to the formation of *m/z* 91, $C_7H_7^+$, requires 1.7 eV, and a rearrangement that produces $C_7H_8^{+}$ (*m/z* 92) requires 1.1 eV. By examination of the relative abundances of these two fragment ions in a CAD spectrum, a qualitative estimation of average energy deposition during the collisional activation is obtained. For the present experiments, the molecular ion of *n*-butylbenzene was apex isolated, then activated via application of a resonant voltage (161 kHz) across the end-cap electrodes. The resulting fragment ions were stored for variable amounts of time to allow collisional cooling. At 0.1 mtorr helium pressure, *n*-butylbenzene ions were resonantly activated at $q_z = 0.4$ for 5 ms, then the abundance of the resulting fragment ions were monitored as a function of cooling time prior to detection. The results are shown in Figure 4, along with the relative distribution of fragment ions for a standard 1 mtorr experiment (at $q_z = 0.4$, activation at 1 $V_{(p-p)}$ for 5 ms).

Several points are noteworthy. First, and most important, the ratio of the abundance of *m/z* 92 to 91 stabilizes at 2.5:1 for the 0.1 mtorr experiment, whereas the ratio is 5:1 for a standard 1 mtorr experiment (as shown on the right-hand side of the plot). This result confirms that the higher energy direct cleavage process is more favorable at lower pressures. Overall, the average internal energy deposition of CAD experiments is higher as the buffer gas pressure is reduced. Second, although at short cooling times the ion signal is weak and not analytically useful, at longer cooling times (i.e., >20 ms), the signal is as intense as the signal measured for a conventional CAD experiment at 1 mtorr. Once again, this latter result indicates that many more ions are stored than are detected at low pressures, but an added delay time maximizes the signal. The effectiveness of collisional cooling suggests that higher internal energy deposition CAD experiments are viable in the quadrupole ion trap and holds promise for collisional activation experiments of larger biomolecules that typically require greater energy deposition to induce fragmentation. It should be noted, however, that for cases in which competing ion/molecule reactions are problematic at longer delay times (such as deprotonation of fragment ions by analyte neutrals present in the trap), this analytical capability for higher CAD energy deposition would be defeated. The use of a pulsed valve for introduction of analytes and an ultra-high

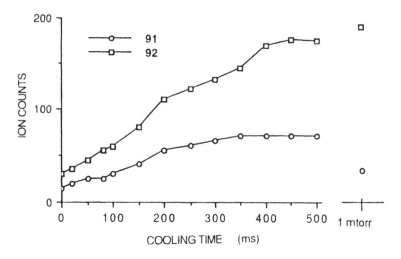

FIGURE 4

Effects of collisional cooling on the detection of fragment ions of n-butylbenzene as a function of time: $q_z = 0.4$, activation time = 5 ms, activation voltage = 1 $V_{(p-p)}$, pressure = 0.1 mtorr. At the right-hand side is shown the fragment ion distribution for CAD of n-butyl-benzene ions acquired at 1.0 mtorr under the standard conditions of $q_z = 0.4$, activation time = 5 ms, and activation voltage = 1 $V_{(p-p)}$. (From Wu, U.-F., and Brodbelt, J.S. *Int. J. Mass Spectrom. Ion Processes*, 115, 67–83, 1992. With permission.)

vacuum chamber would allow attainment of lower base pressures and reduce the interference from competitive ion/molecule reactions.

VI. CONCLUSIONS

Collisional cooling plays a significant role in the efficient storage and detection of ions; however, it is not always necessary to operate the quadrupole ion trap with 1 mtorr helium buffer gas. Lower pressures (<0.5 mtorr) are suitable when an appropriate collisional cooling period is added to the scanning program. Ions may be focused to the center of the trap by undergoing multiple collisions with helium, and this focusing then enhances their ejection for detection. Cooling rates are directly proportional to the helium pressure which is used to focus collisionally the ion cloud. Additionally, as the depth of the trapping well (related to q_z) is increased, more ions may be trapped effectively upon collisional cooling. When the average kinetic energy of an ion is augmented by resonance excitation with a supplementary RF voltage, the cooling effects are even more dramatic. For CAD experiments, in general the average internal energy deposition is somewhat higher at lower helium pressures. The loss of sensitivity observed when operating at low helium pressures (<0.5 mtorr) can be largely offset by adding a cooling period to collapse the fragment ion cloud for more efficient detection. This latter point promises to be of analytical importance when higher internal energy deposition CAD experiments are desired, i.e., in the structural characterization of biomolecules. Such low helium pressure/long cooling time experiments may provide an effective alternative to the standard 1 mtorr operation of the quadrupole ion trap if the occurrence of competing reactions (such as proton transfer or charge exchange of analyte ions and background neutrals) which causes loss of analyte signal is not overwhelming at long times. The nature of the buffer gas used for collisional cooling is also of primary importance. For example, it has been shown that heavier gases are more efficient targets for collisional cooling and collisional activation of ions, but these gases ultimately result in overall poorer trapping efficiencies. This parameter has been the subject of several recent investigations, not described herein.

REFERENCES

1. Stafford, G.C., Jr.; Kelley, P.E.; Syka, J.E.P.; Reynolds, W.E.; Todd, J.F.J., *Int. J. Mass Spectrom. Ion Processes.* 1984, *60*, 85.
2. Holzscheiter, M.H., *Phys. Scripta.* 1988, *T22*, 73.
3. (a) Schaaf, A.H.; Schmeling, U.; Werth, G., *Appl. Phys.* 1981, *25*, 249. (b) André, J.; Schermann, J.P., *Phys. Lett.* 1973, *45A*, 139. (c) Gronowska, J.; Paradisi, C.; Traldi, P.; Vettori, U., *Rapid Commun. Mass Spectrom.* 1990, *4*, 306. (d) Dawson, P.H., *Int. J. Mass Spectrom. Ion Phys.* 1977, *24*, 447. (e) Neuhauser, W.; Hohenstatt, M.; Toschek, P.; Dehmelt, H.G., *Phys. Rev. Lett.* 1978, *41*, 233. (f) Knight, R.D.; Prior, M.H., *J. Appl. Phys.* 1979, *50*, 3044. (g) March, R.E.; McMahon, A.W.; Londry, F.A.; Alfred, R.L.; Todd, J.F.J.; Vedel, F., *Int. J. Mass Spectrom. Ion Processes.* 1989, *95*, 119. (h) March, R.E.; McMahon, A.W.; Allinson, E.T.; Londry, F.A.; Alfred, R.L.; Todd, J.F.J.; Vedel, F., *Int. J. Mass Spectrom. Ion Processes.* 1990, *99*, 109. (i) McLuckey, S.A.; Glish, G.L.; Asano, K.G.; Bartmess, J.E., *Int. J. Mass Spectrom. Ion Processes.* 1991, *109*, 171.
4. Wu, H.-F.; Brodbelt, J.S., *Int. J. Mass Spectrom. Ion Processes.* 1992, *115*, 67.
5. Chen, J.H.; Hays, J.D.; Dunbar, R.C., *J. Phys. Chem.* 1984, *88*, 4759.
6. (a) Vedel, F.; André, J.; Vedel, M.; Brincourt, G., *Phys. Rev.* 1983, *A27*, 2321. (b) Vedel, F.; André, J., *Phys. Rev.* 1984, *A29*, 2098. (c) Vedel, F.; André, J., *Int. J. Mass Spectrom. Ion Processes.* 1985, *65*, 1.
7. (a) Todd, J.F.J.; Lawson, G.; Bonner, R.F., in *Quadrupole Mass Spectrometry and Its Applications.* Dawson, P.H., Ed., Elsevier: Amsterdam, 1976; p. 181. (b) Major, F.G.; Dehmelt, H.G. *Phys. Rev.* 1968, *179*, 91.
8. Atkins, P.W., *Physical Chemistry.* W.H. Freeman: San Francisco, 1982.
9. Reiser, H.-P.; Kaiser, R.E.; Savickas, P.J.; Cooks, R.G., *Int. J. Mass Spectrom. Ion Processes.* 1991, *106*, 237.
10. Louris, J.N.; Cooks, R.G.; Syka, J.E.P.; Kelley, P.E.; Stafford, G.C., Jr., Todd, J.F.J., *Anal. Chem.* 1987, *59*, 1677.
11. (a) Griffiths, I.W.; Harris, F.M.; Mukhtar, E.S.; Beynon, J.H., *Int. J. Mass Spectrom. Ion Phys.* 1981, *43*, 83. (b) Chen, J.H.; Hayes, J.D.; Dunbar, R.C., *J. Phys. Chem.* 1984, *88*, 4759. (c) McLuckey, S.A.; Ouwerkerk, C.E.D.; Boerboom, A.J.H.; Kistemaker, P.G., *Int. J. Mass Spectrom. Ion Phys.* 1984, *59*, 85. (d) Nacson, S.; Harrison, A.G., *Int. J. Mass Spectrom. Ion Processes.* 1985, *63*, 325.

Chapter 6

ION TRAJECTORY SIMULATIONS

Randall K. Julian, Jr., R. Graham Cooks,
Raymond E. March, and Frank A. Londry

CONTENTS

0-8493-4452-2/95/$0.00+$.50

I. INTRODUCTION

It is possible to calculate the trajectory of a charged species in the quadrupole ion trap by numerical integration of the Mathieu equation. However, it is not necessary to follow the motion of the ion continuously;

integration of the Mathieu equation offers the advantage that the details of the trajectory of an ion can be calculated directly at any given point in time with great accuracy. The majority of such simulation studies has been carried out with the radiofrequency (RF) drive potential applied to the ring electrode and with the end-cap electrodes grounded. In this case, and for the case of resonance excitation in the quadrupolar mode, ion trajectories may be calculated directly by integration of the Mathieu equation.

When a buffer gas is added to the ion trap so that collisions occur between ions and buffer gas atoms, the trajectory between successive collisions may be calculated directly by integration of the Mathieu equation. For the calculation of trajectories of ions subjected to resonance excitation in the dipolar and monopolar modes, integration of the Mathieu equation is no longer valid, and a field interpolation or other method must be used. Approximation methods have been applied also to the calculation of ion trajectories.

In the calculation of the trajectory of an ion within a quadrupole ion trap of known r_0, it is necessary to define the initial conditions of ion position and velocity, i.e., $x, y, z, \dot{x}, \dot{y}, \dot{z}$, the RF drive amplitude V of phase ξ_0 and radial frequency Ω, and the mass m of the assumed singly charged ion. Once the initial conditions have been selected, the ionic parameters $a_z, q_z, \beta_z, C_{2n}, A, B$, and u_{max} may be calculated as outlined in Chapter 2.

A. Early Simulation Applications

The first application of numerical methods to the analysis of single ion trajectories in the quadrupole ion trap was reported some 25 years ago by Dawson and Whetten.[1] At that time, application was limited to the use of the ion trap as a storage device for specific ions, not as a mass spectrometer.[2,3] These early studies were extended by Dawson and Whetten[4] to an examination of ion ejection in which a range of ion masses was confined simultaneously. The application of Monte Carlo methods permitted the investigation of ion motion, with the ions undergoing momentum-dissipating collisions and charge exchange; these studies indicated that ions migrate to the center of the ion trap with an accompanying diminution in ion kinetic energy, and from which the ions can be extracted with increased efficiency.[5,6] The use of Monte Carlo methods was extended to the study of an ensemble of two ion species.[7]

André and Schermann[8] earlier studied the spatial and velocity distribution of a population of ions stored in an ion trap and subjected to collisions with rare gas atoms. A mathematical formalism of temporal invariance, developed by Vedel et al.,[9] has been applied to the study of the statistical spatial and energy distributions of trapped ions subjected to collisional cooling. These studies were extended to the space charge case by Vedel and André.[10,11] Early applications of numerical and analytical methods to the analysis of trapped ion behavior have been reviewed previously.[12]

B. Recent Simulation Applications

Representative trajectories for trapped ions of three different masses, with working points on the q_z axis and in a collision-free and resonance-free system, have been calculated by Louris et al.[13] by integration of the Mathieu equation. Lunney et al.[14] have applied finite element analysis to the calculation of fields within the ion trap from which trajectories were calculated. Their interpolation method lacked sufficient accuracy and has been superseded by a field modeling process, developed by Lunney and Moore,[15] in which the field was reconstructed using a summation of multipole potentials. Further simulation studies of the transfer of beams of charged particles in and out of an ion trap are discussed in Chapter 16.

Some previously unpublished trajectory plots by Bexon and Todd and by Fies, Jr., for ions stored within the ion trap and subjected to collisions with helium buffer gas, appeared in a recent publication by Todd.[16] Bexon and Todd calculated one-dimensional (axial) trajectories of (mass/charge ratio) m/z 25 ions showing the effect of regular collisions with helium atoms; the motion of m/z 69 ions in the r–z space plane both with and without helium buffer gas was calculated by Fies, Jr.

Pedder and Yost[17] have developed the HYPERION program for the calculation of ion trajectories; this program uses a modified form of the Mathieu equation together with a fourth-order Runge-Kutta algorithm. A parallel capacitor model was used for dipolar excitation; the simulated final kinetic energy as a function of tickle frequency at constant fluence compared well with experimental loss in parent ion intensity as a function of tickle frequency. In the absence of a tickle voltage, HYPERION reduces to the unperturbed Mathieu equation for trajectory simulation and calculation of ion kinetic energy.

An ion trap simulation program (ITSIM) has been used with great success by Cooks and co-workers.[18] The features of the program permit simulation of external ion injection, bath gas dampening, and resonance excitation. The equations used to calculate the development of the trajectory of an ion are a modified form of the conventional equations of ion motion for an ion trap. Plots of unperturbed ion motion and ion axial excursions by resonance excitation have appeared recently.[19] Two further simulations have been carried out with ITSIM: (1) rapid radial excitation to instability upon the application of a short duration, high voltage direct current (DC) pulse to the end-cap electrodes[20] is compatible with the observed fragmentation pattern induced by collisions with a surface (ring electrode); and (2) calculated ion kinetic energies upon ejection agree well with experimental values.[21] The limitations of the single ion simulation method, together with the dramatic improvements in computer speed, prompted Cooks and co-workers[22,23] to develop multiparticle simulations using a numerical quadrupole simulation (NQS) program. A copy of the revised ITSIM program[22] has been made available to the research community. The application of ITSIM and NQS programs is described below.

A Taylor-McLaurin expansion in the solution of the trajectory equations, used in the phase space formalism, has been proposed for the calculation of ion trajectories in the ion trap;[24] however, no example of this method has been developed.

Single ion simulations have been carried out by Franzen and co-workers[25,26] in order to investigate the performance of ion traps constructed with nonlinear fields superimposed on the main quadrupole field. These simulations are described in detail in Chapter 3.

Time-of-flight simulation studies of Xe^+ have been carried out by Vedel and Vedel,[27] and more recently from the same laboratory by Rebatel.[28]

A series of simulation programs for ion trajectory calculation has been developed by March et al. A relatively simple computer program, MA, which employed integration of the Mathieu equation is used for the calculation of unperturbed trajectories and to provide a standard to which their other programs reduce when specialty features are eliminated. A specific program for quadrupole resonance, SPQR, was prepared;[29,30] in the absence of quadrupolar excitation, the SPQR program reduces to the MA program. A field interpolation method (FIM) has been developed for the simulation of ion trajectories for examination of resonance excitation applied in monopolar and dipolar modes.[31] Revised versions of SPQR and FIM have been prepared that employ continuously ramped DC and RF potentials; the field interpolation with ramps method is designated as FIRM.[32] The facility to vary continuously the DC and RF potentials permits ready examination of processes such as ion isolation, where the working point of a given ion species is moved within the stability diagram. The application of these programs is given below.

PART A
MULTIPARTICLE SIMULATIONS OF THE PERFORMANCE OF THE QUADRUPOLE ION TRAP MASS SPECTROMETER

Randall K. Julian, Jr. and R. Graham Cooks

II. INTRODUCTION TO MULTIPARTICLE SIMULATIONS

Simulations that compute the trajectory of a single ion in the quadrupole ion trap have been used for some time to analyze the performance of the instrument as a mass spectrometer[1] and as a storage device.[5-7] Recently, single ion simulations have been used to investigate resonance excitation[29] and ion isolation methods.[30] Single ion simulations were also used by Wang and Franzen[25] to investigate the performance of ion traps constructed with nonlinear fields superimposed on the main quadrupole field.

A. Scientific and Design Goals of the Simulation

Due to the sensitivity of the ion trajectories to initial ion conditions, several runs of a single ion program are usually required to relate the single trajectory observed to the behavior of an ensemble of ions. Further, because of the large number of ions required, production of a mass spectrum which can be compared to experiment, is very difficult to obtain. Finally, it is not possible to model ion–ion interactions while simulating single ions moving within a field. These basic limitations of the single ion simulation method, together with dramatic improvements in computer speed, prompted our attempt to develop multiparticle simulations.[22,23]

The scientific goal of multiparticle simulations is to develop a model of performance of the ion trap that can be verified against experiment. Such a simulation can be used to study existing experiments and suggest new experiments and operating modes. To reach these scientific goals, design goals were set at the beginning of the project to ensure eventual success.

The design goals of the multiparticle simulations are

1. To develop capabilities to plot interactively the trajectories and energies of ensembles of ions with various m/z values and initial conditions
2. To duplicate the instrumental operating conditions including RF, alternating current (AC), and DC scan functions
3. To allow for ion neutral and ion–ion interactions
4. To generate mass spectra and kinetic energy data which can be compared to experimental data

Two separated simulations were written to meet all of these goals. The first, ITSIM, generates interactive graphic displays for small numbers of ions (1 to 1000) on a personal computer (PC). The second, NQS, generates mass spectra and kinetic energy data for a large number of ions (1000 to 20,000), and relies on vector and parallel processor systems. These programs are both based on the same simulation methodology but differ in their exact implementation because the interactive PC and the massively parallel supercomputer lie at opposite ends of the computer hardware spectrum. Both programs are written in C for portability and to allow the use of advanced data structures, which are not available in FORTRAN.

B. Theory

The integration of the equations of motion is performed via a Taylor series approximation using a small integration step size (1/100 of an RF period). Because the Mathieu equation is a second-order differential equation, its integration is performed in two steps. First, the velocity term is updated:

$$\frac{du}{dt}(t_{n+1}) = \frac{du}{dt}(t_n) + \Delta t \frac{d}{dt}\left(\frac{du}{dt}\right)(t_n) + \ldots + \frac{(\Delta t)^s}{s!}\frac{d^s}{dt^s}\left(\frac{du}{dt}\right)(t_n) \tag{1}$$

and then the position term is updated:

$$u(t) = u(t_n) + \Delta t \frac{du}{dt}(t_n) + \frac{(\Delta t)^2}{2!}\frac{d^2u}{dt^2}(t_n) + \ldots + \frac{(\Delta t)^s}{s!}\frac{d^su}{dt^s}(t_n) \tag{2}$$

where u is a coordinate (x, y, z), $u(t_n)$ is the displacement of the ion from the origin along a given axis at time t_n, and $du/dt(t_n)$ is the velocity component along the same axis at time t_n. The Δt term represents the time increment $t_{n+1} - t_n$. The current form of the simulation computes Equation 1 to $s = 1$ and Equation 2 to $s = 2$; obviously, more terms could be added at the expense of computational speed. The program is begun with an initial data file containing an ion population which has been created by the user. The initial ion population supplies the program values for m/z, initial ion position, $u(t_0)$, and velocity, $du/dt(t_0)$. Due to the statistical nature of an ion population, positions and velocities are generated from a user-specified distribution function. For example, ion positions can be selected to follow a Gaussian distribution along the x-, y-, and z-axis with a selected mean and standard deviation, causing the program to position ions randomly to achieve the specified distribution.

The acceleration terms in Equations 1 and 2 are given by the Mathieu equation, which can be written for the x and y coordinates in terms of a quadrupole electric field E_{qu}. A generalized form of the quadrupole electric field equation[33] for the grounded end-cap electrode operating mode is given in Equation 3. Normally the negative cosine is used for the periodic function in Equation 3; however, a positive sine is used here because it is desirable to begin the simulation with a zero RF amplitude. The acceleration is given by Equation 4.

$$E_{qu} = -2\frac{U + V \sin(\Omega t_n + \delta)u}{(r_0^2 + 2z_0^2)} \tag{3}$$

$$\frac{d^2u}{dt^2}(t_n) = \frac{eE_{qu}}{m} \tag{4}$$

When one introduces a dipole field along the z-axis, the acceleration term in the z-direction includes the electric field terms for both the quadrupole, E_{qz}, and the dipole, E_{dz}, fields:

$$\frac{d^2z}{dt^2}(t_n) = \frac{eE_{qu}}{m} + \frac{eE_{dz}}{m} \tag{5}$$

$$E_{qz} = 4\frac{U + V\sin(\Omega t_n + \delta)z}{\left(r_0^2 + 2z_0^2\right)} \tag{6}$$

$$E_{dz} = \frac{V_{aux}\sin(\Omega_{aux}t + \delta_{aux})}{2z_0} \tag{7}$$

To avoid confusing charge z with the z coordinate, m is defined as the m/z ratio, because the ion mass and charge are not uncoupled within the program. V is the zero-to-peak amplitude $(0 - p)$ of the RF voltage applied to the ring electrode at a frequency Ω, and a phase shift δ; U is the amplitude of the DC voltage applied to the ring electrode; V_{aux} is the amplitude of the voltage applied to the end-cap electrodes to create a dipole field, the end-caps are 180° out-of-phase from one another at a frequency Ω_{aux} and a phase shift δ_{aux}; δ and δ_{aux} are used to maintain a continuous function during frequency scans; additionally $\delta - \delta_{aux}$ can be used to specify an initial RF/AC phase relationship.

III. SIMULATION METHODOLOGY

A. Program Overview

The method chosen for simulating multiple particles moving in the quadrupole ion trap involves an initial ion data set which represents the initial conditions of the ions prior to the simulation. This data set is processed by a trajectory routine that tracks stable ions; records ejection time, position, and velocity; and saves a final data set at the end of the simulation. In the case of ITSIM, a graphical output of the trajectories is displayed in either position or phase space. A simulated experiment is defined in much the same way a real ion trap experiment is: in terms of operating parameter scan functions. These operating parameters correspond to variables in the equations of motion given above. Table 1 supplies the operating parameters used by the program, giving the variable name in the equations of motion, the units and the criteria used by the scheduler routine which controls the simulated clock, and schedules the execution of update routines based on the time. The scan function is a

TABLE 1

Operating Parameters for the Quadrupole ITMS™ and Related Variables in the Equation of Motion

Operating parameter	Parameter in equations of motion	Units	Scheduler update criteria
RF amplitude	V	V_{0-p}	Update only at RF zero crossings
RF frequency	Ω	Hz	Update only at RF phase minima or maxima
AC amplitude	V_{aux}	V_{0-p}	Update only at AC zero crossings
AC frequency	Ω_{aux}	Hz	Update only at AC phase minima or maxima
DC amplitude	U	V_{DC}	Update every clock tick

Note: Also given are the criteria for updating the value of the parameter by the scheduler routine.

description of the values that the operating parameters have as a function of time. Figure 1 gives a typical scan function for a resonance ejection mass instability experiment. A separate parameter update routine adjusts the value of each parameter according to the scan function. The trajectory update routine is executed every clock tick to update the positions and velocities of the ions with the current values of the operating parameters. Because a Taylor series approximation to the solution of the Mathieu equation is being used, the accuracy of the integration is dependent on the size of the time increment. In general, time increments of 0.01 of the RF period have proven to give satisfactory accuracy without causing the program to run prohibitively slowly.[23] When the RF scan function ends, the simulation is considered complete, the ion data set is saved, and the program exits. In the case of NQS, a histogram of ion count vs. ejection time is performed to generate a mass spectrum. Histogram analysis of ion count vs. ion kinetic energy is performed to generate an ejection kinetic energy spectrum. These analyses are also available for ITSIM; however, the limited number of ions that can be simulated on this PC-based program still makes it more difficult to generate spectra with statistically meaningful peak shapes or ejection positions.

B. Describing the Experiment To Be Simulated

As described in the overview, the simulation is performed according to user-specified scan functions which define how the operating parameters will change over time and when the simulation is to end. This technique is based on the commercial ion trap mass spectrometer (ITMS™,

Finnigan), where the scan function is called a microscan. A microscan is
constructed by linking together segments (called scan tables by Finnigan)
in which various potentials are either constant or are ramped. For each
operating parameter in Table 1, there is an ASCII file containing a list of
scan tables that describe how that parameter will be controlled. The scan
table includes a start and stop value, a step size, and start and stop times.
The scheduler routine calls the update routine for a parameter as indicated in Table 1, which uses the scan table data to determine whether it
is time to change the value of the parameter and determines when the
scan function for this parameter has terminated. Once the experiment to

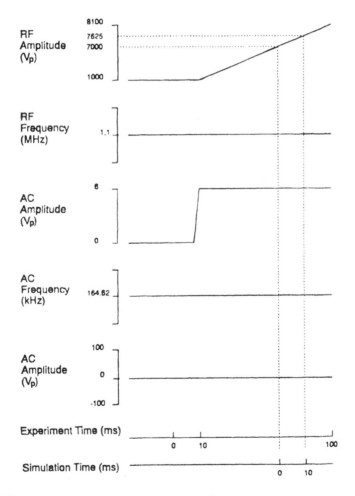

FIGURE 1
Typical scan function, RF amplitude, RF frequency, AC amplitude, AC frequency, and DC
amplitude.

be simulated has been described in terms of scan tables, the ion initial conditions must be specified.

C. Specifying Ion Initial Conditions

The simulation begins with the collection of ions represented in a binary data file created by an ion distribution program (IONGEN, CREATION). Table 2 gives the basic data structure used to hold the ion data. The ion distribution program follows user-specified parameters to select the values for the data in Table 2. The user specifies ions in groups, which are any number of ions with the same m/z value and the same statistical distribution for position and velocity. Any number of groups can be created with any combination of m/z values and position and velocity distributions. The position distributions are generated using a probability density function to determine the probability of an ion occurring with a randomly selected position for each coordinate. The rejection method is then used either to retain or to reject the value for position. The rejection method is an established algorithm based on generating two random numbers for each desired value. A trial value is generated from a rectangular distribution by the pseudo-random number generator. The probability density function is then used to determine the probability of obtaining the trial value from the selected distribution. A second pseudorandom number is then generated, and if it is larger than the probability of obtaining the trial value, the trial value is rejected and the process is repeated. The accepted values will, therefore, follow the selected distribution. Because parameters to the distribution function are supplied separately for each coordinate, the asymmetry of the quadrupole field in the axial direction can be taken into account. A common position distribution for the ions prior to a mass-selective instability scan is a Gaussian distribution centered about zero with a standard deviation determined by laser tomography experiments.[34] Figure 2 shows a plot of the ion distribution determined by laser tomography for the benzoyl ion (m/z 105), which is the primary fragment from dissociation of acetophenone. The distribution is collected by observing the intensity of the photofragment m/z 77, as a laser is scanned along the z-axis. These data have been fitted to a Gaussian distribution and have a $\sigma \approx 0.5$ mm. Ion velocities are distributed using the same method, although the distribution function also can be chosen to be a superposition of two Gaussians, each with a mean centered on the desired mean velocity. It is important to distribute the velocities in a random fashion to prevent the trajectories from being artificially phase-locked, which is the case when the initial velocity is set to zero. Once each ion has been assigned a position and velocity, the program saves the ion data set in a file for use by the trajectory program.

TABLE 2

Data Structure Used for Ion Data Storage

Variable	Typical representation	Vector	Description
lIonCount	4-Byte integer (unsigned)		Number of ions in the simulation
fRFPhase	4-Byte IEEE float		Phase of the RF signal when the ion data is stored
fACPhase	4-Byte IEEE float		Phase of the AC signal when the ion data is stored
iDetectStat	2-Byte integer	*	Detection status of the ion; indicates whether the ion is trapped or has been ejected to an electrode
fM_Z	4-Byte IEEE float	*	m/z ratio for the ion
fXpos	8-Byte IEEE float	*	Position of the ion, in meters, from the center of the trap along the x-axis
fYpos	8-Byte IEEE float	*	y-axis position of the ion
fZpos	8-Byte IEEE float	*	z-axis position of the ion
fXvel	8-Byte IEEE float	*	x-axis component of the ion velocity, in m s^{-1}
fYvel	8-Byte IEEE float	*	y-axis component of the ion velocity
fZvel	8-Byte IEEE float	*	z-axis component of the ion velocity
fTimeInd	8-Byte IEEE float	*	Ion ejection time indicator in seconds
fEjectParam	4-Byte IEEE float	*	Secondary ejection parameter used to study parameters, not usually saved
ulCollCnt	4-Byte integer (unsigned)	*	Total number of collisions experienced by an ion
fCollMass	4-Byte IEEE float	*	$(m_{ion} - m_{buffer})/(m_{ion} + m_{buffer})$; computed by the program prior to execution

Note: The variable name is given with the typical machine representation. Representations allow control over the precision of the data and the computational time. The 4-byte IEEE float has 8 digits of precision, while the 8-byte IEEE float has 10 significant digits. An asterisk in the Vector column indicates that there is one copy of this variable for each ion being simulated.

D. Describing the Electric Field for the Quadrupole Ion Trap

In order to simulate real experiments, the simulation must be supplied with accurate values for the experimental operating parameters. This requirement is especially true for the main RF drive and electrode shape parameters because small changes in the RF amplitude or the

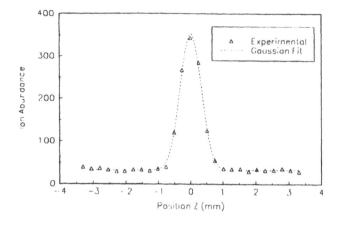

FIGURE 2
Axial ion distribution determined by laser tomography. (Experimental data from Hemberger, P.H. et al., *Chemical Physics Letters*, 191, 405, 1992. With permission.)

electrode shape cause large changes in the electric field strength. The electric field strength has the largest effect on the character of ion motion and on the RF voltage needed for ion ejection. The Finnigan ITMS™ electrodes were measured accurately to obtain values for r_0 and z_0, as well as coordinates (r, z) to describe the surface shape. The most common solution of Laplace's equation for a pure quadrupole field gives ring and end-cap electrodes which have a hyperbolic cross-section:

$$\text{Ring}: \frac{r^2}{r_0^2} - \frac{z^2}{a^2} = 1 \qquad \text{End-cap}: \frac{r^2}{b^2} - \frac{z^2}{z_0^2} = -1 \tag{8}$$

This general solution can be simplified by assigning b to the ring internal radius r_0 and a to the inscribed end-cap radius z_0, which leads to the familiar statement:

$$r_0^2 = 2z_0^2 \tag{9}$$

As discussed by Knight,[33] the simplified relationship given in Equation 9 is not strictly required to create a quadrupole field. As long as the ring and end-cap asymptotes are equal and have a slope of $1/\sqrt{2}$, a pure quadrupole field is generated by hyperbolic electrodes:

$$\frac{a}{r_0} = \frac{z_0}{b} = \frac{1}{\sqrt{2}} \tag{10}$$

The values for a, b, r_0, and z_0 for ideal electrodes based on the simplified geometry of Equation 9 and for the Finnigan ITMS™ are given in Table 3. The value for z_0 was found to be larger than ideal. Coordinates (r, z) were obtained, to the nearest 0.001 cm, from the electrode surface and fitted to Equation 8 to determine a and b. The effect of a pure quadrupole

TABLE 3

Geometric Parameters Used to Describe Quadrupole Ion Trap Electrodes

Geometric parameter	Simplified geometry $(r_0^2 = 2z_0^2)$	ITMS geometry	ITMS dimension measurement error	Geometric definition
r_0	1.0000	1.0000	0.001	Inscribed radius (cm) of the endcap electrode
z_0	0.7071	0.781	0.001	Radius (cm) of the ring electrode
a	0.7071	0.740	0.011	Conjugate vertex of the hyperbola describing the ring electrode (i.e., the value of z_0 which results in a shared ring/end-cap asymptote; see Equation 8)
b	1.0000	1.081	0.003	Conjugate vertex of the hyperbola describing the end-cap electrodes (see Equation 8)

Note: Measured from actual electrode. The values for r_0 and z_0 describe the size of the electrodes while the parameters a and b are used to describe their shape.

field with a larger z_0 is a weaker electric field than would be obtained by ideal electrodes at the same RF voltage. The effect is summarized by the general equation for q_z given by Knight:[33]

$$q_z = -2q_r = \frac{8eV}{m\left(r_0^2 + 2z_0^2\right)\Omega^2} \tag{11}$$

The electrode coordinates and the ideal and actual electric field lines are given in Figure 3. The actual electrode coordinates were used with the Poisson program[35] to compute the actual potentials within the ion trap. Figure 3A shows the equipotential lines for an ion trap derived from the simple relationship in Equation 9. Figure 3B shows the equipotential lines for the actual ITMS™ electrodes. Deviations in the ITMS™ potential from the asymptote given in Equation 10 produce even-order fields superimposed on the quadrupole field. In the ITMS™, higher-order fields are added because the end-cap electrodes-to-center distance z_0 was "stretched" without a corresponding change to the end-cap electrode shape, which changes the asymptote angle indirectly. Directly changing the asymptote angle is, in fact, the method used by Wang and Franzen[25] to deliberately superimpose octopole fields in the nonlinear QUISTOR. The deviation

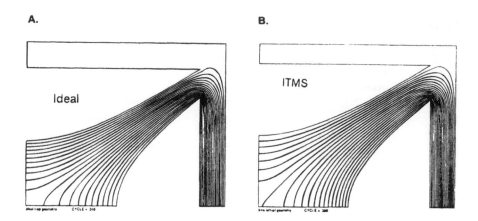

FIGURE 3
Equipotential lines for the quadrupole ion trap from calculations done with the Poisson program. In this example, 1000 V is placed on the ring electrode (right) while the end-cap electrode (top) is grounded. (A) Equipotential lines for the ideal ion trap electrodes created using Equation 10; (B) equipotential lines for the ITMS electrodes used to collect experimental data. The deviations shown produce even-order fields, the strongest of which is the octopole. The octopole field produces nonlinear resonances at $\beta_z = 0.5$ and at $\beta_r + \beta_z = 1$.

shown in Figure 3 creates a small octopole field term in the ITMS™ which is responsible for resonances at $\beta_z = 0.5$ and $\beta_z + \beta_r = 1$.[25] These resonance lines have been observed in the ITMS™ by a number of groups at large ion displacement[36–38] and long storage times.[39] For the present simulation study, a pure quadrupole field, using the actual values for r_0 and z_0 was used. However, more recent ITSIM data often consider higher order field components.

E.   Analyzing Simulation Results

Beyond the graphical trajectory and phase-space plots available (ITSIM), the final state-space of the entire ion population, including ejection time, is available at the end of the simulation for further analysis. The most common analysis is an ejection time histogram which represents the mass spectrum produced by ions ejected via the exit end-cap. A histogram also can be generated from a subset of the total data to account for end-cap hole size or to analyze ions ejected in nondetectable directions, that is, to the ring or injection end-cap. The data can be analyzed in terms of the ion final kinetic energy to produce kinetic energy distributions for various operating conditions. The reason mass spectra and kinetic energy distributions are generated by the simulation is to provide a point of comparison with the experimental data. This comparison allows validation of the simulation so that new experiments may be suggested and existing experimental data can be studied in terms of ion motion. Finally, the original data set can be compared to the final data set so the initial conditions of the ion can be compared to the ejection time, location (position), or velocity. The final data set has the same structure as the input data set so that after ejected ions are removed, the data can be used as the input to another simulation. To ensure the RF and AC phase remains continuous between simulations, the phases of both signals are stored in the data file.

IV. SIMULATIONS INVOLVING SEVERAL PARTICLES

The primary goal of modeling the behavior of several ions (<1000) in a quadrupole ion trap is visualization of ensembles of ions to give insight into the character of the trajectories (frequency and magnitude) and to allow a direct comparison between ions with different m/z values. Examining the phase-space plots of particles has been a valuable tool in analyzing the dynamics of the ion trap for some time.[40,41] A plot showing the phase-space dynamics of several particles adds to the ability to determine the ranges of position and kinetic energy within the ion cloud as a function of time under various operating conditions. Visualization of trajectories is performed using two PC-based programs, one for position space (ITSIM), and one for phase space (ITSIMAN). Memory limits for DOS-based 80×86 PC machines set the limit for the number of ions that may be simulated with these programs, while speed considerations limit the total simulation time over which the trajectories may be accurately integrated. The interactive nature of the PC-based programs make them ideal for testing the models which are used to describe the quadrupole ion trap in larger simulations.

A. Components of the Model of Ion Motion

The components that comprise the model of ion motion can be stated in terms of potentials in the second-order differential equation describing the acceleration of an ion:

$$\frac{d^2z}{dt^2} = -\frac{Cdz}{dt} + \frac{4e(U - V\cos\Omega t)z}{m(r_0^2 + 2z_0^2)} + \frac{e}{2mz_0}z_0 V_{aux}\cos\Omega_{aux}t + \frac{e^2}{4\pi\in m}\frac{1}{\sum_i (r - r_i)^2}$$

(12)

where V, Ω, V_{aux}, Ω_{aux}, and U are operating parameters given in Table 1 and r_0 and z_0 are geometric parameters given in Table 3. This equation implements a dissipative term for collisions with a neutral buffer gas, which reduces the kinetic energy of the ion according to a cooling factor C. Space-charge is also implemented as a Coloumbic repulsive potential. In this term, ε is the permittivity of vacuum and $\mathbf{r}_i$ represents the positions of all other ions. Note that the space-charge term involves all three coordinates with $\mathbf{r} = (x, y, z)$, coupling the motion of ions in all three directions. Equation 12 represents the most important features of ion motion, at least in a rudimentary fashion. The first improvement that can be made is replacing the viscous dampening term (Cdz/dt) with random collisions using a statistical model for determining the probability of a collision. This method, which is currently implemented in both the small- (ITSIM) and the large-scale (NQS) programs, is based on a simple hard-sphere collision model. The collision diameter (D_{12}) is used along with the number density of the buffer gas (N) and velocity of the ion (v_x, v_y, v_z) to compute the probability of a collision (P) which can be expressed in terms of the collision frequency $v(t)$as:

$$P = vdt = C\sqrt{v_x^2 + v_y^2 + v_z^2} \tag{13}$$

where

$$C = \pi ND_{12}^2 \, dt \tag{14}$$

If a collision occurs, energy is removed from the ion by reducing its velocity in a random fashion, varying from no energy loss (fly-by) to the maximum possible for an elastic collision according to the following equation:

$$\Delta v_u = 1 - \left(1 - \frac{4m_{buffer}}{m}\right)^{1/2} k \tag{15}$$

where K is a random number varying from 0 to 1, representing a kind of impact parameter. This model is an improvement over the viscous drag type models because the energy is removed at random times corresponding to random ion positions, rather than at every time increment.

B.   Examples of Trajectories and Phase-Space Plots

The small-scale simulation is primarily used to visualize ion trajectories, in space as a function of time, and as phase-space plots. Trajectory plots are observed as position traces tracking the past position of an ion, or animation of multiple particles within the trap can be displayed. Phase-space plots can be plotted directly or simplified by constructing Poincaré sections.[42] Figure 4 gives an example of these three types of information for a collection of stable ions of m/z 202 at $q_z = 0.4$. The top trace shows

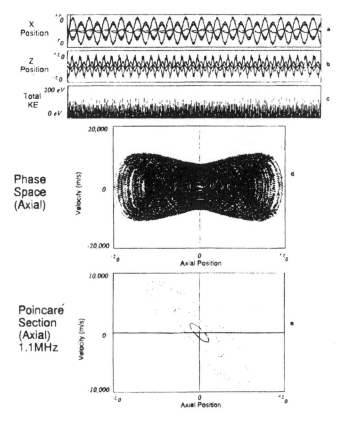

FIGURE 4

Simulation of stable motion for five ions of m/z 202 at $q_z = 0.4$ (RF = 1110 V_{0-p}). (a) Trajectory along the x-axis (plot range $\pm r_0$; (b) trajectory along the z-axis (plot range $\pm z_0$); (c) total kinetic energy (plot range 0 to 200 eV); (d) phase-space plot (axial velocity vs. axial position); (e) Poincare section of (d) strobed at 1.1 MHz showing an attractor.

the x-position, the second trace shows the z-position, and the third shows the total kinetic energy. The fourth plot is an axial (z) phase-space plot and the last is a Poincaré section of the phase-space taken at the RF drive frequency of 1.1 MHz, sampled at a zero phase angle. From the trajectory plots, the character of the ion motion is visible, while examination of the phase-space plot and the Poincaré section gives more information on the overall stability and periodic nature of the entire collection of ions.

C. Simulations

Three ion trap experiments are examined using simulations of collections of particles: (1) the mass-selective instability scan, conventionally used to obtain mass spectra; (2) resonance ejection of ions, which is used to enhance resolution and to extend the mass range of the instrument; (3) a new experiment involving the application of DC pulses to the end-cap electrodes in an attempt to force the ions further from the center or to force collisions with the various electrode surfaces. Because collections of ions are observed in these simulations, a flow[42] along a solution surface is sampled instead of a single path on that surface, and this gives information on ranges of values such as energy, position, and ejection conditions.

1. Mass-Selective Instability Scan

The most important development in the field of quadrupole ITMS since the invention of the ion trap is the mass-selective instability scan developed by Stafford and co-workers at Finnigan.[43] This scan is used to generate mass spectra from the instrument by mass selectively ejecting ions to an external detector. All other developments such as the ability to perform multiple stages of mass analysis,[13,44] the extension of the mass range,[45] the ability to obtain high resolution spectra,[46] and the incredible sensitivity improvements[47] have this scan at the core of the experiment. It is, therefore, critical to understand the ion ejection process, which is controlled by the dynamic effects of the scan on ion motion. Figure 5 shows a simulation of ion motion during the mass-selective instability scan. In this simulation, the amplitude of the RF voltage applied to the ring electrode is ramped, forcing ions to have trajectories that are mathematically unbounded. When the amplitude of the ion trajectory exceeds the dimensions of the electrode in the z-direction, ions are ejected. Those ions that eject via the exit end-cap electrode strike the electron multiplier and are detected. In the instrument, this signal is associated with the RF voltage, which can be converted to a mass scale to create a mass spectrum. Note in Figure 5, that ions which have the same mass have the same frequency of motion but are phase shifted from each other. The trajectories of ions with a higher m/z ratio are still stable mathematically and the ions remain in the trap while those of the lighter ions become unstable and the ions are ejected. The mass limit of an ion trap operating in this mode is set by the maximum voltage that can be applied to the ring electrode. From Figure 5 it is also evident that buffer gas collisions dampen the amplitude of the ion motion, allowing some ions to remain within the electrode boundary despite the mathematical instability of the trajectories.

2. Resonance Ejection

By coupling an external field to the ion motion in the z-direction, ion trajections that are stable mathematically can be resonantly excited, and ions can be ejected at lower RF voltages than is possible with the stan-

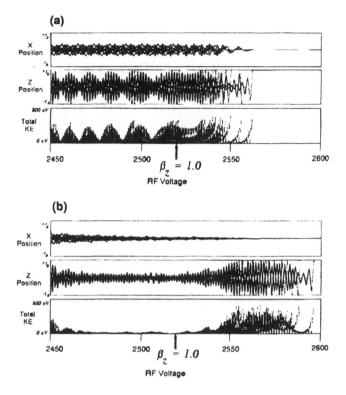

FIGURE 5
Simulation of mass-selective instability scan (20 ions trapped). (a) Ejection of m/z 202 at $\beta_z = 1$ without buffer gas; (b) ejection of m/z 202 with helium buffer gas at a partial pressure of 1.0 mtorr.

dard instability scan, thus increasing the mass range of the instrument. The addition of a dipole field in the z-direction changes the equations of motion from the Mathieu equation, which can be solved analytically, to a forced Mathieu-type equation for which there is no closed form solution. Numerical integration of Newton's laws of motion is one approximate method available to solve the trajectories of the ions in this type of field.

By approximating the potential applied between the end-cap electrodes as a pure dipole, it is possible to simulate the motion of ions during resonance ejection. Figure 6 shows a simulation in which ions are ejected via resonance ejection. An auxiliary field is applied that matches the ion frequency in the z-direction for the selected RF amplitude. The amplitude of the auxiliary potential depends on the degree of resonance excitation desired. In the presence of helium buffer gas, low amplitude excitation gives ions sufficient velocity to undergo collisions which can

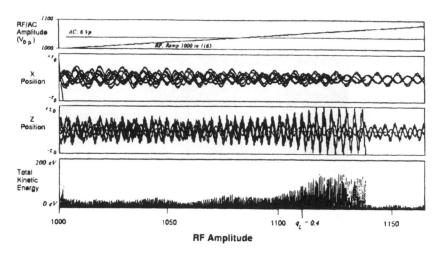

FIGURE 6

Resonance ejection with mass range extension. This figure shows the ejection of 10 ions of m/z 202 at $q_z = 0.4$ using an auxiliary axial frequency of 160 kHz at an amplitude of 6 $V_{(0-p)}$. Helium buffer gas is present at a partial pressure of 1.0 mtorr. A higher mass ion at m/z 269 is not ejected because its motion is not in resonance with the auxiliary dipole.

result in an increase in the internal energy of the ion and subsequent ion fragmentation. Higher amplitude excitations result in direct ion ejection. Due to the energy absorption profile of the ion, the amplitude of motion will increase before it reaches the exact resonance frequency, although the actual ejection may not occur for several secular cycles. The exact ejection time depends on the initial maximum displacement of an ion prior to the application of the dipole and on the buffer gas. Buffer pressure and auxiliary field amplitude can be adjusted to maximize mass assignment accuracy by controlling the initial conditions of the ions prior to excitation.

3. Simulation of DC Pulse Experiments

Resonance excitation is used routinely to cause ions to undergo activating collisions with a buffer gas which leads to their fragmentation. This technique is somewhat limited by the fact that very high excitation amplitudes can eject an ion without fragmenting it. A new experiment was devised to improve internal energy deposition by attempting to cause surface-induced dissociation (SID)[48] via fast DC pulses.[49] A DC pulse is directed along the z-axis and, depending on the polarity, it can cause either axial[50] or radial excitation. It was observed that the amount of fragmentation caused by the axial DC pulse was dependent on the pulse amplitude.[51] Multiparticle simulation was used to determine the translational energy achieved by ions when fast DC pulses were applied to one end-cap electrode. To measure internal energy deposition, n-butylbenzene was used as a thermometer molecule by comparing induced dissociation. As discussed in Reference 50 and references therein, the intensity ratio of m/z 91/92 is used as an indication of the amount of internal energy deposited in this molecule. Using the normal resonant excitation method, the highest value for this ratio achieved in an ion trap was 4.[13] Using axial DC pulses, a value of 20 was achieved[50] at large DC pulse amplitudes. Figure 7 shows the mass spectra collected after n-butylbenzene is subjected to 3 μs of duration axial DC pulses of increasing amplitude. The pulses were applied when the parent ion (m/z 134) had a q_z value of 0.5. As can be seen in Figure 7, the m/z 91/92 intensity ratio increases with increasing pulse amplitude. The ITSIM program was modified to model the DC dipole pulse so that its effects on ion motion could be studied. The results are shown as phase-space plots in Figure 8. As expected, ions do not appear to be undergoing resonant excitation due to the frequency content of the pulse, but rather are responding to the steep, short-lived gradient. In Figure 8a, the 70-V pulse from Figure 7b is shown to excite axially the ions much in the way resonant excitation would, only for a much shorter period of time, resulting in a low m/z 91/92 intensity ratio of 0.7. Figure 8b corresponds to the spectrum in Figure 7c and shows the result of a pulse of 104 V. This pulse also results in a short-lived axial excitation, however, surface collisions covering a broad range of collision energies are also indicated. Figure 8c shows the result of a pulse of 157 V, which corresponds to the spectrum in Figure 7d. Nearly all of the parent ion population is forced to the electrode surface, with a very broad collision energy range, suggesting that both surface and gas collisions are involved. Figure 8d corresponds to the spectrum in Figure 7e, resulting from a pulse of 187 V. Here, a dramatic increase in the m/z 91/92 intensity ratio is observed and is accompanied by a significant reduction in dissociation efficiency. Both facts are explained by the phase-space plot in Figure 8d, which shows the collision energy range as the ions strike the surface. Ions strike the surface with a relatively narrow energy range that is sufficiently high to deposit large amounts of internal energy.

V. SIMULATIONS OF LARGE NUMBERS OF PARTICLES

Thus far the discussion of simulations has involved a very small number of ions compared to the number probably trapped by a real instrument. The above simulations attempted to draw insights into device performance by examining the trajectories and kinetic energy spectra for a small number of ions. The approach taken in the NQS described here is to elucidate how operating conditions affect the final mass spectra by producing simulated spectra rather than trajectory plots. Numerical quadru-

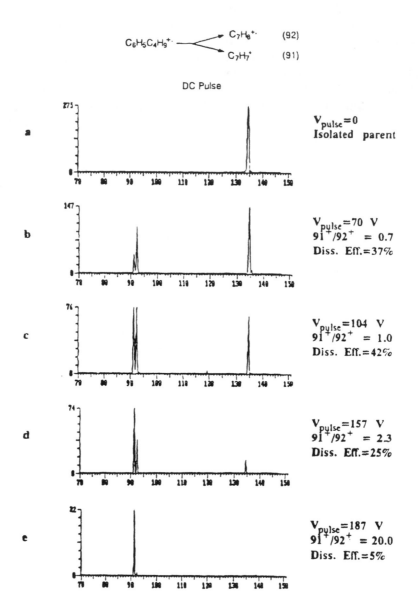

FIGURE 7

Experimental MS/MS spectra of *n*-butylbenzene (*m/z* 134) ionized via electron impact. These data shows fragment peaks C_7H_7· (*m/z* 91) and C_7H_7· (*m/z* 92) caused by application of a 3-μs DC dipole pulse while the parent ion was at $q_z = 0.5$. (a) Pulse off, showing no fragmentation occurring due to isolation or storage; (b) pulse voltage 70 V_{dc}; (c) 104 V_{dc}; (d) 157 V_{dc}; (e) 187 V_{dc}. (From Lammert, S.A. and Cooks, R.G., *Rapid Communications in Mass Spectrometry*, 6, 528, 1992. With permission.)

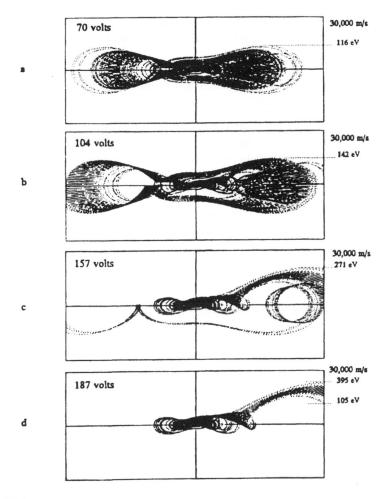

FIGURE 8
Phase-space plots from ITSIMAN of the DC dipole experiments shown in Figure 7. Here, an initial ion population is subjected to the DC pulse to observe the energy ranges achieved. (a) 3 μs, 70 V_{dc} pulse; (b) 104 V_{dc}; (c) 157 V_{dc}; (d) 187 V_{dc}. Abscissa shows axial position; origional shows ion velocity.

pole simulation utilizes parallel and vector supercomputer technology to compute the trajectories for a statistical population of ions using models of ion motion that are as realistic as possible. The final results are mass spectra and kinetic energy distributions which can be compared to actual experiments. The aim is to verify the model against experiment and, ultimately, to identify what effect each component of the model has on the mass spectrum. From this analysis, it will be shown that improved methods of operation are suggested by simulation results. In addition to computing a large number of ion trajectories, the program is designed to allow the specification of any type of scan function for the RF, AC, and DC signals applied to the ion trap electrodes in the same way as are ITSIM and ITSIMAN.

A. Simulations Involving Statistical Ion Populations

Because the simulation is expected to generate a spectrum comparable to experiment, a population of ions that represents a typical experimental population is required. As described in Section III.C, the initial conditions of an ion population can be set following a distribution derived from experiment. Statistically, a small number of initial positions and velocities selected from a simulated distribution may not resemble the experimental distribution, regardless of how well the simulated distribution matches reality. If the simulated distribution is in error, this problem will be apparent only when a large number of values are generated. Simulations involving large numbers of particles are, therefore, more sensitive to the initial condition distribution functions, and care must be taken in choosing them. Simulations involving large numbers of particles also tax the pseudorandom number generator which is used during both the initialization and execution of the simulation. For example, the rejection method used by this simulation to select initial values, depending on the distribution function being used, can easily require 100 random values per ion. Because 10,000 ions typically are generated, 10^6 random numbers could be required. Care must be taken that the random number generator has a period greater than the number of random values needed.

B. Generation of Mass Spectra

To compare the simulation directly to the experimental results from the ion trap mass spectrometer, simulated mass spectra are generated from the final ion data. During the course of the simulation, when the position of an ion exceeds the limits of the electrode dimensions, it is considered ejected. During the simulated time increment that an ion is ejected, the program records the simulated time and electrode at which the ion is lost and freezes the values for the position and velocity of the ion. On vector systems, a system usually exists to prevent a memory location from being updated within a vector, in turn preventing the ejection data from being lost. On massively parallel systems such as Thinking Machines Connection Machine CM-2,[52] an individual processor can be deactivated if the ion for which it is computing data is ejected. Either the simulated scan function will terminate or all of the ions will be ejected to terminate the entire simulation. When the simulation is complete, the final ion data set can be examined for ejected ions. Because the ejection time was recorded for each ejected ion, a histogram can be constructed by counting the number of ions ejected within each time window. The width of the time window can be controlled to simulate the bandwidth of the detector circuit on the actual instrument, if desired. Setting a window of 28 μs results in the number of ions ejected within that window to be integrated at approximately the bandwidth of the ITMS™ pre-amplifier and integrator. The scan start voltage and ramp rate are used then to compute the RF voltage for each time window, which can be converted to mass for direct comparison to the experiment. A more accurate method for direct comparison in turn is obtained by converting the experimental mass scale to an RF voltage and then comparing RF ejection voltages. Application of this method to the ITMS™ is discussed in Section V.D.

C. Implementation of Equations of Motion for Vector and Parallel Processors

The key to calculating the trajectories for a very large number of ions is to take advantage of the fact that the equations of motion apply to each ion independently. Thus, there is a single set of instructions which are to be performed on each ion, i.e., a single instruction path/multiple data path (SIMD). The SIMD architecture is implemented in processing units called vector processors, which require the calculation to be vectorized. Vectorization is achieved by coding Equations 1 and 2 such that the same operations are performed on multiple data items in parallel by the vector processor. The vector processor therefore performs operations on *arrays* of data, and some of its operations differ from traditional *vector* operations. For example, an *array* multiplication (Equation 22) indicates that the first element of the first array is multiplied by the first element of the second array; the result is stored in the first element of the result array. This operation is not to be confused with the standard notion of multiplying two vectors to produce a scalar quantity.

The vectorization of the trajectory calculations consists of separating the ion-independent terms from the ion-dependent terms in Equations 3 and 5. The ion-independent terms, which include the values of E_{qu}, E_{qz}, and E_{dz}, are calculated for each time increment to form field constants, which are quantities that vary with the electric field but are constant for every ion. The field constants used in the velocity calculation are

$$S_{qu} = (\Delta t)\, eE_{qu} \qquad (16)$$

$$S_{qz} = (\Delta t)\, eE_{qz} \qquad (17)$$

$$S_{dz} = (\Delta t)\, eE_{dz} \qquad (18)$$

The field constants used in the position calculation are

$$P_u = \frac{(\Delta t)^2}{2}\, eE_{qu} \qquad (19)$$

$$P_{qz} = \frac{(\Delta t)^2}{2} e E_{qz} \tag{20}$$

$$P_{dz} = \frac{(\Delta t)^2}{2} e E_{dz} \tag{21}$$

The ion positions are stored in arrays x_i, y_i, and z_i. The x- and y-position arrays are collectively called u_i. The ion velocities are stored in arrays $v_{x,i}$, $v_{y,i}$, and $v_{z,i}$. The arrays holding the velocities in the x- and y-directions are collectively called $v_{u,i}$. The ion positions and m/z ratios are combined according to Equation 22 into three vectors, $A_{x,i}$, $A_{y,i}$, and $A_{z,i}$, collectively called $A_{u,i}$. The array B_i holding the inverse of the m/z ratios m_i is computed using Equation 23 at the beginning of the program and is used to speed the computations by eliminating division operations during the trajectory calculation.

$$\mathbf{B}_i = \frac{1}{\mathbf{m}_i} \tag{22}$$

$$\mathbf{A}_{u,i} = \mathbf{B}_i \mathbf{u}_i \tag{23}$$

Combining the ion-independent constants S_u, S_{qz}, and S_{dz} allows the Taylor series in Equation 1 to be rewritten as a vector operation:

$$v_{u,i}(t_{n+1}) = v_{u,i}(t_n) + \mathbf{A}_{u,i} S_u \tag{24}$$

$$v_{z,i}(t_{n+1}) = v_{z,i}(t_n) + \mathbf{A}_{u,i} S_{qz} + \mathbf{B}_i S_{dz} \tag{25}$$

and likewise using P_u, P_{qz}, the series in Equation 2 is also rewritten as a vector operation, first for the x and y coordinates:

$$u_i(t_{n+1}) = u_i(t_n) + V_{u,i}(t_{n+1}) \Delta t + \mathbf{A}_{u,i} P_u \tag{26}$$

and then for the z coordinate:

$$z_i(t_{n+1}) = z_i(t_n) + v_{z,i}(t_{n+1}) \Delta t + \mathbf{A}_{z,i} P_{qz} + \mathbf{B}_i P_{dz} \tag{27}$$

To simulate accurately the spectra produced by the ITMS™ exact values of r_0, z_0, Ω and V must be supplied to Equation 4, 6, and 7. Careful measurements of the electrode spacing were combined with high voltage measurements of V during the mass analysis scan to allow a quantitative comparison between the simulation and experiment.

D. Comparing Simulation Results to Experiment

The ITMS™ data system uses the user-selected mass analysis scan table to generate internal signals that control the RF voltage applied to the ring electrode. The start mass is converted into a step number for the 12-bit RF control digital-to-analog convertor (DAC). The RF signal is then modulated by the RF DAC output prior to amplification. The effects of non-linear components used in the modulation and amplification stages are corrected, to a degree, by a feedback circuit that measures the RF level, compares it to the DAC setpoint, then applies a correction voltage. The end result is a linear RF ramp during mass analysis. The RF voltage can be measured as a function of DAC output voltage to create a calibration curve which can be used to convert the mass axis of the data system to RF voltage. To measure the RF voltage on the ring, a high voltage probe with 1% precision was connected to the RF line outside the vacuum chamber. The capacitance of the probe added to the RF driver circuit required retapping and then fine tuning of the RF coil to balance the LC circuit. The results of the calibration are shown in Figure 9. The RF voltage measured also confirms the ITMS™ electrode measurements. By rearranging Equation 11 to yield m/z (m), then substituting the ITMS parameters from Table 3, the maximum RF voltage needed to achieve a mass range of 650 u using mass-selective instability is calculated to 8105 V_{0-p}. The ITMS RF control DAC generates a 10.01-V output for mass 650 u, which generates 8100 V_{0-p} on the ring electrode as measured by the 1% high voltage probe.

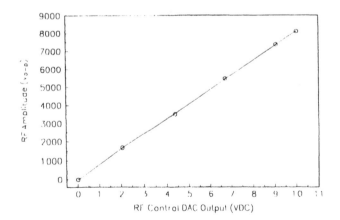

FIGURE 9

Calibration plot showing the linear relationship between the RF control DAC output voltage and the actual RF voltage applied to the ring electrode. By measuring the RF control DAC output when ion ejection occurs, this relationship can be used to establish the RF voltage of ejection and to allow simulation and experiment to be compared.

Using the calibration curve from Figure 9, the mass scale from the experimental data was converted to RF amplitude for direct comparison to the simulation RF amplitude at which ion ejection occurred.

E. Simulations

1. Resonance Ejection Scan

A simulated low resolution, resonant ejection mass spectrum of the protonated substance P molecular ion region is shown in Figure 10A and the experimental spectrum is shown for comparison in Figure 10B. The simulated data were obtained from 10,000 ions distributed initially as described above, computed without buffer gas collisions or space-charge

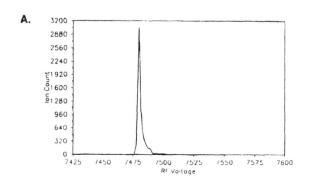

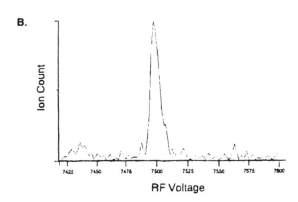

FIGURE 10

(A) Simulated resonant ejection mass spectrum showing the molecular ion region, [M + H]]', of the decapeptide substance P (protonated; 12,500 Da/s). This mass spectrum was generated using the isotope ratios given in Table 4 and the parameters given in Table 5. (B) Experimental mass spectrum of protonated substance P (molecular ion region) collected using resonance ejection and normal RF scan speed; 2.25× mass range extension, 12,500 Da/s. Parameters are given in Table 6.

interactions. The experimental data were collected using the prototype ITMS described earlier.[53] Peptides were dissolved in 1:1 MeOH:H₂O and 1 µl of the solution was mixed with 1 µl of a 1:1 glycerol:thioglycerol matrix. The protonated peptide was desorbed from a gold-plated probe tip with a pulsed beam of Cs ions at 7 kV from a Cs⁺ source which is described elsewhere.[54] A 1-µl sample of substance P in a 37.1 pmol/µl solution was applied to the Cs⁺ probe. Table 4 gives the m/z value and the number of ions simulated for each isotope. In the actual experiment, the mass range of the ITMS must be extended using resonance ejection to eject these ions. Applying an axial frequency of 162.64 kHz produces a 2.25× mass range extension. Table 5 gives the scan conditions used in such an experiment. The standard RF ramp speed produces a mass analysis scan rate of 5555.5 Da s⁻¹; applying the 2.25× extension factor gives the actual scan rate used, 12,500 Da s⁻¹.

During the experiment, helium buffer gas was used to allow the trapping of the injected ions of protonated substance P. In the simula-

TABLE 4

Isotopic Abundances of Protonated Substance P Based on Procomp-Peptide Analysis Program V1.2

m/z	No. of ions
1347.73	4221
1348.73	3385
1349.73	1631
1350.73	587
1351.73	167

TABLE 5

Conditions Used for Measurement at Normal Resolution

Drive Potential (applied to ring)
    RF frequency: 1.1 MHz
    RF scan: 7430–7550 $V_{(0-p)}$
    RF voltage increment: 56.73170 mV/step
    RF voltage step time: 909.0909 ns/step (12,500 µs⁻¹)

Auxiliary potential (applied to end-caps)
    AC frequency: 162.640 kHz
    AC amplitude: 6.0 $V_{(0-p)}$

Ejection conditions
    β_{eject} = 0.29571
    q_{eject} = 0.40355 (2.25× mass range extension)

TABLE 6

Conditions Used for Measurement at High Resolution

Drive potential (applied to ring)
    RF frequency: 1.1 MHz
    RF scan: 7440–7550 $V_{(0-p)}$
    RF voltage increment: 5.673170 mV/step
    RF voltage step time: 909.0909 ns/step (1250 μs^{-1})

Auxiliary potential (applied to end-caps)
    AC frequency: 162.640 kHz
    AC amplitude: 5.0 $V_{(0-p)}$

Ejection conditions
    β_{eject} = 0.29571
    q_{eject} = 0.40355 (2.25× mass range extension)

tion, no buffer gas was used, but its effects on the ion cloud radius at the start of the simulation were accounted for by selecting an initial Gaussian distribution of ions with a standard deviation of 5.0×10^{-4} m, as suggested by laser tomography experiments carried out by Hemberger and co-workers.[34]

2. High Resolution Resonance Ejection Scan

Schwartz and co-workers,[46] as well as other groups, have demonstrated that mass resolution in the ITMS may be improved by decreasing the rate at which the amplitude of the RF voltage is incremented. Experimentally, this is performed by attenuating the RF amplitude control signal during the mass analysis scan by reducing the voltage increment produced by incrementing the RF DAC by a single step. To reach the desired starting voltage, the voltage supplied by the attenuated RF ramp is added to an offset voltage. To simulate this experiment, the same conditions were chosen as for the low resolution experiment except that the RF voltage increment was reduced by a factor of 10, giving a tenfold attenuation in scan speed. Using a voltage step of 5.67317 mV, the scan rate is reduced to 1250 Da s^{-1} at the 2.25× mass range extension. Table 6 gives the operating parameters for the simulation of the high resolution experiment. Note that, as with the low resolution result, no buffer collisions or space charge interactions were computed. Figure 11A shows the simulated mass spectrum for the high resolution experiment and Figure 11B shows the corresponding experimental mass spectrum. Exactly the same ion population was used in both the high and low resolution scans; the only difference between the two experiments was the RF ramp speed. The improved mass resolution evident in the experiments (compare

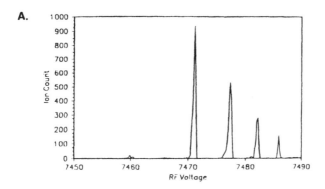

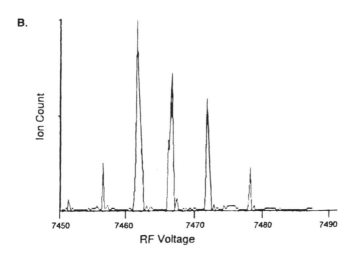

FIGURE 11
(A) Simulated mass spectrum of protonated substance P (molecular ion region; 1250 Da/s) generated using the parameters given in Table 6. (B) Experimental mass spectrum of substance P (molecular ion region; 1250 Da/s, 2.25× mass range) collected using the parameters given in Table 6.

Figures 10B and 11B) is reproduced in the simulation (compare Figures 10A and 11A).

It is important to compare the RF voltages of the simulated and experimental peaks. Using the simplified electrode geometry obtained by substituting Equation 9 into Equation 11, the voltage required to eject m/z 1348 at $q_z = 0.040355$ is calculated to be 6730.3 V_{0-p}. Simulations using this geometry produced a value of 6730.10 V_{0-p}, which is approximately 770 V_{0-p} lower than the experimental result. The simulated mass peak in Figure 10A is ejected 20 V_{0-p} lower than the experimental peak in Figure 10B. In the high resolution simulation shown in Figure 11A, m/z 1347.7 is ejected at 7470 V_{0-p}, 7 V_{0-p} higher than the experimental peak at 7463 V_{0-p} in Figure 11B. Possible reasons for these shifts and their direction are higher-order fields, buffer collisions, and space charge. Each of these effects can be shown to give slight shifts in the positions of peaks in experimental mass spectra. A detailed study of experimental mass shifts associated with resonant ejection is given in the thesis of Reiser.[55]

3. Resonant Ejection Phase Relationships

When the data from a simulated peak is expanded in time, individual ion ejection events become visible as shown in Figure 12A. This effect has also been observed experimentally using broad-band amplification of the electron multiplier signal and a digital storage oscilloscope. Figure 12B suggests that positive ions are only ejected via the detector end-cap electrode during the positive AC phase; i.e., when the injector end-cap is positive and the detector end-cap is negative. Not only must the AC be of the correct phase, but ion ejection via the exit end-cap only occurs when the RF has a negative slope, just prior to reaching the minimum (and vice versa for ejection through the injection end-cap). In all cases, positive ions are ejected only when the RF phase is negative and has a negative slope and when the AC phase is in the appropriate quadrant for the ion polarity and the direction of ejection.

Typically, when resonant ejection is used for mass range extension, the AC frequently is chosen to give a convenient mass range extension factor. Thus, the AC frequency usually does not have an integer relationship to the RF frequency, so if the RF and AC begin in phase, after one AC cycle the phase angle will have rotated to a new, nonzero value related to the difference in frequency. Resonant ejection under these conditions, termed "rotating phase resonant ejection", limits the number of AC cycles during which ions may be ejected because the critical phase relationship cannot occur every AC cycle. This situation is shown in Figure 12A for m/z 1347.7, ejected at 162.640 kHz, giving a 2.25x mass range extension. By choosing the AC frequency to have an integer relation to RF frequency, the phase angle can be fixed, leading to fixed phase resonance ejection. Under fixed phase ejection conditions the initial RF/AC phase angle can be chosen such that the critical RF/AC phase relationship occurs every AC cycle. A simulated fixed phase ejection of the same ion data set at 137.5 kHz is shown in Figure 13A. This spectrum suggests that choosing the AC frequency and initial RF/AC phase angle appropriately should force ions to be ejected on every AC cycle. Further, optimizing the initial RF/AC phase relationship, the conditions required for ion ejection can be optimized as shown in Figure 13B. Experiments utilizing this technique are underway and, in preliminary low resolution results, the amplitude and resolution of ion ejection are improved.

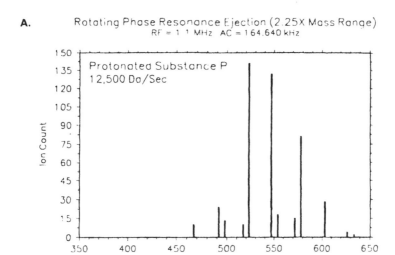

A. Rotating Phase Resonance Ejection (2.25X Mass Range)
RF = 1.1 MHz   AC = 164.640 kHz

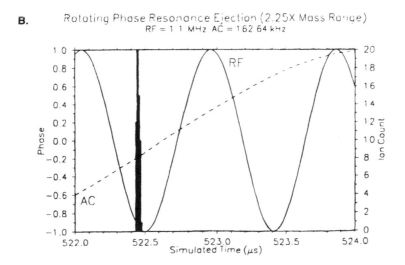

B. Rotating Phase Resonance Ejection (2.25X Mass Range)
RF = 1.1 MHz   AC = 162.64 kHz

FIGURE 12

(A) Rotating phase resonance ejection of m/z 1347.7 at 162.64 kHz. This figure indicates that a critical RF/AC phase relationship is required for ejection because ions are not ejected at every AC cycle; this relationship suggests that controlling the RF/AC phase angle will give some control over the ejection process. Because the critical phase relationship does not occur every AC cycle, ions are ejected over several AC cycles. (B) An expanded time axis allows examination of the relationship among the RF phase, the AC phase, and the number of ions ejected through the exit end-cap for the peak at 522.5 μs from the spectrum in (A).

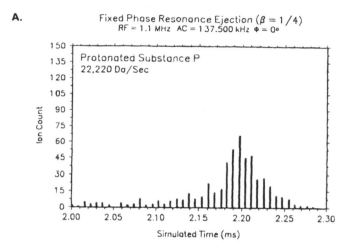

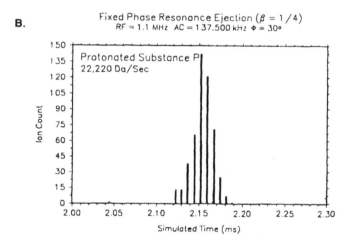

FIGURE 13
(A) Resonance ejection of *m/z* 1347.7 at an AC frequency of 137.5 kHz. Now the RF and AC have a fixed phase relationship. The initial phase angle between the RF and the AC is 0°, resulting in ion ejection every AC cycle; the process is not optimized. (B) By choosing the initial RF/AC phase angle to be 30°, efficient ion ejection occurs every AC cycle over a narrow voltage range, greatly improving resolution.

ACKNOWLEDGMENTS

This work was supported by the National Science Foundation (CHE92-23791) and by Finnigan MAT through the Chemistry Department Industrial Associate Program. The simulation project began as a team effort with Hans-Peter Reiser, the author of the ITSIM program. R.K.J. acknowledges support through a Fellowship provided by Eli Lilly and Company. Computer facilities were provided by the Purdue University Computer Center on the ETA-10P* and the VAX 8800 through the Presidential Computing Reserves. Access to the Connection Machine CM-200 at Los Alamos National Laboratory's Advanced Computing Laboratory was facilitated by Phil Hemberger of Los Alamos. Electric field calculations were performed using the Poisson/Superfish codes obtained from the Los Alamos Accelerator Code Group, which provided the code with funding from the Department of Energy. Finally, thanks to Steve Lammert, Kenny Morand, Kathy Cox, and Jon Williams for their invaluable assistance.

PART B
SIMULATION OF SINGLE PARTICLES IN
THE QUADRUPOLE ION TRAP

Frank A. Londry and Raymond E. March

VI. SCOPE OF THE SIMULATION

The simulation studies described here are concerned with ion isolation by each of two methods, concurrent isolation and consecutive isolation, together with an examination of resonance excitation by each of three modes of irradiation of the ion trap. The performance of the simulation programs has been compared with experiment in two areas: in the determination of relative RF absorption coefficients, wherein quite good agreement was achieved both with SPQR for quadrupolar excitation[31] and in the determination of the induction fluence required for ion ejection, where fluence is defined as the product of a supplementary tickle voltage amplitude, zero-to-peak, and the duration of excitation.[56] In this latter area, agreement is now satisfactory.

A basic premise of the simulation studies carried out in this laboratory is that the detailed inspection of relatively few, though precisely calculated, ion trajectories that display a variety of behaviors can be both quite informative and indicative of the behavior of an ensemble of ions.

VII. MASS-SELECTIVE ION ISOLATION

The ion trap is particularly well suited for carrying out tandem mass spectrometry, i.e., repeated mass-selective analyses performed consecutively in time. An essential step in this process is the isolation, or mass-selective storage, of a single ion species. Once ion isolation has been performed, the *purified* species is available for resonance excitation with concomitant collision-induced dissociation or for reaction with neutral species in the trap, or both. The product ions of these processes stored within the trap may be subjected subsequently to mass-selective analysis. This sequence of operations may be repeated several times and is described as (MS)^n where n is the number of mass-selective stages. An excellent example of the utility of this method has been furnished by Glish et al.,[57] who used (MS)^4 in the investigation of ion structures and ion composition.

During the process of ion isolation, the trajectories of unwanted ion species become unstable when the working point (*vide infra*) reaches a boundary of the stability diagram; there are three ways in which this condition may be achieved, as discussed below. As ion isolation is now used routinely with the ITMS™, it was of interest to simulate ion trajectories during this process and to examine ion kinetic energies, both for those ions whose trajectories remain stable during ion isolation and those for which they became unstable. The program used here was SPQR, which was modified for the work described here by the incorporation of RF or DC voltage ramps or both. This feature of the program permits calculation of ion trajectories while the working point is being moved both within and beyond the stability region.

A. Concurrent Ion Isolation

In this method, which is known also as apex isolation, the working point of the ion to be isolated is moved to the upper apex of the stability diagram, whereupon ions of higher m/z ratio are lost radially and, concurrently, ions of lower m/z ratio are lost axially.

1. Working Point Movement in the Stability Diagram

For a stability diagram such as is shown in Figure 14, in which the ordinate and abscissa are expressed in terms of a_z and q_z, respectively, specific values of these parameters define a working point in (a_z, q_z) space which may or may not lie within the boundaries of the stability diagram. Because a_z and q_z are directly proportional to the DC and RF potentials, respectively, the working point may be moved in linear fashion by variation of the DC or RF potentials or both. The boundaries of the stability diagram are defined by $\beta_r = 0$ and 1, and $\beta_z = 0$ and 1; the $\beta_r = 0$ and $\beta_z = 1$ boundaries are identified in Figure 14; the $\beta_r = 0$ and $\beta_z = 0$ boundaries emerge from the origin and intersect with the $\beta_z = 1$ and $\beta_r = 1$ boundaries, respectively. The respective coordinates (a_z, q_z), of the intersections of these boundaries of the stability diagram are (0.149998, 0.780909) and (−0.671359, 1.239780). The coordinates of the intersection of the $\beta_r = 1$ and $\beta_z = 1$ boundaries are (−0.543965, 1.351220).

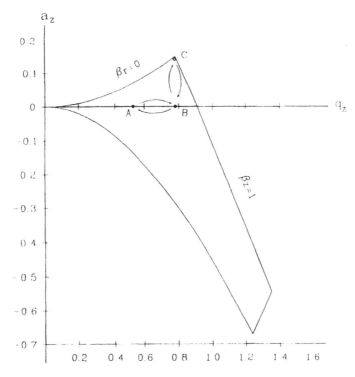

FIGURE 14
Stability diagram for the quadrupole ion trap showing the changes in location of the working point for a selected ion species undergoing ion isolation; ions of lower and higher m/z ratio are ejected concurrently. (From March, R.E. et al., *Int J. Mass Spectrom. Ion Processes*, *112*, 247, 1992. With permission.)

The "trajectory" of the working point for the ion species undergoing isolation is shown schematically in Figure 14, starting at point A. The working point is moved initially from A to B. The location of B is critical in that it must be located directly below the upper apex, i.e., the intersection of the $\beta_r = 0$ and $\beta_z = 1$ boundaries of the stability diagram. The working point is then moved from B to C. The location of C is equally critical in that it must lie close to the upper apex. The working point is returned to B and then to A in two successive stages.

The sequence of changes in RF and DC potentials associated with ion isolation is shown schematically in Figure 15.

2. Isolation of m/z 146

a. Trajectory Calculations for m/z 144, 146, and 148

The trajectories of single representative singly charged ions of three ion species, m/z 144, 146, and 148, were simulated to demonstrate isola-

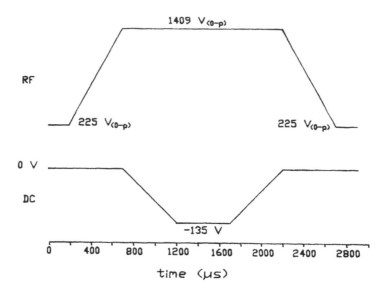

FIGURE 15

A scan function for ion isolation using RF and DC voltages. The ordinate shows the relative voltage amplitudes, though not to scale, and the abscissa shows the duration of the simulation. (From March, R.E. et al., *Int J. Mass Spectrom. Ion Processes*, 112, 247, 1992. With permission.)

tion of m/z 146. Initially, the RF voltage amplitude was set at 225 $V_{(0-p)}$ such that the values for q_z were 0.1228, 0.1245, and 0.1262 for m/z 148, 146, and 144, respectively. The trajectory of each ion was calculated separately, starting at zero degrees phase angle of the RF voltage and then once per RF cycle for 200 μs. The instantaneous ion kinetic energy was calculated once each RF cycle, i.e., in time steps of 9.091×10^{-7} s, and is shown for m/z 146 in Figure 16A; there are 220 datum points in this figure.

The final 40 μs of this time period were calculated in time steps of 40 ns and show the temporal variations of ion kinetic energy (Figure 16B) in much greater detail. There are 1000 datum points in this figure. It is seen that the maxima in ion kinetic energy of 0.16 eV in Figure 16B exceed those of the low resolution data of Figure 16A.

The radial and axial excursions during the final 40 μs are shown in Figure 16C and D, in which two major frequency components are evident. The lower amplitude component oscillates at the drive frequency of 1.1 MHz, and during the time period of 40 μs shown in Figure 16D, 44 maxima due to the drive frequency can be clearly identified. The frequency component of greater amplitude appears to be common to both radial and axial motion, but this is not so. For radial motion, for which the value of r is always positive, one cycle corresponds to the time interval between every second maximum. The radial fundamental secular

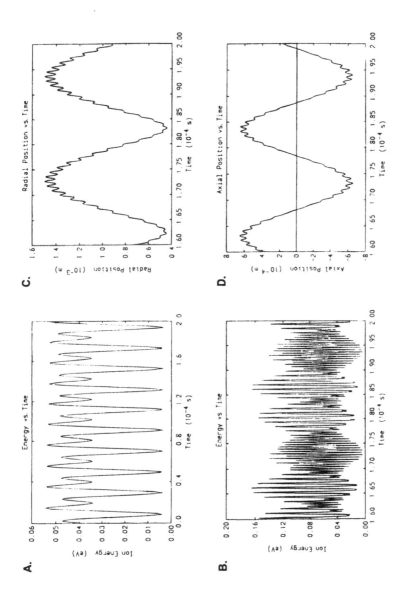

FIGURE 16

Trajectory simulation of m/z 146 at $a_z = 0$, $q_z = 0.1245$. (A) kinetic energy variation over the time period 0 to 200 μs in time steps of 9.091×10^{-7} s; (B) kinetic energy variation over the time period 160 to 200 μs in time steps of 40 ns; (C) radial excursions over the time period 160 to 200 μs; (D) axial excursions over the time period 160 to 200 μs. (From March, R.E. et al., *Int J. Mass Spectrom. Ion Processes, 112*, 247, 1992. With permission.)

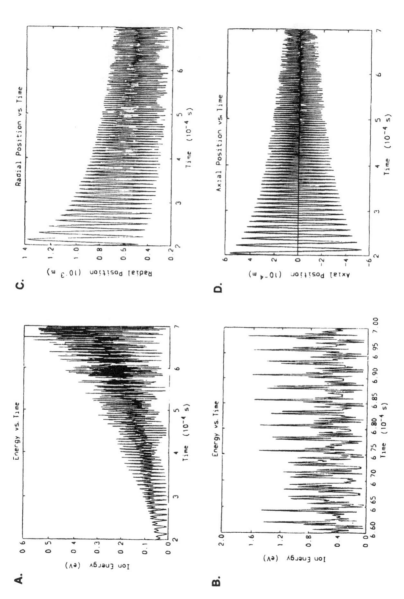

FIGURE 17

Trajectory simulation of m/z 146 during application of the RF ramp so that q_z is increased from 0.1245 to 0.7797. (A) Kinetic energy variation over the time period 200 to 700 μs in time steps of 9.091×10^{-7} s; (B) kinetic energy variation over the time period 660 to 700 μs in time steps of 40 ns; (C) radial excursions over the time period 200 to 700 μs; (D) axial excursions over the time period 200 to 700 μs. (From March, R.E. et al., *Int J. Mass Spectrom. Ion Processes*, 112, 247, 1992. With permission.)

frequency is 24.230 kHz, while that for axial motion is 48.571 kHz. The temporal variation of axial ion motion shown in Figure 16D consists of 1.94 cycles of the axial fundamental secular frequency.

b. Application of Radiofrequency and Direct Current Voltage Ramps

The working point of m/z 146 was moved from point A in Figure 14 to point B by ramping the RF potential from 225 $V_{(0-p)}$ to 1409 $V_{(0-p)}$ at the rate of 2.368×10^6 $V_{(0-p)}$ s^{-1} for 500 μs. In Figure 17A is shown the temporal variation of the kinetic energy of m/z 146 during this period. The progressive though irregular increase in kinetic energy is seen clearly here. It should be noted that the time axis, which is linear with respect to q_z as the RF voltage ramp is constant, is cumulative in that it shows the time elapsed in the simulation, 200 to 700 μs. The fine structure of kinetic energy variation during the final 40 μs of the RF voltage ramp is shown in Figure 17B. The radial and axial excursions during this period, which are shown in Figures 4C and D, indicate that the trajectory of the ion becomes more tightly focused near the center of the ion trap during the RF voltage ramp. Parenthetically, this observation is supportive of the onion model of ion storage; here, the ions of lower m/z ratio yet of higher q_z value are more tightly focused at the center than are ions of (relatively) higher m/z ratio, thus leading to a model wherein ions are arranged in the trap as the layers of an onion. Clearly, such an arrangement is not necessarily conducive to interference-free manipulation of ion trajectories, for example, during mass-selective ion ejection in order of ascending m/z ratio.

At the end of this period, point B, the values of q_z for ions of m/z 144, 146, and 148 were 0.7905, 0.7797, and 0.7692, respectively. A continuous DC voltage ramp of -2.700×10^5 Vs^{-1} was then applied over a period of 500 μs such that at the end of this period and 1200 μs into the simulation, a DC potential of -135 V was imposed on the ring electrode. At this juncture, the working point for m/z 146 was located at point C. The value of a_z for ions of m/z 146 was 0.1494, while those for ions of m/z 144 and 148 were 0.1515 and 0.1474, respectively. The working points for m/z 144 and 148 now lay outside of the stability diagram, while that for m/z 146 remained inside, so that the m/z 146 ion species was isolated within the ion trap. The trajectories of ions of m/z 144 and 148 became unstable, and these ions left the ion trap as shown below.

c. Fate of m/z 144

The fate of the m/z 144 ions during the DC ramp is shown in Figure 18A to F. The temporal variations of ion kinetic energy, radial and axial excursions, are shown in Figure 18A to C. The increase in both kinetic energy and axial excursions were modest until the final 10 μs of the sim-

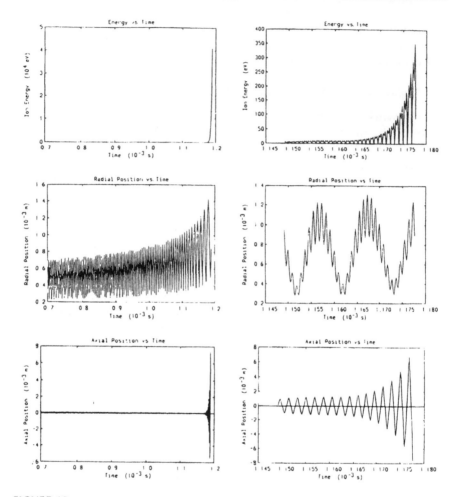

FIGURE 18

Trajectory simulation of m/z 144 with $q_z = 0.7905$, during application of the DC ramp so that a_z is increased from zero to $0.1494 \times 146/144$ (the final working point lies outside the stability diagram). (A) Kinetic energy variation over the time period 700 to 1177 μs (time of ejection), in time steps of 9.091×10^7 s; (B) radial excursions over the time period 700 to 1177 μs; (C) axial excursions over the time period 700 to 1177 μs; (D) kinetic energy variation over the time period 1147 to 1177 μs in time steps of 40 ns; (E) radial excursions over the time period 1147 to 1177 μs; (F) axial excursions over the time period 1147 to 1177 μs. (From March, R.E. et al., *Int J. Mass Spectrom. Ion Processes*, 112, 247, 1992. With permission.)

ulation. Ion kinetic energy reached a value of ≈350 eV shortly before the axial excursion exceeded the value of z_0 and, after 1.177 ms of the simulation, the ion was lost from the trap. Radial excursions increased by a factor of about 2 during the application of the DC pulse. The variations of ion kinetic energy and radial and axial excursions during the final 30 μs of the trajectory of m/z 144 are shown in Figure 18D to F, respectively.

d.   Fate of m/z 148

The trajectory parameters of kinetic energy and radial and axial excursions of the m/z 148 ions during the DC ramp resemble closely those of m/z 144 ions. At the end of the DC voltage ramp the ion remained in the ion trap although the trajectory was already exhibiting signs of incipient instability. The plateau value of the DC voltage, –135 V, was maintained for a period of 500 μs and the trajectory of m/z 148 remained confined to the ion trap for only 10.88 μs further. In Figures 19A to C are shown the variations of trajectory parameters of kinetic energy and radial and axial excursions of the m/z 148 ions during the DC ramp. The variations of these parameters during the final 10.88 μs are shown in Figure 19D to F; at the end of this period, the radial excursion of the m/z 148 ion exceeded r_0 and the ion was lost. Thus, as expected, the ion of lower m/z ratio was lost axially and that of higher m/z ratio was lost radially.

e.   Fate of m/z 146

The DC plateau was maintained for 500 μs, and the variation of trajectory parameters for m/z 146 during this period are shown in Figure 20A to C; the variations of these parameters during the final 40 μs are shown in Figures 20D to F. It is seen that the trajectory of m/z 146 is quite stable during this period. The kinetic energy variation shown in Figure 20A has a maximum value of 4.7 eV, but the kinetic energy was calculated only once per RF cycle; when the kinetic energy was calculated in time steps of 40 ns, it was seen that kinetic energy maxima reached 22 eV. While such energies are maintained for brief periods, collisions during these periods could lead to loss of ions. The axial component of motion (Figures 20A and B) shows pronounced beating as is discussed in Chapter 3. Upon removal of the DC voltage and lowering of the RF voltage, the working point of m/z 146 is returned to point A. The m/z 146 ion has been isolated and is now available for further examination.

B.   Isolation of Ions Having the Highest *m/z* Ratio

A special case of ion isolation is the removal of all ions having m/z ratios lower than that of a chosen species, for example, isolation of the molecular ion from all of the fragment ions. This case is similar to that described above where the RF voltage amplitude is ramped. Here, the RF voltage is increased until the working points of the ions having lower m/z ratios pass through the $\beta_z = 1$ boundary of the stability diagram. This case is not considered further here.

C.   Consecutive Isolation

In this method, unwanted ions are removed in two consecutive stages; ions of m/z ratio lower than that of the selected ion species are ejected at

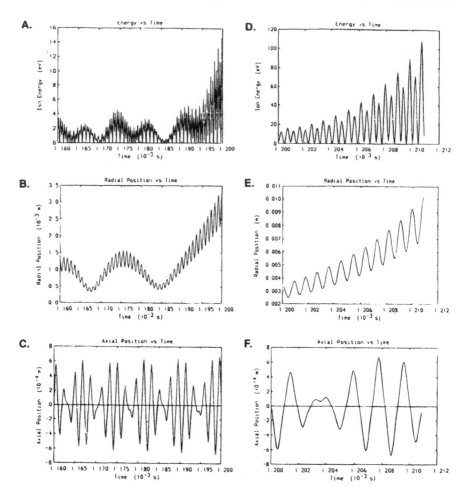

FIGURE 19

Trajectory simulation of m/z 148 with $q_r = 0.7692$, during application of the DC ramp and immediately following it so that a_z is increased from zero to $0.1494 \times 146/148$ (the final working point lies only just outside the stability diagram). (A) Kinetic energy variation over the time period 1160 to 1200 μs in time steps of 40 ns; (B) radial excursions over the time period 1160 to 1200 μs; (C) axial excursions over the time period 1160 to 1200 μs; (D) kinetic energy variation over the time period 1200 to 1210.9 μs (time of ejection) in time steps of 40 ns; (E) radial excursions over the time period 1200 to 1210.9 μs; (F) axial excursions over the time period 1200 to 1210.9 μs. (From March, R.E. et al., *Int J. Mass Spectrom. Ion Processes*, 112, 247, 1992. With permission.)

the $\beta_z = 1$ boundary of the stability diagram, while those of higher m/z ratio are ejected at the $\beta_z = 0$ boundary. At first sight, it would appear that little is to be gained by consecutive isolation in that this process is both more cumbersome and more time consuming than concurrent isolation. Upon closer examination, it is found that the consequences of mov-

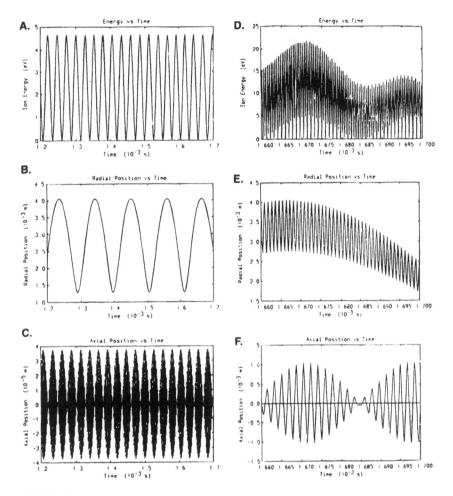

FIGURE 20

Trajectory simulation of m/z 146 during the plateau, or sorting period. (A) Kinetic energy variation over the time period 1200 to 1700 μs in time steps of 9.091×10^{-7} s; (B) radial excursions over the same time period; (C) axial excursions over the same time period; (D) kinetic energy variation over the time period 1660 to 1700 μs in time steps of 40 ns; (E) radial excursions over the same time period; (F) axial excursions over the same time period. (From March, R.E. et al., *Int J. Mass Spectrom. Ion Processes*, 112, 247, 1992. With permission.)

ing the working point of the ion species undergoing isolation to the vicinity of two stability boundaries in turn are profound. Under these circumstances, excitation of the ion species by the trapping field in the presence of helium buffer gas can lead to ion isolation, ion loss, and ion fragmentation. Thus, in this simulation study, attention is focused on the variation of ion kinetic energy during the isolation process, and on the effects of ion/atom collisions.

The findings of this simulation study will challenge some of the common perceptions of ion behavior; in particular, the general understanding that all collisions lead to a loss in ion momentum (with a concomitant enhancement of trajectory stability by reduction in ion excursion from the center of the device) will need to be reexamined. The role of collisions during prolonged collisional focusing near the center of the ion trap also requires reconsideration.

1. Origin of the Method

The scan function and movement of the working points within the stability diagram for this consecutive process of ion isolation are shown in Figure 21; they correspond to those employed originally by Traldi and co-workers.[58,59] The scan function corresponds to the removal of unwanted ion species in two consecutive stages. The order of ion ejection carried out by the Traldi group is that of ions of low m/z followed by ions of high m/z. While in theory the order of ion ejection should be of no import, in practice, the order is critical.

2. System Undergoing Isolation

The consecutive method of ion isolation has been applied[58] to an ion-structure investigation of [M-H]$\cdot$ ions, m/z 207, formed by hydride loss from chalcone (2-propen-1-ona-1,3-diaphenyl). As the scan function for the consecutive method was reported in detail, this particular ion isolation process was chosen for simulation. Although the electron impact mass spectrum of chalcone shows ions at m/z 207 and 208 but not at m/z 206, the simulation included a hypothetical ion of m/z 206 in order to illustrate isolation from neighboring ions differing in mass by 1 u. A helium pressure of 0.7×10^{-3} torr was assumed in the simulation study, in agreement with the reported experimental condition.[28]

3. Order of Ion Ejection

In the early stages of the development of this consecutive method of ion isolation, it was observed that selected ion species could indeed be isolated when the $\beta_z = 0$ boundary was visited first: here, ions of higher m/z were ejected; upon visiting the proximity of the $\beta_z = 1$ boundary, ions of lower m/z were ejected, and the selected ion species remained isolated in the ion trap ready for photodissociation. When the order of ion ejection was reversed, ions of lower and higher m/z were ejected as expected at the $\beta_z = 1$ and $\beta_z = 0$ boundaries, respectively; however, at the conclusion of the process, ions of lower m/z were detected in the subsequent mass analysis scan.[60]

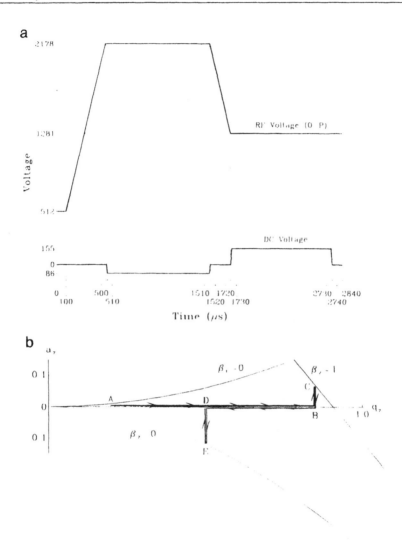

FIGURE 21
(a) A scan function for consecutive ion isolation using RF and DC voltages; (b) stability diagram for the quadrupole ion trap showing the changes in location of the working point, ABCBDED, for an ion undergoing consecutive ion isolation. (From March, R.E. et al., *Int J. Mass Spectrom. Ion Processes*, 125, 9, 1993. With permission.)

It has been reported[61] that while the working point of the ion undergoing isolation is in the vicinity of the $\beta_z = 0$ boundary, the selected ion species becomes kinetically excited, and after collision(s) with bath gas atoms, some ions dissociate. Fragment ions formed thus have working points that lie within the stability boundaries and are stored. This type of behavior has been described as boundary-activated dissociation (BAD). Thus, it would appear that when collision-induced dissociation (CID) of the isolated ion is envisaged, that consecutive ion isolation obviates the need for resonance excitation with a supplementary RF tickle voltage oscillating at a specific frequency. A detailed discussion of boundary activation is presented in Chapter 7.

4. Collisions of Ions with Buffer Gas Atoms

A critical aspect of consecutive ion isolation is the observation of ion fragmentation when the working point of the ion undergoing isolation is moved to the vicinity of the $\beta_z = 0$ boundary. As fragmentation occurs from the conversion of ion translational energy to internal energy through collisions of the ion with buffer gas atoms, collisions were considered in the simulation of this isolation method. Although the consideration of collisions per se does not yield information on the fragmentation processes, the collisional behavior of ions undergoing isolation is of interest, particularly the temporal variation of ion kinetic energy under collision conditions.

Collisions between an ion and buffer gas atoms are simulated using the Langevin theory.[62] As an ion approaches an inert gas atom, the charged ion induces an electric dipole moment in the neutral, giving rise to a net attractive force between the two. There is a certain critical value of the impact parameter, b, such that when $b < b_0$, a short-lived complex will be formed and a transfer of energy will occur. According to the Langevin theory and assuming SI units, the collision cross-section for a singly charged, point-particle ion with an atom, or molecule with no permanent dipole moment, is given by[63]

$$\sigma = \pi b_0^2 = \frac{2\pi e}{4\pi \,\epsilon_0 \, v} \sqrt{\frac{\alpha_e}{\mu}} \tag{28}$$

where e is the electronic charge, ϵ_0 is the permittivity of free space, v is the relative velocity of the collision partners, α_e is the electric polarizability of the gas atom, and μ is the reduced mass of the system. A detailed description of an individual collision is given in the Appendix. 6.

The probability of a collision per unit time can be calculated from the collision cross-section and the number density of collision-gas atoms, n, as

$$P = nv\sigma \tag{29}$$

Using the ideal gas law to approximate n and substituting for σ from Equation 28

$$P = \frac{e}{2\,\epsilon_0} \sqrt{\frac{\alpha_e}{\mu}} \frac{p}{kT} \tag{30}$$

where p and T are the pressure and temperature of the collision-gas, respectively, and k is the Boltzmann constant. Conveniently, this quantity is independent of the relative velocity of the collision partners and needs to be calculated only once, at the beginning of a run.

Random collisions were implemented as follows:

1. At each time step in the calculation a random number between 0 and 1 was generated. When this number was less than or equal to the probability of a collision occurring over a period of one time step, a collision was deemed to have occurred. (As a helium pressure of 0.7×10^{-3} torr was assumed and the phase-space coordinates of an ion were recalculated 20 times each RF cycle, a collision would occur on average once every 1800 time steps.)

2. The magnitude of the velocity of each collision-gas atom was chosen randomly from a Maxwell-Boltzmann distribution of velocities.

3. The collision-gas atom was given a random direction in the laboratory reference frame.

4. The velocity of the center of mass was determined.

5. The magnitude of the velocity of the ion in the center of mass reference frame was calculated. (For elastic collisions, in the center of mass frame, only the directions of the velocities of the collision partners are changed by the collision. The magnitudes of the velocities remain unchanged.)

6. The scattered ion was given a random direction in the center of mass frame. This step was equivalent to choosing a random value for the impact parameter, which varied between zero and the maximum value for which capture would occur.

7. The new velocity of the scattered ion was transformed to the laboratory reference frame.

A detailed description of the system of computer programs used to simulate single-ion trajectories in an ion trap, such as those presented here, is given elsewhere.[64]

5. Description of the Path Traced by the Working Point

The "trajectory" of the working point for the ion undergoing isolation, m/z 207, is shown schematically in Figure 21b, starting at point A for which $q_z = 0.2$. The working point is moved from A to B which has a q_z value of 0.85. The location of B is not particularly critical; this value appears to have been chosen as it lies about midway between the q_z value of the upper apex and that at which the q_z-axis intercepts the $\beta_z = 1$ boundary. The working point is then moved from B to C; the position of C is determined by both the value chosen for B and the negative DC potential applied to the ring electrode. The position of C ($a_z = 0.0673$, $q_z = 0.85$)

can be defined in terms of a quantity Δm_1 which is the difference in atomic mass units between the mass of the selected singly charged ion (with its working point located at C) and that of a hypothetical ion of lower mass lying on the $\beta_z = 1$ boundary. Δm then defines the separation, in atomic mass units, of a given working point from a boundary of the stability diagram. In this study, $\Delta m_1 = 0.3$ u such that the working point of the ion of m/z 206.7 lay on the $\beta_z = 1$ boundary. After a sorting time of 1 ms at C, the working point is returned to B and then moved to point D on the q_z axis with $q_z = 0.50$. The working point is moved to point E in the vicinity of the $\beta_z = 0$ boundary and then, after a second sorting time of 1 ms, is returned to point D. The locations of C and E are critical in that they determine the proximity of the working points of the selected ion, m/z 207, to the $\beta_z = 1$ and $\beta_z = 0$ boundaries, respectively, and the degrees to which the working points for m/z 206 and m/z 208 lie beyond the respective boundaries.

Point E $(a_z = -0.1213, q_z = 0.50)$ is defined in terms of a quantity Δm_h, which is the difference in atomic mass units between the mass of the selected ion and that of a hypothetical ion of *higher* mass lying on the $\beta_z = 0$ boundary; in this case, m/z 207.8, and $\Delta m_h = 0.8$ u.

6. Scan Function for Consecutive Isolation

The selected ion was maintained at point A on the q_z-axis $(q_z = 0.200)$ for 100 μs. The RF voltage was ramped from 512.4 $V_{(0-p)}$ to 2177.8 V at the rate of 4.163×10^6 Vs^{-1}, to move from A to B in 400 μs. A DC voltage of −86.20 V was applied at the rate of -8.620×10^6 V s^{-1}, to move from B to C in 10 μs. Point C was maintained for 1000 μs. The DC and RF voltages were reduced in succession at the rates of 8.620×10^6 V s^{-1} and -4.484×10^6 V s^{-1}, respectively, so as to reattain point B in 10 μs and then arrive at point D in 200 μs. Point D corresponded to an RF level of 1281 V. A DC voltage of 155.4 V was applied at the rate of 15.54×10^6 V s^{-1} to move from D to E in 10 μs. Point E was maintained for 1000 μs; the DC voltage was then reduced to zero at the same rate so as to regain point D in 10 μs. After a further period of 100 μs, the simulation was terminated. Thus, simulation of the process of consecutive ion isolation was treated in ten stages. The total duration of the simulated process was 2.840 ms.

7. Ion Initial Positions and Velocities

The initial position (x, y, z), velocities $(\dot{x}, \dot{y}, \dot{z})$, and RF phase for each of the ten ions were chosen at random; the distance from the center of the ion trap to the initial position of each ion, R, was restricted to $0.01 <$ R < 1 mm, while the initial ion kinetic energy, T, was restricted to $0.05 <$

$T < 0.3$ eV. However, these restrictions on the initial position and velocity were found to be inadequate. Examination of the calculated trajectories for m/z 207 ions revealed that excursions from the trap center of some of the trajectories exceeded the physical dimensions of the ion trap when the working point was moved to the vicinity of a stability boundary. The absorption of power from the RF field increases with the distance from the center of the ion trap, and although the trajectories remained mathematically stable, the amplitudes of the ion's secular excursions increased such that some of the ions collided with the end-cap electrodes.

The simulation process was repeated for m/z 207 ions, using the same ten ions as before, but the ions were subjected to an additional appropriate period of collisional cooling so as to produce well-cooled ions.

8. Determination of the Cooling Process

The trajectory of a single ion, with a working point on the q_z-axis at $q_z = 0.2$, was calculated for 50 µs in a collision-free system to verify trajectory stability. The ion had relatively large secular amplitudes, both radial and axial, and consequently was highly energetic. During the first collision-free 50 µs of ion motion, the ion exhibited a maximum kinetic energy of 33 eV and maximum radial and axial excursions of 7.7 and 6.2 mm, respectively. Thus, the ion was energy rich and its trajectory filled the major portion of the ion trap. The trajectory was calculated for 2000 µs further with an ambient helium pressure of 1.5×10^{-2} torr; this relatively high pressure of helium was chosen to permit foreshortening of the computational time. The motion of the ion was not unduly restricted at this pressure as the probability of a collision was 0.012 for each time step (4.545×10^{-8} s); i.e., on average, there were 83 time steps between successive collisions. Following a collisional period of 2000 µs at an ambient helium pressure of 1.5×10^{-2} torr, the fundamental secular frequencies of the trajectories were retained, the ions remain confined within the ion trap, and the ion kinetic energy had been greatly reduced.

The temporal variations of ion kinetic energy and radial and axial excursions during the 2000 µs are shown in Figure 22. Maximum dampening of ion kinetic energy and radial and axial excursions were achieved after ca. 1000 µs. A high-pressure cooling time of 1000 µs was judged to be adequate for the formation of a thoroughly cooled ion.

Each of the ten original ions was subjected to a cooling period of 1000 µs, with an ambient helium pressure of 1.5×10^{-2} torr. Under these conditions, each of the ions suffered ca. 250 collisions in 1000 µs or, on average, one collision every 88 time steps or 4.5 RF cycles. The coordinates and velocities of the cooled ions, together with the RF phase, were taken as initial conditions for the subsequent simulation of the ion isolation process.

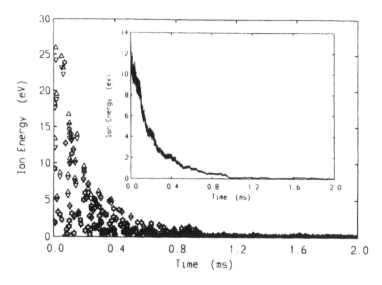

FIGURE 22

Trajectory simulation of m/z 207, with collisions corresponding to a pressure of helium of 1.5×10^{-2} torr, at the working point ($a_z = 0$, $q_z = 0.2$), over a time period of 1000 µs. Upright and inverted triangles correspond, respectively, to ion kinetic energies before and after each of 510 collisions (the inset to this figure shows the average kinetic energy over the same time period). (From March, R.E. et al., *Int J. Mass Spectrom. Ion Processes*, 125, 9, 1993. With permission.)

9. Simulation Results

The trajectories of ten cooled ions of m/z 207 were calculated in ten stages for the path ABCBDED shown in Figure 21. Random collisions with helium were considered at a collision frequency appropriate to a pressure of 7×10^{-4} torr. The simulation was repeated for ten cool ions in a collision-free system. The entire simulation was repeated for ten warm ions, i.e., ions with the same initial positions and velocities as the cooled ions but not subjected to the cooling procedure.

Trajectories for the ten cool subject ions were calculated up to and including the plateau at point C for both collisional and collision-free systems. Maximum trajectory excursions and kinetic energy maxima at point C are shown in Tables 7 and 8, respectively, and compared with the behavior of warm ions under the same conditions. The maximum excursions of the cool ions, under all conditions, were smaller than those of the warm ions by at least a factor of 2. For example, the average of the maximum radial excursions for the collision-free cool ions is 4.2×10^{-4} m, exactly half of that for the warm ions, while the average of the maximum axial excursions for the cool ions is almost one fourth that of the warm ions. For the collisonal systems, virtually the same ratios of averages as above were obtained for the maximum radial and axial excursions. A striking feature of the difference in behavior of the cool ions compared with that of the warm ions was that none of the cool ions exceeded the physical dimensions of the ion trap, whereas collisions caused warm ions A_3 and A_9 to exceed the dimensions. Maximum radial excursions for both cool and warm ions are virtually unaffected by collisions, but due to the proximity of the working point to the $\beta_z = 1$ boundary, maximum axial excursions for both cool and warm ions are increased by the action of collisions. The effects of collisions are seen also in Table 8, wherein the kinetic energy maxima for warm and cool ions at point C are presented. While the average maxima of kinetic energy for cool ions both with and without collisions are less than those for warm ions, collisions generally lead to increased kinetic energy.

The entire simulation was continued to the working point E. Again, the maximum radial excursions for cool ions, as shown in Table 9, were about half those of warm ions, and both types of ions were affected by collisions. The average maximum axial excursion for cool ions was approximately one sixth that for warm ions in the collision-free system, and about one half that for warm ions in the collisional system. In only a single case, A_4, did the axial excursion of a cool ion exceed the physical dimensions of the ion trap. A detailed description of the final collision of ion A_4 is given in the Appendix. The average maximum kinetic energy for cool ions both with and without collisions are again less than those for warm ions as shown in Table 10. While collisions produced a small decrease in the average maximum kinetic energy for warm ions, the average maximum kinetic energy for cool ions was approximately doubled by the action of collisions.

10. Conclusions

The trajectories were calculated for each of ten warm and ten cool ions of m/z 206, 207, and 208 which had been subjected to a consecutive ion isolation process under collisional and collision-free conditions. Ejection of all unwanted ions (m/z 206 and 208) was accomplished readily. Examination of the trajectories of warm m/z 207 ions revealed a variety of behavior; under all conditions, and a high degree of ion loss. Small variations in the positions of working points C and E brought little change in ion loss. Collisional precooling of the ions of m/z 207 had a profound effect upon trajectory behavior in that ion loss was reduced (albeit, for a small sample of ions) in both collisional and collision-free conditions. While nine of ten cool ions remained in the trap under collisional conditions, in general, collisions that occur while the working point of an ion is near a stability boundary lead to increased kinetic energy.

It is clearly of importance to consider the ion kinetic energy maxima of an ion species undergoing isolation, particularly during the period in

TABLE 7

Maximum Radial and Axial Trajectory Excursions at the Working Point C for "Warm" and "Cool" m/z 207 Ions Both With and Without Collisions with Helium Atoms

Ion	Warm[a]				Cool[b]			
	$P_{He} = 0$ (/10^{-4} m)		$P_{He} = 0.7$ mtorr (/10^{-4} m)		$P_{He} = 0$ (/10^{-4} m)		$P_{He} = 0.7$ mtorr (/10^{-4} m)	
	r^c	z^d	r^c	z^d	r^c	z^d	r^c	z^d
A_1	9.5	20.0	7.0	25.0	3.1	7.0	3.5	5.2
A_2	11.0	15.0	11.0	15.0	4.7	4.5	4.5	10.0
A_3	11.0	40.0	8.0	72.0[e]	3.0	9.0	2.9	3.8
A_4	16.0	10.0	13.0	60.0	5.0	4.8	4.5	32.0
A_5	2.4	33.0	4.0	50.0	5.3	15.0	4.2	20.0
A_6	3.0	9.0	2.7	9.0	3.0	5.5	3.4	10.0
A_7	3.7	25.0	5.0	72.0[e]	5.0	3.5	5.0	3.2
A_8	6.5	35.0	2.8	9.0	5.5	7.5	6.2	4.0
A_9	9.0	47.0	11.0	72.0[e]	3.5	3.8	2.8	5.0
A_{10}	12.0	22.0	12.0	40.0	4.0	2.5	3.7	10.0
Range (/10^{-4} m)	2.4–16.0	9.0–47.0	2.7–13.0	9.0–72.0	3.0–5.5	2.5–15.0	2.8–6.2	3.2–32.0
Average (/10^{-4} m)	8.4	26	7.6	42.7	4.2	6.3	4.1	10

[a] Initial position, R, restricted to $0.01 < R < 1$ mm; initial kinetic energy, T, restricted to $0.05 < T < 0.3$ eV.
[b] Ions created initially at random and subjected to ca. 250 collisions with helium atoms.
[c] Maximum radial excursion.
[d] Maximum axial excursion.
[e] Axial excursions exceed the physical dimensions of the ion trap.

Source: From March, R.E. et al., *Int J. Mass Spectrom. Ion Processes*, 125, 9, 1993. With permission.

TABLE 8

Kinetic Energy Maxima at the Working Point C for "Warm" and "Cool" m/z 207 Ions Both With and Without Collisions with Helium Atoms

	Warm[a]		Cool[b]	
	$P_{He} = 0$ (/eV)	$P_{He} = 0.7$ mtorr (/eV)	$P_{He} = 0$ (/eV)	$P_{He} = 0.7$ mtorr (/eV)
Ion	E^c	E^c	E^c	E^c
A_1	37.0	51.0	4.0	2.6
A_2	23.0	21.0	2.0	9.0
A_3	150.0	250.0[d]	7.0	1.3
A_4	14.0	7.0	2.3	90.0
A_5	100.0	210.0	19.0	34.0
A_6	7.0	7.5	2.8	8.0
A_7	57.0	420.0[d]	1.6	1.6
A_8	112.0	7.3	5.0	2.3
A_9	195.0	410.0[d]	1.4	2.1
A_{10}	45.0	45.0	0.8	10.0
Range (/eV)	7.0–195.0	7.0–420.0	0.8–19.0	1.3–90.0
Average (/eV)	74	150	4.6	16

[a,b] As in Table 1.
[c] Maximum kinetic energy.
[d] Axial excursions exceed the physical dimensions of the ion trap.

Source: From March, R.E. et al., *Int J.Mass Spectrom. Ion Processes, 125*, 9, 1993. With permission.

which ions of other species are being ejected. When the working point of the ion undergoing isolation is moved to the vicinity of a boundary of the stability diagram, ion kinetic energy is enhanced, and the degree of enhancement can be moderated by the collisional history of the ion.

It appears that the removal, at point C, of ions lower in mass by 1 u can be accomplished rapidly. Furthermore, when the ions are sufficiently well cooled, their kinetic energies are moderated so that their trajectories do not exceed the physical dimensions of the ion trap, with the result that the first stage of consecutive ion isolation is accomplished efficiently.

For the second stage of ion isolation, at point E, where ions greater in mass by 1 u are again removed rapidly, it should be possible to select the operating conditions so as to optimize ion kinetic energies for BAD with minimal loss of ions undergoing isolation. Ion kinetic energy maxima attained in a collisional system will increase the probability of internal excitation of the ion, due to conversion of ion kinetic energy in collisions with helium atoms, and of dissociation. These same kinetic energy maxima can be accompanied by large excursions which result in ion losses.

The search for a compromise is clearly justified here, as the possibility of ejection of higher m/z ions concurrent with dissociation of the mass-

TABLE 9

Maximum Radial and Axial Trajectory Excursions at the Working Point E for "Warm" and "Cool" m/z 207 Ions Both With and Without Collisions with Helium Atoms

	Warm[a]				Cool[b]			
	$P_{He} = 0$ (/10⁻⁴ m)		$P_{He} = 0.7$ mtorr(/10⁻⁴ m)		$P_{He} = 0$ (/10⁻⁴ m)		$P_{He} = 0.7$ mtorr (/10⁻⁴ m)	
Ion	r^c	z^d	r^c	z^d	r^c	z^d	r^c	z^d
A_1	8.0	55.0	5.7	72.0*	2.5	2.0	2.2	22.0
A_2	9.0	27.0	9.0	16.0	4.0	8.0	4.0	9.0
A_3	8.2	72.0*	—	—	2.5	4.0	2.3	6.0
A_4	13.0	6.0	10.0	23.0	4.3	10.0	3.7	72.0*
A_5	1.8	72.0*	3.4	72.0*	4.5	7.0	4.2	60.0
A_6	2.5	25.0	2.2	25.0	2.6	15.0	2.8	27.0
A_7	3.1	72.0*	—	—	4.3	1.1	3.6	32.0
A_8	5.7	72.0*	2.9	19.0	4.5	20.0	5.0	20.0
A_9	7.5	72.0*	—	—	2.7	15.0	1.4	20.0
A_{10}	10.0	37.0	9.5	72.0*	3.2	6.0	3.4	6.2
Range (/10⁻⁴ m)	1.8–13.0	6.0–72.0	2.2–10.0	16.0–72.0	2.5–4.5	1.1–20.0	1.4–5.0	6.2–72.0
Average (/10⁻⁴ m)	6.9	51	6.1	43	3.5	8.8	3.3	27

[a,b]As in Table 1.

Source: From March, R.E. et al., *Int J.Mass Spectrom. Ion Processes, 112, 247, 1992.* With permission.

TABLE 10

Kinetic Energy Maxima at the Working Point E for "Warm" and
"Cool" m/z 207 Ions Both With and Without Collisions with
Helium Atoms

Ion	Warm[a]		Cool[b]	
	$P_{He} = 0$ E^c	$P_{He} = 0.7$ mtorr E^c	$P_{He} = 0$ E^c	$P_{He} = 0.7$ mtorr E^c
A_1	55.0	110.0[d]	8.0	11.0
A_2	15.0	6.0	1.4	1.8
A_3	140.0[d]	—	37.0	0.8
A_4	3.4	12.0	2.3	118.0[d]
A_5	130.0[d]	130.0[d]	99.0	70.0
A_6	14.0	12.0	4.6	15.0
A_7	110.0[d]	—	0.3	20.0
A_8	115.0[d]	7.0	8.0	9.0
A_9	110.0[d]	—	4.2	8.0
A_{10}	27.0	110.0[d]	0.8	80.0
Range (/eV)	3.4–140.0	6.0–130.0	0.3–99.0	0.8–118.0
Average (/eV)	72	55	14	34

[a] See footnote a to Table 1.
[b] See footnote b to Table 1.
[c] See footnote c to Table 2.
[d] See footnote e to Table 1.

Source: From March, R.E. et al., *Int J.Mass Spectrom. Ion Processes, 125,* 9, 1993. With permission.

selected ions and retention of fragment ions will be of considerable value in that the need for specific resonance excitation will be obviated.

VIII. RESONANCE EXCITATION

A. Introduction

The secular motion of an ion can be excited along the axis of a quadrupole ion trap by the resonant absorption of power from an auxiliary generator, as was employed initially by Paul et al.[65] and Fischer[66] as a method of trapped ion detection. The onset of energy absorption from an auxiliary oscillator by charged particles has been observed visually.[67] With the development of commercial ion traps, resonance excitation of ions stored in a quadrupole ion trap is now being employed on a much wider scale.[68] The supplementary or auxiliary RF potential oscillating at a selected frequency has been referred to as a "tickle" as opposed to the main RF potential, which is known as the "drive" potential.

Resonance excitation can be used to eject ions from an ion trap or to enhance stored ion kinetic energy. Ions having a specific m/z ratio can be ejected from the ion trap upon irradiation at a specific tickle frequency and with the RF drive amplitude held constant, provided the magnitude of the tickle potential amplitude and the duration of irradiation are sufficient. When the magnitude of the tickle potential and the duration of irradiation are moderated, the rate of enhancement of ion kinetic energy by resonance excitation also is moderated. In this case, enhanced ion kinetic energy can open endothermic ion/molecule reaction pathways, or ion kinetic energy can be transferred to internal energy, through collisions with helium buffer gas, and accumulated in the resonantly excited species to the point of ion fragmentation. Collision induced dissociation by resonance excitation is very commonly used in MS/MS. When the RF drive amplitude is ramped during resonance excitation, several or all ion species can be ejected mass selectively as each species comes into resonance with the chosen tickle frequency. In this RF scanning mode, the mass range of the ion trap can be increased by an order of magnitude or so.[53]

B. Application of an Auxiliary Potential

Resonance excitation can be wrought with the application of an auxiliary potential oscillating at a frequency which is usually chosen to coincide with one of the frequency components of the secular motion of a subject ion. There are three such arrangements for applying an RF potential to ion trap electrodes in order to induce resonance excitation: in-phase potentials may be applied to each end-cap electrode (quadrupolar); potentials 180° out-of-phase may be applied to each end-cap electrode (dipolar); and a potential may be applied to one end-cap electrode while the other is grounded (monopolar).

The secular frequency, $\omega_{u,n}$, was defined in Chapter 2 as the angular frequency of order n for motion in the direction u ($u = r, z$), such that

$$\omega_{u,n} = \left(n + \frac{1}{2}\beta_u \right)\Omega, \quad 0 \leq n < \infty \tag{28}$$

and

$$\omega_{u,n} = -\left(n + \frac{1}{2}\beta_u \right)\Omega, \quad -\infty < n < 0 \tag{29}$$

When $n = 0$, the fundamental angular frequency $\omega_{u,o}$ in either the r- or z-direction, is given by

$$\omega_{u,0} = \frac{1}{2}\beta_u\Omega \tag{30}$$

Therefore, ions with the same β_r or β_z values will have identical secular frequencies. The forms of ion trajectories have the general appearance of a Lissajous curve composed of two frequency components ω_r and ω_z of the secular motion, with a superimposed micromotion of frequency $\Omega/2\pi$ (Hz), where

$$\omega_{r,0} = \beta_r\Omega/2 \qquad \omega_{z,0} = \beta_z\Omega/2 \tag{31}$$

The higher-order frequencies correspond to $n = \pm1, \pm2$, etc.

C. Quadrupolar Resonance Excitation

Resonance excitation in the quadrupolar mode is effected when a potential is applied in phase to the end-cap electrodes. The application of a potential ϕ to the ring electrode is entirely equivalent to applying a potential of $-\phi$ to the end-cap electrodes. With the drive and the auxiliary potentials given as ϕ^d and ϕ^a, respectively, the potential within the device is given by

$$\phi_{(r,z)} = \frac{\left(\phi^d - \phi^a\right)\left(r^2 - 2z^2\right)}{2r_0^2} + \frac{\left(\phi^d + \phi^a\right)}{2} \tag{32}$$

The resulting equations of motion are given by

$$\frac{d^2u}{dt^2} = \frac{\alpha e}{mr_0^2}\left[U + V\cos(\Omega t + \gamma) - V_1\cos(\omega t + \gamma_1)\right]u \tag{33}$$

where $\alpha = 1$ for $u = x, y$ and $\alpha = -2$ for $u = z$. V_1 and ω are, respectively, the amplitude and angular frequency of the auxiliary potential; γ and γ_1 are, respectively, the initial phase angles of the drive and auxiliary potentials. Equation 33 can be solved by numerical integration. Ion trajectory calculations were carried out on microcomputers using the Livermore Solver for Ordinary Differential Equations.

1. Simulation Results

a. Ion Irradiation in the Presence of a Buffer Gas

In a typical simulation experiment, a trapped ion with $a_z = 0$ and $q_z = 0.1955$ was subjected to collisions with neon atoms, then subjected to irradiation at the fundamental axial secular frequency, $\beta_z\Omega/2$. Ion energy was reduced progressively under the influence of momentum-dissipating collisions with neon. A tickle voltage of 3.5 $V_{(0-p)}$ was applied for 0.3

ms; ion energy increased rapidly due to resonant absorption and, despite some ten collisions with neon, the ion was ejected after 121 μs of irradiation. While radial and axial excursions were diminished prior to irradiation, ion motion was excited radially upon irradiation and the ion was ejected radially. Radial ejection under these conditions is characteristic of quadrupolar excitation.

b. Ion Irradiation Under Collision-Free Conditions

Ions of m/z 100 were irradiated at $\beta_z \Omega/2$ in the absence of buffer gas, and the zero-to-peak amplitude of the tickle voltage was varied from 0.125 to 10 V; the ion kinetic energy was calculated to the time of ion ejection. If we consider the tickle voltage as a measure of the irradiation intensity, then the product of amplitude and irradiation duration can be considered as a fluence. The fluence required for ejection was found to be constant at ~0.36 V ms^{-1} over the range of tickle voltages. Under these conditions, it can be concluded that the irradiation period required for ion ejection varies inversely with tickle voltage.

c. Phase Variation

In a collision-free system, the phases γ and γ_1 of the drive potential, ϕ^d, and the auxiliary potential, ϕ^a, respectively, were varied in turn in steps of $\pi/5$ rad. Ion axial and radial motions each exhibited composite wave behavior such that the envelopes encompassing the temporal variations of ion motions showed low-frequency beating at ~4.7 kHz. Ion ejection times varied smoothly as each of γ and γ_1 were varied.

d. Variation of Tickle Frequency

The trajectories of four ions of m/z 134 were calculated in turn at a number of working points on the q_z-axis, such that $0.0985 \leq q_z \leq 0.8813$, while they were irradiated at a number of frequencies. This laborious experiment resembled the point-by-point determination of an absorption spectrum. Each simulation was calculated for 10 μs or up to the time of ion ejection; when ejection occurred, the final kinetic energy, time to ejection, and direction of ejection were noted. At the $\beta_z \Omega/2$ frequency, three ions at $q_z = 0.0985$ were ejected axially and one radially within the time allotted. As the working point was moved along the q_z-axis to $q_z = .4910$, radial ejection became predominant, while the time to ejection and the final kinetic energy increased. Strong absorption was observed over a narrow range of q_z values, $0.7799 \leq q_z \leq 0.7901$; this region is intersected by the iso-β_z line, $\beta_z = 2/3$, at which nonlinear resonances can be observed. Generally weak absorption was observed with irradiation at $\beta_r \Omega/2$ with ejection occurring only at low and high q_z values.

Excitation effects were observed at four new frequencies, $\beta_z\Omega$, $(1 \pm \beta_z)\Omega$, and $\beta_z\Omega$; absorption at $\beta_z\Omega$ and $\beta_z\Omega$ was particularly strong. A theoretical treatment of quadrupolar excitation which predicts absorption at these four frequencies was presented recently.[69] Excitation at $\beta_z\Omega$ and $\beta_z\Omega$ is known as parametric resonance.

e.   Ion Kinetic Energy Upon Ejection

Resonant irradiation of stored ions leads to enhancement of ion kinetic energy (averaged over an RF cycle, for example) so that in the limit the trapping potential is overcome and the ion is lost from the ion trap. Hence, the kinetic energy of an ion upon ejection from the trap can be interpreted as a measure of the trapping potential well depth.[70,71] Trajectories of four subject ions were calculated to the point of ejection, and the kinetic energy over the final RF cycle was noted; the simulations were carried out over a variety of q_z values (on the q_z-axis), and with resonant excitation at each of the six irradiation frequencies $\beta_z\Omega/2$, $\beta_r\Omega/2$, $\beta_z\Omega$, $(1 \pm \beta_z)\Omega$, and $\beta_z\Omega$.

The variation of average ion kinetic energies upon ejection as a function of q_z and the relatively small spread of these kinetic energies at each value of q_z were interpreted in terms of the potential well depth created by the storage field. When these data are compared with an estimate of the axial potential well depth, $\overline{D}_z$, derived originally by Major and Dehmelt[72] and developed by Lawson and co-workers,[73,74]

$$\overline{D}_z = q_z V / 8 \tag{34}$$

the agreement is quite remarkable.

f.   Resonance Excitation at Working Points with Negative a_z Values

Three ions were subjected individually to radial and axial parametric resonance in three separate regions lying below the q_z-axis of the stability diagram (i.e., with negative a_z values); the three regions are defined as $\beta_z\Omega = \beta_r\Omega$, $\beta_r\Omega < \beta_z\Omega$, and $\beta_z\Omega < \beta_r\Omega$. In all regions, axial parametric resonance invariably induced axial ion ejection, while radial parametric resonance induced radial ion ejection except in the vicinity of those boundaries of the stability diagram where $\beta_z = 0$ and 1, here axial ion ejection was induced.

g.   Conclusion

Quadrupolar excitation of trapped ions is novel in that it can channel energy into the radial or axial component of ion motion and can permit selection of the direction of ion ejection. The concept of a constant fluence required for ion ejection at a given value of q_z and for a given resonant frequency may also be applicable to both dipolar and monopolar irradiation.

IX. INTERACTION OF SIMULATION AND EXPERIMENT

As a general rule, simulation tends to follow experiment, and the field of ITMS is no exception. While the speed and versatility of PCs have increased enormously in recent years, the ease with which simulations may be carried out now has not necessarily been accompanied by enhanced confidence in the results. Yet, on occasion, due to the increased complexity of computerized instrumentation and to the increased separation of the experimentalist from the experimental apparatus, simulation results may indicate that assumptions concerning experimental conditions are not justified. The authors recount such an example.

A. Resonance Excitation: the Fluence Required for Ion Ejection

It was discussed earlier that the product of tickle voltage zero-to-peak amplitude and duration of irradiation can be considered as a fluence, and that simulation studies had shown that the fluence required for ion ejection is sensibly constant for a collision-free system. An opportunity was presented for a relatively facile comparison of simulation with experiment with respect to the minimum, or induction, fluence required for ion ejection under collisional conditions.

1. Experimental System

The experimental results of the resonance excitation of the isolated molecular ion, m/z 146, $[C_6H_4^{35}Cl_2]^+$ from 1,2-dichlorobenzene were reported earlier,[75] so this system was selected for the exploration of the variation of induction fluence under resonance excitation conditions.[76] The sequence of operations followed in the experimental study was that of ionization, pause, concurrent isolation of m/z 146, reduction of the RF voltage to obtain a q_z value appropriate for resonance excitation and mass analysis by the method of axial modulation. The initial pause provided a sorting period during which ions in unstable trajectories left the ion trap. There were similar pauses following each stage of the ion isolation process. Resonance excitation of m/z 146 in the dipolar mode induced ion loss by both ion ejection from the ion trap and ion dissociation to form product ions of m/z 111 as in

$$\left[C_6H_4^{35}Cl_2\right]^{+\cdot} + \phi \rightarrow \left[C_6H_4^{35}Cl\right]^+ + {}^{35}Cl^{\cdot} \qquad (35)$$

where ϕ represents the resonance excitation. The method of dynamically programmed scans[74] was used to follow the above sequence. For a given dynamically programmed microscan, the amplitude of the tickle voltage was fixed at (for example) 1350 mV$_{(0-p)}$ and the initial duration of irradiation was 100 µs; following irradiation the ions were mass analyzed and the data were stored. Several microscans were summed so as to obtain a macroscan with good signal-to-noise ratio. The duration of irradiation was incremented in steps of 50 µs to 10.05 ms. The data can be plotted in the form of the logarithm of ion signal intensity vs. irradiation time. Such a kinetic plot shows the diminution of the signal intensities for both m/z 146 and the total number of ions (the decrease in the total number of ions is due to ejection of some of the m/z 146 ions), the appearance of ions of m/z 111 and, at high radiation fluences, the appearance of m/z 75 due to loss of HCl from excited $[C_6H_4{}^{35}Cl]^+$ (m/z 111). Of particular interest here is the induction period of ca. 600 µs which precedes the loss of m/z 146 ions either by dissociation or ejection.

A series of dynamically programmed scans was carried out in which the tickle voltage amplitude was varied over the range 250 to 1900 mV; the induction period was noted for each value of the amplitude. The values of the induction fluence obtained with tickle voltage amplitudes in the range 0.6 to 1.5 V$_{(0-p)}$ are sensibly constant. At high tickle voltage amplitudes, the induction fluence appears to increase slightly for which we present an explanation (*vide infra*).

2. Simulation of Resonance Excitation Under Collisional Conditions

In order to simulate the experimentally observed variation of the induction fluence the trajectories for each of ten ions of m/z 146 were calculated for the conditions of the above experiment. The initial stage of the simulation consisted of a collisional cooling episode, similar to that described earlier, of 1 ms at 10^{-2} torr. Ions thus cooled survived the ion isolation process. The scan function for ion creation, isolation, and resonance excitation was followed as closely as possible, and an ambient pressure of 10^{-3} torr of helium was assumed. Following the ion isolation process and re-establishment of the working point on the q_z-axis at $q_z = 0.25$, the coordinates and velocities of each of the ten ions were recorded. Each ion was subjected to resonance excitation at precisely the fundamental axial secular frequency and with a tickle voltage of fixed amplitude. The trajectory of each ion was calculated to the point of ion ejection, and the duration of irradiation required for ejection was noted. The coordinates and velocities of each of the ten ions at $q_z = 0.25$ were recaptured, and the process of resonance excitation repeated at a different tickle voltage amplitude.

For each tickle voltage amplitude, one may then consider the ten ions as a hypothetical ensemble within the ion trap so that the calculated trajectories yield the time of (or fluence required for) ejection of each ion from the ensemble. A kinetic plot may then be constructed of the percentage of ions remaining in the ion trap as a function of the duration of irradiation for a given tickle voltage amplitude. Extrapolation of each plot to 100% yields the duration of irradiation immediately prior to ejection of the first ion from a more numerous, yet fictitious, ensemble of ions. An approximation to this value could have been obtained by inspection of the trajectories and selection of the shortest time required for ejection. However, this procedure could have led to overestimation of the irradiation time required for incipient ejection of ions. The plots are linear and the data points show remarkably little scatter. The induction fluences for these simulations, at a pressure of 10^{-3} torr of helium, are in good agreement with experimental results.

3. Discrepancy

For the tickle voltage amplitude range of 0.6 to 1.5 $V_{(0-p)}$, the induction fluences obtained by simulation are ca. one fifth of the experimental values. The origin of this quite considerable discrepancy was thought to lie in the uncertainty associated with the ambient pressure in the experimental work, as this pressure is not measured directly. The simulation induction fluences were smaller than the experimental values, so it was proposed that the ambient pressure in the experimental work was greater than envisaged. Thus, the entire simulation was repeated for each of the pressures of helium 5×10^{-3} torr and 10×10^{-3} torr. Yet the induction fluence as a function of tickle voltage amplitude at each of these elevated pressures of buffer gas was virtually unchanged. It did not seem reasonable that the experimental fluences for ejection exceeded those of the simulation model by a factor of 5.

4. Explanation

At this juncture, it was decided to change the compound and to check the scan function of the ITMS™ instrument on an oscilloscope. The new compound examined was n-butylbenzene and, to our relief and surprise, the experimental fluences required to eject the molecular ion, m/z 134, were lower than those required for ejection of m/z 146 from 1,2-dichlorobenzene. In addition, examination of the scan function with an oscilloscope showed two possible sources of experimental error. First, there was an appreciable discrepancy between the actual irradiation time and input values; the finite duration of an interrupt increases the duration of an operation compared to the programmed duration. Second, there was a finite risetime for the resonant tickle voltage to reach the programmed amplitude, with the result that the experimental determination overestimates the required fluence. Both observations were of great value in the pursuit of physicochemical investigations with the commercial version of the ion trap.

Almost simultaneously with these observations, it was announced[54] that the ion trap used in the Finnigan MAT ITMS™ and Ion Trap Detector (ITD™) differed significantly from the ideal case with respect to the separation between the end-cap electrodes. The ion trap in the ITMS™ and ITD™ had been "stretched" so that the value of z_0 was increased by 11%, with no corresponding modification to the shapes of the electrodes which would be required in order to maintain a purely quadrupolar geometry. An analysis of the incidence and effects of nonlinear resonances in the stretched ion trap is given in Chapter 3; the story associated with the development of this ion trap is recounted in Chapter 4. An earlier examination of the characteristics of this ion trap is given elsewhere.[77]

The revelation concerning the stretched ion trap in the ITMS™ and ITD™ instruments provided a basis for rationalization of the variation in induction fluence between experiment and simulation. The effect of this significant increase in z_0 is to introduce higher-order (multipole) terms to the electric potential in the trapping region, as discussed in Chapter 3. Higher-order multipoles give rise to nonlinear forces within the ion trap; consequently, the secular frequency of ion motion becomes dependent upon the ion displacement from the center of the trap. Thus, as an ion moves away from the center, it experiences a shift in its secular frequency.

Under conditions of resonance excitation, wherein ions become kinetically heated and move away from the center, they will experience shifts in their secular frequencies. When the nonlinear forces are sufficiently strong, the secular frequency of an excited ion can fall out of phase with the fixed frequency of the excitation applied to the end-cap electrodes before it is ejected; the ion is refocused subsequently toward the center of the ion trap. Consequently, there is a pronounced diminution in absorption leading to an increase in the required induction fluence for ion ejection.

5. Resolution of the Discrepancy

Further simulation studies of ions in a stretched ion trap were carried out using the FIM program described earlier[31] together with frequency analysis of calculated trajectories. It was shown that in the stretched ion trap the fundamental secular frequency of an ion confined near the center of the ion trap is less than that obtained by both simulation and theory for an ion confined in a pure quadrupole ion trap. Following resonance excitation of the ion in the stretched ion trap so that trajectory excursions reached ca. $0.9\ z_0$, frequency analysis of the trajectory showed that the fundamental frequency increased by ca. 4%. Thus, ions in the stretched trap can fall off resonance as they begin to experience nonlinear forces away from the center of the ion trap.

While appropriately corrected experimental induction fluences for the loss of m/z 146 from 1,2-dichlorobenzene remain greater than the simulation results, the discrepancy has been reduced. The experimental induction fluences for the loss of m/z 134, M$^+$ of n-butylbenzene, at each of $q_z = 0.25$ and 0.45 were obtained with corrected tickle durations and are in quite good agreement with the simulation results for m/z 146. It is not expected that the simulation results for m/z 134 will differ substantially from those for m/z 146. It remains for the discrepancy between the simulation results and the experimental data for m/z 146 to be explained.

X. CONCLUSIONS

From the simulation results of the two aspects of MS/MS described here, i.e., mass-selective ion isolation and mass-selective resonance excitation, it is evident that the experimental conditions during each operation should be arranged for optimum control of ion kinetic energy. In the former operation, it is highly desirable that ion kinetic energy be minimized by collisional cooling prior to the movement of a working point to the vicinity of a stability boundary; otherwise, ion losses at the trap electrodes will occur. Preparation of the ions by collisional cooling in this fashion is analogous to optimizing the entry conditions for ions into a quadrupole mass filter when maximum ion transmission is to be achieved. In mass-selective resonance excitation with the commercial ion trap, it may be necessary to retune the tickle frequency during excitation in order to enhance the acquisition of kinetic energy during the tickle excitation process. The simulation results have demonstrated the quite considerable power of mathematical modeling of ion behavior in the ion trap. Simulation studies have been shown to be both complementary to and synergistic with experimentation. These studies have indicated that the instrumental scan function should be monitored when the temporal variations of experimental parameters are to be known with precision; these studies have indicated further that the resonant frequency of ions changes with the magnitude of axial displacement of ions from the center of the ion trap.

APPENDIX

Presented here are the dynamics of the final collision suffered by a cool ion of m/z 207 at the working point E (ion A_4 in Table 11) during the process of consecutive isolation. Following this collision, A_4 was lost even-

TABLE 11

Pre- and Postcollision Parameters (kinetic energy and velocity) for a
"Cool" Ion of m/z 207 (A_4) in Collision with a Helium Atom at the
Working Point E

Parameter	Precollision	Postcollision	Loss (percent)
Total E	0.168 eV	0.139 eV	17.3
Radial E	0.00103 eV	0.000675 eV	34.5
Azimuthal E	0.00198 eV	0.00695 eV	−251.0
Axial E	0.165 eV	0.131 eV	20.6
Radial v	−30.97 ms⁻¹	−25.08 ms⁻¹	19.0
Azimuthal v	−42.94 ms⁻¹	−80.50 ms⁻¹	−87.5
Axial v	−392.2 ms⁻¹	−350.0 ms⁻¹	10.8
v	395.0 ms⁻¹	360.0 ms⁻¹	29.0

Note: The percentage loss in the value of each parameter is given also.

Source: From March, R.E. et al., *Int J. Mass Spectrom. Ion Processes*, 125, 9, 1993.
With permission.

tually at an end-cap electrode. Energy and momentum were conserved
during the simulated collision.

The final collision for this cool ion occurred at 1.91324 ms of the simulation. From the values of the pre- and post-collision parameters shown
in Table 11 it is seen that as a result of this collision, the total ion kinetic
was reduced by virtue of the reductions in ion radial and axial kinetic
energies. However, azimuthal kinetic energy was increased by some
250%, although the absolute increase (ca. 5 meV) was minute. The effect
of these changes in component kinetic energies was that the trajectory of
the ion eventually exceeded the physical dimensions of the ion trap, even
though the changes in component kinetic energies were minor. The behavior of the ion prior to and subsequent to the collision, in addition to
the immediate changes in component kinetic energies wrought by the
collision, is shown in the accompanying figures. In some figures, not only
is the behavior of the ion subsequent to the collision shown, but also the
behavior of the ion had the collision not occurred, so as to illustrate the
marked effect of the field on the ion subsequent to the minor collisional
changes in ion motion.

In Figure 23(a) is shown the temporal variation of the ion kinetic energy over a simulation period of 100 μs, from 1.86 to 1.96 ms, where a
collision occurred at 1.91324 ms. The upright and inverted filled triangles, which are overlapped in this figure, represent kinetic energy prior
to and subsequent to the collision, respectively. While the change in kinetic energy due to the collision alone was minute compared with the ordinate scale of Figure 23(a), the subsequent action of the field upon the
ion had a marked effect in that ion kinetic energy maxima were increased
subsequently threefold. In Figure 23(b) the time and energy scales have

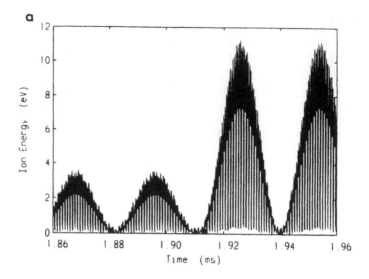

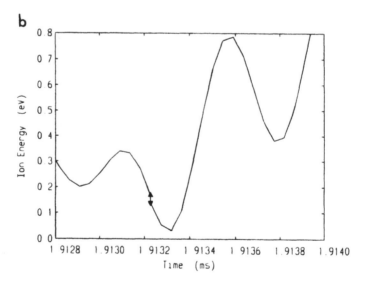

FIGURE 23

Temporal variation of ion kinetic energy of the cool ion A_4; (a), over a simulation period of 100 µs, from 1.86 to 1.96 ms; (b) over a simulation period of 1.2 µs, from 1.9128 to 1.9140 ms. Note that a collision occurred at 1.91324 ms. The upright and inverted filled triangles represent kinetic energy prior to and subsequent to the collision, respectively. The dotted line in (b) shows the temporal variation of ion kinetic energy had the collision not occurred. (From March, R.E. et al., *Int J. Mass Spectrom. Ion Processes*, 125, 9, 1993. With permission.)

been diminished to show the loss in kinetic energy as a result of the collision. The solid line shows ion kinetic energy prior to, at, and following the collision, while the dotted line shows ion kinetic energy had the collision not occurred.

In Figures 24(a) and (b) are shown the radial and azimuthal kinetic energies, respectively, over a simulation period of 10 μs. The loss in radial kinetic energy is minute again on the scale of Figure 24(a) so that the filled triangles are overlapped. The dotted line in this figure shows that the collision had little effect upon subsequent radial kinetic energy. The gain in azimuthal kinetic energy as a result of the collision is seen clearly in Figure 24(b) and, in comparison with the dotted line, is seen to persist following the collision.

The temporal variation of ion axial kinetic energy shown in Figure 25(a) covers a simulation period of 100 μs. The time of collision is shown by the overlapping filled triangles. The temporal variation shown here is quite similar to that of ion kinetic energy shown in Figure 25(a), as the bulk of the ion kinetic energy resides in axial motion. However, the variations in axial kinetic energy minima and maxima are more smooth than those for total ion kinetic energy. A magnified presentation of the variation in axial kinetic energy is shown in Figure 25(b). Here, the postcollision effect of the field upon axial kinetic energy is seen to be marked in comparison to that had the collision not occurred, as indicated by the dotted line.

The temporal variations of each of radial and axial position throughout a simulation period of 100 μs are shown in Figures 26(a) and (b), respectively. Although a collision occurred at 1.91324 ms, the variation in radial position following the collision is virtually identical to that prior to the collision, as shown in Figure 26(a). The variation in axial position is, however, changed markedly following the collision in that axial excursions from the center of the ion trap are twice those calculated prior to the collision.

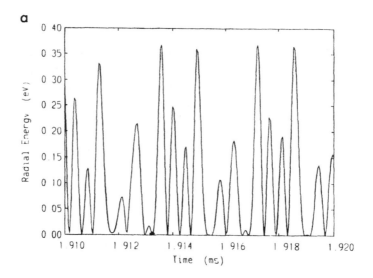

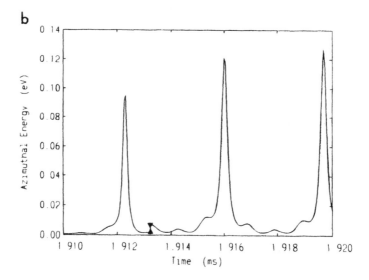

FIGURE 24

Temporal variation of radial and azimuthal components of ion kinetic energy of the cool ion A_4; (a) radial kinetic energy over a simulation period of 10 μs, from 1.910 to 1.920 ms; (b) azimuthal kinetic energy over the same simulation period. In both of these figures, the filled triangles depict collisional kinetic energy values, while the dotted lines show the behavior of the component energy had the collision not occurred. (From March, R.E. et al., *Int J. Mass Spectrom. Ion Processes, 125,* 9, 1993. With permission.)

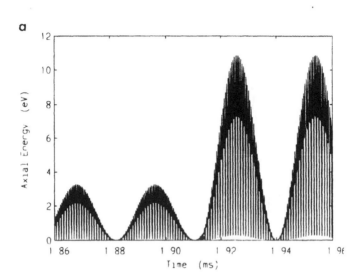

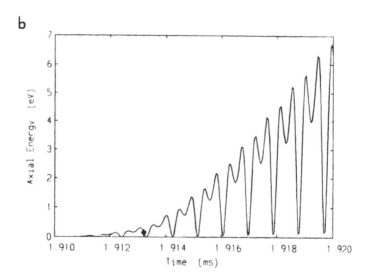

FIGURE 25

Temporal variation of ion axial kinetic energy of the cool ion A_4. (a) Over a simulation period of 100 μs, from 1.86 to 1.96 ms; (b) over a simulation period of 10 μs, from 1.910 to 1.920 ms. The filled triangles in (a) and (b) and the dotted line in (b) have the same significance as in other figures. (From March, R.E. et al., *Int J. Mass Spectrom. Ion Processes*, 125, 9, 1993. With permission.)

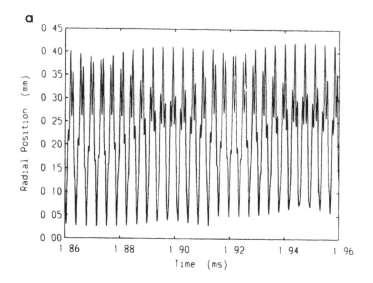

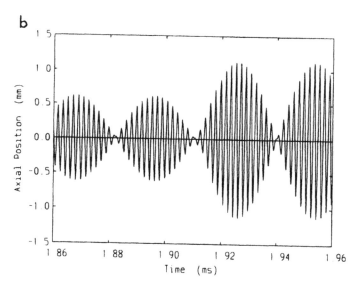

FIGURE 26

Tremporal variation of radial and axial position of the cool ion A_4 with respect to the center of the ion trap. (a) Radial position over a simulation period of 100 µs, from 1.86 to 1.96 ms; (b) axial position over a simulation period of 100 µs, from 1.86 to 1.96 ms. Note that a collision occurred at 1.91324 ms. (From March, R.E. et al., *Int J. Mass Spectrom. Ion Processes*, *125*, 9, 1993. With permission.)

REFERENCES

1. Dawson, P.H.; Whetten, N.R. *J. Vac. Sci. Technol.* 1968, *5*, 1.
2. Dawson, P.H.; Whetten, N.R. *Int. J. Mass Spectrom. Ion Phys.* 1969, *2*, 45.
3. Dawson, P.H.; Whetten, N.R. *Int. J. Mass Spectrom. Ion Phys.* 1975, *16*, 269.
4. Dawson, P.H.; Whetten, N.R. *Int. J. Mass Spectrom. Ion Phys.* 1974, *14*, 339.
5. Bonner, R.F.; March, R.E.; Durup, J. *Int. J. Mass Spectrom. Ion Phys.* 1976, *22*, 17.
6. Bonner, R.F.; March, R.E. *Int. J. Mass Spectrom. Ion Phys.* 1977, *25*, 411.
7. Doran, M.C.; Fulford, J.E.; Hughes, R.J.; Morita, Y.; March, R.E. *Int. J. Mass Spectrom. Ion Phys.* 1980, *33*, 139.
8. André, J.; Schermann, J.P. *Phys. Lett.* 1973, *A45*, 139.
9. Vedel, F.; André, J.; Vedel, M.; Brincourt, G. *Phys. Rev.* 1983, *A27*, 2321.
10. Vedel, F.; André, J. *Phys. Rev.* 1984, *A29*, 2098.
11. Vedel, F.; André, J. *Int. J. Mass Spectrom. Ion Processes.* 1985, *65*, 1.
12. March, R.E.; Hughes, R.J. *Quadrupole Storage Mass Spectrometry*, Chemical Analysis Series, Vol. 102. Wiley: New York, 1989.
13. Louris, J.N.; Cooks, R.G.; Syka, J.E.P.; Kelley, P.E.; Stafford, G.C., Jr.; Todd, J.F.J. *Anal. Chem.* 1987, *59*, 1677.
14. Lunney, M.D.; Webb, J.P.; Moore, R.B. *J. Appl. Phys.* 1989, *65*, 2883.
15. Lunney, M.D.; Moore, R.B. *I.E.E.E. Trans. Magnet.* 1991, *27*, 4174.
16. Todd, J.F.J. *Mass Spectrom. Rev.* 1991, *10*, 3.
17. Pedder, R.E.; Yost, R.A. *Proc. 36th ASMS Conf. Mass Spectrom. Allied Topics.* San Francisco, 1988; p. 632.
18. Julian, R.K., Jr.; Reiser, H.-P.; Cooks, R.G. personal communication, 1991.
19. Cooks, R.G.; Glish, G.L.; McLuckey, S.A.; Kaiser, R.E., Jr. *Chem. Eng. News.* 1991, *69*, 26.
20. Tucker, D.B.; Hameister, C.H.; Bradshaw, S.C.; Hoekman, D.J.; Weber-Grabau, M. *Proc. 36th ASMS Conf. Mass Spectrom. Allied Topics*, San Francisco, 1988; p. 628.
21. Reiser, H.-P.; Kaiser, R.E., Jr.; Savickas, P.J.; Cooks, R.G. *Int. J. Mass Spectrom. Ion Processes.* 1991, *106*, 237.
22. Reiser, H.-P.; Julian, R.K., Jr.; Cooks, R.G. *Int. J. Mass Spectrom. Ion Processes.* 1992, *121*, 49.
23. Julian, R.K., Jr.; Reiser, H.-P.; Cooks, R.G. *Int. J. Mass Spectrom. Ion Processes*, 1993, *123*, 85.
24. Baril, M.; Le, R.; Marchand, P. *Int. J. Mass Spectrom. Ion Processes.* 1990, *98*, 87.
25. Wang, Y.; Franzen, J. *Int. J. Mass Spectrom. Ion Processes.* 1992, *112*, 167.
26. Wang, Y.; Franzen, J.; Wanczek, K.-P. *Int. J. Mass Spectrom. Ion Processes*, 1993, *124*, 125.
27. Vedel, F.; Vedel, M. *Proc. 11th Int. Mass Spectrom. Conf.* Published in *Advances in Mass Spectrometry*. Heyden & Son: London, 1988; p. 244.
28. Rebatel, I. Doctoral thesis, University of Provence, Provence, Italy, 1993.
29. March, R.E.; McMahon, A.W.; Londry, F.A.; Alfred, R.L.; Todd, J.F.J.; Vedel, F. *Int. J. Mass Spectrom. Ion Processes.* 1989, *95*, 119.
30. March, R.E.; McMahon, A.W.; Allinson, E.T.; Londry, F.A.; Alfred, R.L.; Todd, J.F.J.; Vedel, F. *Int. J. Mass Spectrom. Ion Processes.* 1990, *99*, 109.
31. March, R.E.; Londry, F.A.; Alfred, R.L.; Todd, J.F.J.; Penman, A.D.; Vedel, F.; Vedel, M. *Int. J. Mass Spectrom. Ion Processes.* 1991, *110*, 159.
32. March, R.E.; Londry, F.A.; Alfred, R.L.; Franklin, A.M.; Todd, J.F.J. *Int. J. Mass Spectrom. Ion Processes.* 1992, *112*, 247.
33. Knight, R.D. *Int. J. Mass Spectrom. Ion Phys.* 1983, *51*, 127.
34. Hemberger, P.H.; Nogar, N.S.; Williams, J.D.; Cooks, R.G.; Syka, J.E.P. *Chem. Phys. Lett.* 1992, *191*, 405.
35. Reference Manual for the Poisson/Superfish Group of Codes. (LA-UR-87-126), January 1, 1987 Version, Los Alamos Accelerator Code Group, Los Alamos National Laboratory, Los Alamos, NM.

36. Guidugli, F.; Traldi, P. *Rapid Commun. Mass Spectrom.* 1991, *5*, 343.
37. Morand, K.L.; Lammert, S.A.; Cooks, R.G. *Rapid Commun. Mass Spectrom.* 1991, *5*, 491.
38. Guidugli, F.; Traldi, P.; Franklin, A.M.; Langford, M.L.; Murrel, J.; Todd, J.F.J. *Rapid Commun. Mass Spectrom.* 1992, *6*, 229.
39. Eades, D.M.; Kleintop, B.K.; Yost, R.A. *Proc. 40th ASMS Conf. Mass Spectrom. Allied Topics.* Washington, D.C., 1992; p. 1290.
40. Dawson, P.H.; Lambert, C. *Int. J. Mass Spectrom. Ion Phys.* 1975, *16*, 269.
41. Dawson, P.H. *Quadrupole Mass Spectrometry and its Applications.* Elsevier: Amsterdam, 1976.
42. Moon, F.C. *Chaotic Vibrations.* John Wiley & Sons: New York, 1987 (for a treatment of parametrically driven oscillators which have the Mathieu form, see pp. 79–81).
43. Kelley, P.E.; Stafford, G.C.; Stevens, D.R. *Method of Mass Analyzing a Sample by Use of a Quadrupole Ion Trap.* U.S. Patent 4,540,884; 10 September 1985.
44. Louris, J.L.; Brodbelt-Lustig, J.S.; Cooks, R.G.; Glish, G.L.; Van Berkel, G.J.; McLuckey, S.A. *Int. J. Mass Spectrom. Ion Processes.* 1990, *96*, 117.
45. Kaiser, R.E., Jr.; Cooks, R.G.; Stafford, G.C.; Syka, J.E.P.; Hemberger, P.H. *Int. J. Mass Spectrom. Ion Processes.* 1990, *106*, 79.
46. Schwartz, J.C.; Syka, J.E.P.; Jardine, I. *J. Am. Soc. Mass Spectrom.* 1991, *2*, 198.
47. Kaiser, R.E., Jr.; Cooks, R.G.; Syka, J.E.P.; Stafford, G.C. *Rapid Commun. Mass Spectrom.* 1990, *4*, 30.
48. Mabud, M.A.; DeKrey, M.J.; Cooks, R.G. *Int. J. Mass Spectrom. Ion Processes.* 1985, *67*, 285.
49. Lammert, S.A.; Cooks, R.G. *J. Am. Soc. Mass Spectrom.* 1991, *2*, 487.
50. Lammert, S.A.; Cooks, R.G. *Rapid Commun. Mass Spectrom.* 1992, *6*, 528.
51. Lammert, S.A. Ph.D. thesis, Purdue University, West Lafayette, IN, 1992.
52. Connection Machine User's Guide, Version 6.1, October 1991, Thinking Machines Corporation, Cambridge, MA.
53. Kaiser, R.E., Jr.; Louris, J.N.; Amy, J.W.; Cooks, R.G. *Rapid Commun. Mass Spectrom.* 1989, *3*, 225.
54. Louris, J.N.; Stafford, G.C.; Syka, J.E.P.; Taylor, D. *Proc. 40th ASMS Conf. Mass Spectrom. Allied Topics.* Washington, D.C., 1992; p. 1003.
55. Reiser, H.-P. Ph.D. thesis, University of Geissen, Geissen, Germany, 1992.
56. March, R.E.; Weir, M.R.; Tkaczyk, M.; Londry, F.A.; Alfred, R.L.; Franklin, A. M.; Langford, M.L.; Todd, J.F.J. *Org. Mass Spectrom,* 1993, *28*, 499.
57. Glish, G.L.; McLuckey, S.A.; Asano, K.G. *J. Am. Soc. Mass Spectrom.* 1990, *1*, 166.
58. Gronowska, J.; Paradisi, C.; Traldi, P.; Vettori, U. *Rapid Commun. Mass Spectrom.* 1990, *4*, 306.
59. Ardanaz, C.E.; Traldi, P.; Vettori, U.; Kavka, J.; Guidugli, F. *Rapid Commun. Mass Spectrom.* 1991, *5*, 5.
60. Todd, J.F.J. Personal communication, 1992.
61. Paradisi, C.; Todd, J.F.J.; Traldi, P.; Vettori, U. *Org. Mass Spectrom.* 1992, *27*, 251.
62. Gioumousis, G.; Stevenson, D.P. *J. Chem. Phys.* 1958, *29*, 294.
63. Su, T.; Bowers, M.T. *J. Chem. Phys.* 1973, *58*, 3027.
64. Londry, F.A.; Alfred, R.L.; March, R.E. *J. Am. Soc. Mass Spectrom,* 1993, *4(9)*, 687.
65. Paul, W.; Osberghaus, O.; Fischer, E. *Forschungsberichte des Wirtschaft und Verkehrministeriums Nordrhein Westfalen.* No. 415, Westdeutscher Verlag: Cologne, 1958.
66. Fischer, E. *Z. Phys.* 1959, *156*, 1.
67. Jungmann, K.; Hoffnagle, J.; DeVoe, R.G.; Brewer, R.G. *Phys. Rev. A.* 1987, *36*, 3451.
68. Paul, W.; Reinhard, H.P.; Von Zahn, U. *Z. Phys.* 1958, *152*, 143. Fulford, J.E.; Hoa, D.-N.; Hughes, R.J.; March, R.E.; Bonner, R.F.; Wong, G.J. *J. Vac. Sci. Technol.* 1980, *17*, 829. Armitage, M.A.; Fulford, J.E.; Hoa, D.-N.; Hughes, R.J.; March, R.E. *Can. J. Chem.* 1979, *57*, 2108.

69. Alfred, R.L.; Londry, F.A.; March, R.E. *Int. J. Mass Spectrom. Ion Processes*, 1993, *125*, 171.
70. March, R.E.; Fulford, J.E. *Int. J. Mass Spectrom. Ion Phys.* 1979, *30*, 39.
71. Mather, R.E.; Todd, J.F.J. *Int. J. Mass Spectrom. Ion Phys.* 1979, *31*, 1.
72. Major, F.G.; Dehmelt, H.G. *Phys. Rev.* 1968, *179*, 91.
73. Dawson, P.H., Ed. *Quadrupole Mass Spectrometry and Its Applications*. Elsevier: Amsterdam, 1976; Chap. 4.
74. Lawson, G.; Todd, J.F.J.; Bonner, R.F. *Dyn. Mass Spectrom.* 1975, *4*, 39.
75. Penman, A.D.; Todd, J.F.J.; Thorner, D.A.; Smith, R.D. *Rapid Commun. Mass Spectrom.* 1990, *4*, 415.
76. Todd, J.F.J.; Penman, A.D.; Franklin, A.M.; March, R.E. *2nd Eur. Meet. Tandem Mass Spectrometry*, Warwick University, Coventry, U.K., July 19 to 22, 1992.
77. March, R.E.; Londry, F.A.; Alfred, R.L. *Org. Mass Spectrom.* 1992, *27*, 1151.

Chapter 7

BOUNDARY EXCITATION

Pietro Traldi, Silvia Catinella, Raymond E. March, and Colin S. Creaser

CONTENTS

0-8493-4452-2/95/$0.00+$.50
© 1995 by CRC Press, Inc.

I. INTRODUCTION

The quadrupole ion trap is a highly effective device for performing tandem mass spectrometry (MS/MS) experiments.[1,2] In Volume I, Chapters 1, 5, and 8, and Volume II, Chapters 1, 2, 4, and 5, much evidence is presented to demonstrate the utility of the simple and relatively inexpensive ion trap as a collision chamber for the determination of ion structures and for the analytical verification of compounds. In most cases, the results obtained in the form of collisional mass spectra are directly comparable with those obtained by more complicated and more expensive sector instruments.

A. Resonant Excitation

The vast majority of ion trap MS/MS data has been obtained with the application of an auxiliary radiofrequency (RF) potential, or "tickle"[3] voltage, across the end-cap electrodes and oscillating at a resonant frequency of a chosen and isolated ion species. However, the tickle frequency must be carefully tuned to the secular frequency of the ion if high fragmentation efficiencies are to be obtained. This tuning procedure is experimentally difficult for transient species, such as those eluting from a chromatographic inlet. A general procedure for ion activation which does not require such careful tuning, therefore, would be useful for ion trap MS/MS. Tuned irradiation of an ion species leads to resonant excitation, and, upon subsequent collisions with the buffer gas, some of the enhanced kinetic energy of the ions is converted to internal energy within the ions.[4] The ions so activated then undergo decomposition processes and the product ions of these processes form the collison mass spectrum, or secondary mass spectrum, or MS/MS mass spectrum.

The intimate mechanism of the action of the supplementary RF field has been the object of both theoretical[5] and experimental[6] investigations. It was found that the weak point in tickle activation lies in the step-by-step deposition of energy in the selected species which favored those decomposition pathways having the lowest critical, or activation, energies.[5] Thus, as a crude generalization and overstating the case somewhat, tickle activation sometimes yielded daughter ion mass spectra with few components and a subsequent loss of information. For this reason, other potential methodologies for nonspecific approaches to ion activation have been proposed, such as the use of random noise, broad-band frequencies, sweep frequency, and pulsed axial activation.[7-9] The alternative methods of photodissociation (PD)[10] and laser PD,[1,11] and the use of fast direct current (DC) pulses[12] which induce interactions with surfaces,[13] seem to be the most promising. Laser PD may be carried out using either a focused laser beam or a fiberoptic interface.[11] Photodissociation can be a highly efficient process in the ion trap and has been demonstrated for the gas chromatography-tandem mass spectrometry (GC/MS/MS) analysis of isomeric butylbenzenes,[14] but requires that the precursor ion absorbs energy at the laser wavelength used. The latter prerequisite limits the range of compounds amenable to the technique. It must be emphasized, however, that these other methodologies which could supplant tickle activation require supplementary equipment such as a laser or a fast DC pulse power supply.

B. Boundary Activation

A further ion activation method was proposed in 1991, and it is readily accessible with the commercial version of the ion trap; this method requires neither instrumental modification nor additional dedicated devices.[15] This technique has aroused considerable interest within the ion trapping community.[16-20] The method is based on the interaction of a selected ion species with the high fields derived from the RF drive potential; this interaction leads to pronounced acceleration of the ion species whereupon the ion is dissociated subsequent to collisions with the buffer gas which cause the ions to be activated. The interaction of the ions with the high fields is brought about experimentally by choosing a_z and q_z values such that the working point of the ion species is located close to the boundary of the stability diagram; it is for this reason that this method has been described as boundary activation, or boundary-activated dissociation (BAD). Boundary-activated dissociation is, therefore, a mass-selective collisional process which does not require the careful tuning of a tickle voltage.

A description of the experimental and theoretical aspects of boundary activation is presented below in two parts. In Part A, attention is focused on four aspects of this novel technique: the basic BAD process, the efficiency of the process, a discussion of internal energy distribution, and an application of the technique to the analysis of a sample of diesel oil. In Part B, a detailed discussion is presented of the experimental aspects of BAD, together with an account of the initial studies of the effects of precooling ions prior to BAD, and an explication of the use of boundary activation for discriminating between consecutive and competing collisionally induced decompositions.

PART A
BOUNDARY-EFFECT ACTIVATED DISSOCIATION
IN TANDEM MASS SPECTROMETRY

Colin S. Creaser

II. BOUNDARY-EFFECT ACTIVATION OF IONS

A. Basic Process

The basic processes of a BAD experiment are illustrated using the stability diagram shown in Figure 1.[19,20] The working point of an isolated population of precursor ions is brought to a convenient q_z value on the $a_z = 0$ line (typically around $q_z = 0.5$) indicated by point A in Figure 1. A short DC voltage pulse (≈ 50 ms) is then applied to the ring electrode, which moves the working point of the ion to the upper or lower edges ($\beta_r = 0$ or $\beta_z = 0$, respectively) of the stability diagram. For example, application of a positive DC pulse moves the working point to the lower boundary indicated by point B in Figure 1. At this boundary, the amplitude of the ion oscillations increased in the z-direction (i.e., toward the end-cap electrodes), while moving the working point to the upper boundary by applying a negative DC pulse, induces an increase in amplitude in the r-direction (i.e., toward the ring electrode). There appears to be little difference in ion activation using either approach. The increase in the amplitude of oscillation induces collisions with the buffer gas and fragmentation of ions within the boundary region. Removal of the DC voltage returns the working points of residual precursor ions and product ions, whose q_z values fall between points B and C in Figure 1, to the $a_z = 0$ line. The product ion spectrum may then be recorded using a normal mass-instability scan. Ions whose working points lie beyond point C will have q_z values >0.908 and will, therefore, be lost from the trap when the DC pulse is removed. This limit determines the low-mass cutoff for the product spectrum.

B. Efficiency of Boundary-Effect Activated Dissociation

The efficiency of the process depends on the precise position of the working point of the ions within the boundary region and the energy required for fragmentation. This condition is illustrated in Figure 2 for the BAD activation of the molecular ion of acetophenone ($C_6H_5.CO.CH_3$, *m/z* 120). As the amplitude of the DC voltage is increased and the working

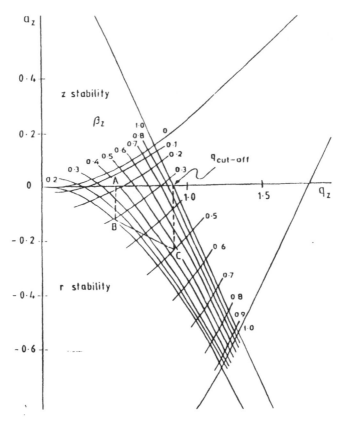

FIGURE 1
Stability diagram for boundary-effect activated dissociation. (From Creaser, C.S. and O'Neill, K.E., *Org. Mass Spectrom.*, 28, 564, 1993. With permission.)

point of the m/z 120 ion approaches the boundary region, the intensity of this ion falls to zero as a result of ion scattering and collisional dissociation. The intensity of the $[M-CH_3]^+$ product ion (m/z 105), however, rises as the voltage is increased, reaching a maximum at +113 V before declining. The intensity of the higher energy product $[C_6H_5]^+$ (m/z 77) is observed only as the internal energy deposition reaches a level sufficient to induce its formation, which occurs as the working point is moved closer to the stability limit. Finally, all ions are lost from the trap when the magnitude of the DC voltage is large enough to move the working point of the precursor ion outside the stable region.

The overall efficiency of BAD fragmentation in MS/MS experiments is compound dependent, as it is in collision-activated dissociation (CAD), but fragmentation efficiencies in the range 9 to <90% have been reported. Generally, fragmentation efficiencies are expected to be slightly lower in

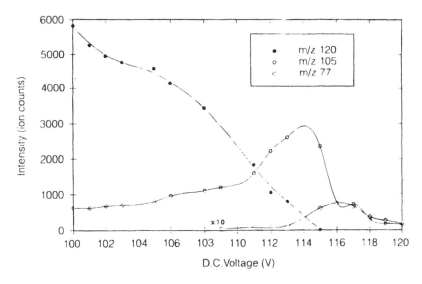

FIGURE 2
Variation of acetophenone-derived precursor and product ion intensities with applied DC voltage.

BAD than in CAD because of the increased probability of precursor ion loss for ions with working points at the edge of the stability region.

C. Internal Energy Distribution

The internal energy distribution of the precursor ions subjected to BAD has been estimated by the activation of compounds with well-characterized fragmentation processes and by comparison of BAD with CAD product ion spectra for model compounds.[16-20] The results of these experiments suggest that the typical internal energy deposition under BAD is slightly lower than that observed for optimized CAD conditions using a tickle voltage; however, some evidence exists of the possibility of much higher energy deposition in some cases.[18,19] The energy deposited by BAD is sufficient to yield products for most precursor ions that have been studied, but some ions such as the molecular ions of polycyclic aromatic hydrocarbons (PAHs), which are amenable to CAD, are not dissociated by BAD.

III. APPLICATION

The advantage of BAD for routine analysis is the ability to determine the a_z value for a given value of q_z, at which any precursor ion is brought to the edge of the stability region. This advantage is applicable to all ions regardless of mass. The a_z value may be determined experimentally from the calibration curve of the applied DC voltages at which maximum BAD-induced fragmentation is observed for precursor ions of a range of masses and structures, shown in Figure 3. The lower and upper horizontal lines for each point represent the onset of fragmentation and the point at which all ions are lost. The slope of the regression line through these points yields an a_z value of −0.1474, from Equation 1,

$$\text{slope} = -\frac{a_z r_0^2 \Omega^2}{8e} \tag{1}$$

which allows the DC voltage required to induce BAD fragmentation to be calculated for any selected ion within the calibration range without the necessity for further tuning.

The application of BAD to the GC/MS/MS analysis of compounds yielding ions of nominal m/z 134 has been demonstrated in a sample of diesel oil.[20] Molecular ions of alkyl benzenes ($C_{10}H_{14}$, accurate mass 134.1095 u) and benzothiophene (C_8H_6S, accurate mass 134.0190 u) were dissociated successfully in a single chromatographic run using the applied DC voltage predicted from the calibration graph. The resulting product ion chromatogram (in the range m/z 74 to 130) and MS/MS spectra of diethylbenzene and benzothiophene are shown in Figure 4.

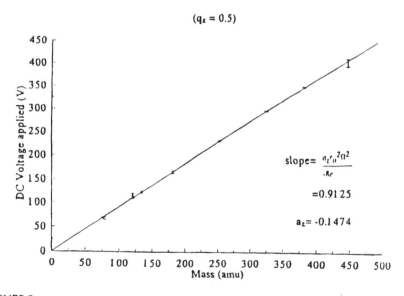

FIGURE 3

Relationship between applied DC voltage and mass for optimum boundary-effect activated dissociation. (From Creaser, C.S. and O'Neill, K.E., *Org. Mass Spectrom.*, 28, 564, 1993. With permission.)

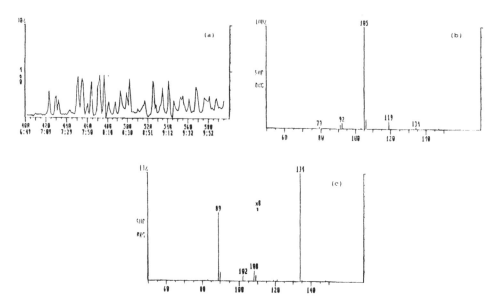

FIGURE 4

GC/MS/MS analysis of diesel fuel using BAD activation. (a) Product ion chromatogram for m/z 134 (range m/z 74 to 130), (b) MS/MS spectrum of diethylbenzene, (c) MS/MS spectrum of benzothiophene. (From Creaser, C.S. and O'Neill, K.E., *Org. Mass Spectrom.*, 28, 564, 1993. With permission.)

IV. CONCLUSION

Fragmentation efficiencies varied from <90% for ethylbenzene to <10% for benzothiophene. Boundary-effect activation thus offers an alternative strategy to tickle-induced CAD and PD, when dealing with complex multicomponent systems.

PART B
APPLICATIONS OF BOUNDARY-ACTIVATED DISSOCIATION

Pietro Traldi, Silvia Catinella, and Raymond E. March

V. EXPERIMENTAL ASPECTS OF BOUNDARY ACTIVATION

The "boundary effect" for collisional activation was first described by Paradisi et al. in a study devoted to the investigation of effects of storing ions under conditions that differed from those usually employed, i.e., under RF-only conditions, such that the working point was located on the q_z-axis.

A. Boundary Activation of the Molecular Ion of 2-(2'-Hydroxybenzoyl)-Benzoic Acid

The molecular ion, M$^+$, of 2-(2'-hydroxybenzoyl)-benzoic acid (**1**), was chosen as a model ion for this study as it had been used similarly in a previous collisional-activation study based on resonant excitation in the ion trap mass spectrometer (ITMS).[6] This compound, **1**, gives rise to the simple fragmentation pattern reported in Scheme 1, and consists of the primary loss of water followed by further losses of H and CO; it had been shown earlier[6] that the loss of water is a low energy process, hence the employment of **1** as a sensitive probe for minor uptakes of energy is valid.

The scan function employed in this investigation is shown in Figure 5. The ionization pulse (a) is followed by the two-step ion isolation phase (b, c). Once these steps have been completed, the DC and RF voltages were adjusted so as to define the a_z and q_z values for the selected ion species which was then stored for a predetermined time (e).

Different sets of experiments were performed using either helium or argon as buffer gas. Extensive fragmentations were observed while the working point was located in the proximity of either the upper ($\beta_r = 0$) or the lower ($\beta_z = 0$) left boundaries of the stability diagram; these fragmentations were related to the peculiar conditions that the ions experience when the working point is so positioned (Figure 6). It can be envisaged that along the borders of the stability diagram the stability and

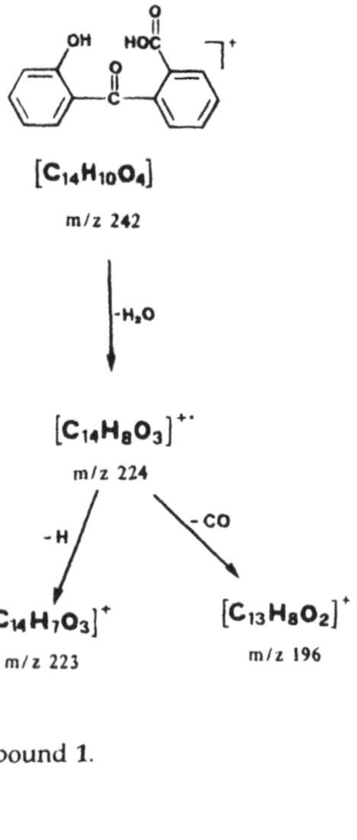

SCHEME 1
Fragmentation pattern of compound 1.

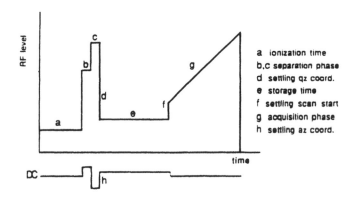

FIGURE 5
Scan function employed in the investigation of compound **1**. (From Paradisi, C. et al., *Org. Mass Spectrom.*, 27, 252, 1992. With permission.)

instability regions are separated by a quasi-stable region in which the ions experience a large field. The interaction of the ions with this field leads to marked increases of ion kinetic energy which leads, in turn, to energy deposition in the ion species as a result of collisions with buffer gas.

1.   Quasi-Stable Region

When this quasi-stable intermediate region was probed using the molecular ion of **1**, a clear discrepancy was found between the theoretical and experimental values of the a_z parameter for the $\beta_r = 0$ and $\beta_z = 0$ boundaries of the stability diagram.

2.   Stretched Ion Trap

Until the summer of 1992, most researchers working with the Finnigan MAT ITMS™ had assumed that they were working with the Paul quadrupole ion trap. However, at the 40th Annual Conference on Mass Spectrometry and Allied Topics, John Louris of Finnigan MAT revealed that the commercially available ion trap has a stretched geometry in the z-direction. It had been assumed that the separation of the end-cap electrodes, $2z_0$, was such that the value for z_0 was close to the theoretical value of 0.707 cm (i.e., when the inscribed radius of the ring electrode, r_0, is equal to 1.00 cm). The actual value for z_0 is 0.783 cm.[21] The origins of the stretched geometry are discussed at length in Chapter 4, and the ramifications of increasing the separation of the end-cap electrodes are discussed in Chapter 3. Thus, the discrepancy in the values for the a_z parameter for the $\beta_r = 0$ and $\beta_z = 0$ boundaries of the stability diagram is due possibly

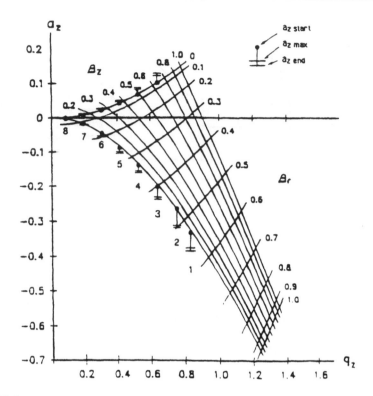

FIGURE 6

Values of a_z for each q_z coordinate, superimposed on the stability diagram, and representing points at which significant variations of ion abundance occur. (From Paradisi, C. et al., *Org. Mass Spectrom.*, 27, 252, 1992. With permission.)

to the difference in geometry between the theoretical trap and the commercial version of the trap.

3. Stretched Ion Trap Stability Diagram

Recently, Johnson et al.[22] have determined experimentally the boundaries of the stability diagram for the stretched version and have presented an equation for the conversion of the Mathieu a_u and q_u parameters to DC and RF voltages for the ITMS™. As can be seen from Figure 7, quite marked differences are apparent between the stability diagram for the pure quadrupole ion trap (Paul ion trap) and that for the ITMS™. It is worth noting that the left side of the stability diagram is in good agreement with the previous findings of Paradisi et al., as reported in Figure 6.

4. Collision-Induced Dissociation with Boundary Activation

Furthermore, the work of Paradisi et al. proved that when the working point of a selected ion species is placed close to either the $\beta_r = 0$ or $\beta_z = 0$ boundaries of the stability diagram, collision-induced dissociation

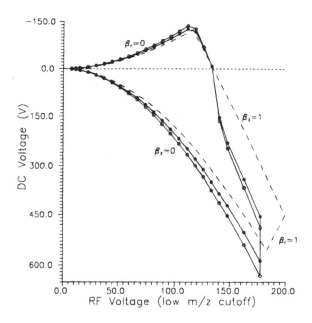

FIGURE 7

Comparison of the experimental and the theoritical stability diagrams for a pure quadrupole ion trap and for a stretched ion trap. ● experimental, DC calibrated; ○, theoretical approximation, stretched ion trap; - - -, theoritical, pure quadrupole ion trap. (From Johnson,
J.V. et al., *Rapid Comm. Mass Spectrom.*, 6, 763, 1992. With permission.)

(CID) takes place; hence, MS/MS experiments can be performed in the
ion trap without the requirement of applying a resonant tickle voltage
across the two end-cap electrodes. The results obtained here are supportive
of the general interpretation of the action of a resonant tickle voltage: the
effective ion acceleration in tickle experiments is not due to resonance of
the ions with the supplementary RF field, rather it originates from the
fields derived from the RF drive potential. The action of the supplementary or tickle voltage is simply to move the trapped ion cloud into regions within the ion trap in which the main RF field is greater. Collisional
activation by the boundary effect is based on the same physical phenomenon;
in this case, the main RF field is accessed by the action of DC voltages of
appropriate magnitudes.

B.   Daughter Ion Mass Spectra of a Pyrrolquinoline Derivative

In a paper by Curcuruto et al.[18] it was shown that the boundary effect leads not only to effective daughter ion mass spectra, but that the
extent of energy deposition by this method is greater than that achieved
by a resonant tickle voltage. The model compound chosen was a
pyrrolquinoline derivative (compound 2, 2,6-dimethyl-9-methoxy-4H-

pyrrole[3,2,1-*ij*]quinolin-4-one), which had been studied previously under both electron impact (EI) and high-energy collision conditions.[23]

1. Electron Impact and Mass-Analyzed Ion Kinetic Energy Spectra

The related mass spectrum and fragmentation pattern of compound **2** are reported in Table 1 and Scheme 2. It can be seen that the molecular ion at *m/z* 227 is present in high abundance, while primary methyl loss is responsible for the formation of the ionic species at *m/z* 212, which leads to the base peak of the mass spectrum. Minor fragments are present at *m/z* 184, 167, 154, and 141. The mass-analyzed ion kinetic energy (MIKE) spectrometry[24] measurements of the relevant ions of **2** are reported in Table 2.

2. Resonant Excitation

The components of the daughter ion mass spectra of the molecular ion of **2** obtained under a variety of conditions are shown in Table 3. Spectrum a, which consists of but a single fragment ion species, was obtained with an ITMS™ where $M^{+\cdot}$ was selected by the two-step isolation procedure; the ion was irradiated at $q_z = 0.9$ with a tickle voltage of 195 mV for 22 ms at an ambient helium pressure of 1.1×10^{-4} torr. The single component, *m/z* 212, of spectrum a was generated by the primary $CH_3^{\cdot}$ loss. When the amplitude of the tickle voltage was increased further to 300 mV, no effect was observed. In this case, the loss of $CH_3^{\cdot}$ represents the process having the lowest activation energy, and no other decomposition pathways are accessed. This result constitutes further evidence for the general understanding of tickle voltage activation. The energy deposition occurs in a step by step fashion, thus favoring the decomposition process having the lowest critical energy.

TABLE 1

Ion Relative Abundances of the EI Mass Spectrum of Compound 2

Ionic species	*m/z*	Relative abundance
$M^{\cdot}$	227	19
$[M-H]^{\cdot}$	226	98
$[M-CH_3]^{\cdot}$	212	41
$[(M-H)-CH_3]^{\cdot}$	211	100
$[M-CO]^{\cdot}$	199	2
$[M-C_2H_3O]^{\cdot}$	184	12
$[M-C_2H_4O]^{\cdot}$	183	11
$[(M-H)-C_2H_4O]^{\cdot}$	182	16
$[M-C_2H_2O-CH_3]^{\cdot}$	170	3
$[M-C_2H_3O-CH_3]^{\cdot}$	169	1
$[M-C_2H_4-CO]^{\cdot}$	155	12
$[M-C_2H_4O-CHO]^{\cdot}$	154	16

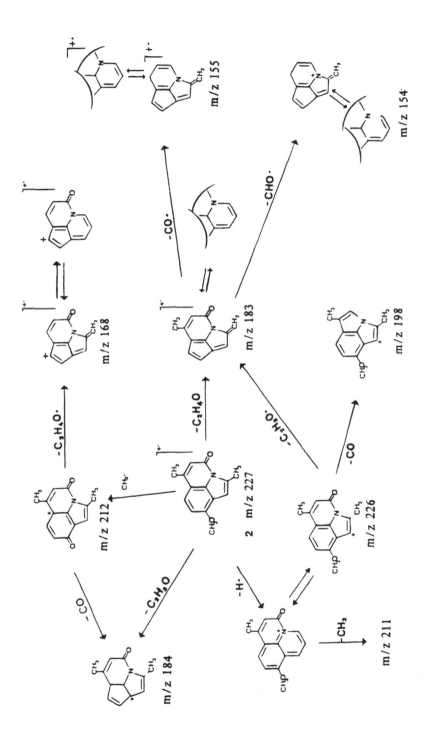

SCHEME 2
Fragmentation pattern of compound **2**. (From Fontana, S. et al., *Org. Mass Spectrom*, 27, 1256, 1992. With permission.)

TABLE 2

Ion Relative Abundances of the MIKE Spectrum of Compound **2**

Parent ion	m/z	Absolute abundances
M$^+$ (m/z 227)		
	226	10
	212	64
	199	—
	198	20
	184	2
	183	2
	170	—
	169	—
	168	1
	155	1
	154	—

TABLE 3

Ion Relative Abundances of the Components of Daughter Ion Mass Spectra of the Molecular Ion of Compound **2** Obtained under a Variety of Conditions

m/z	Relative abundances			
	a	b	c	d
212	100	92	100	97
198	—	3.4	—	1
184	—	0.8	—	2
183	—	0.4	—	—

^{a} = Tickle voltage activation in an ITMS.
^{b} = boundary-effect activation following application of a DC voltage in an ITMS.
^{c} = CID in a TSQ instrument; helium collision gas, ion kinetic energy of 5 eV.
^{d} = As for c, but with an ion kinetic energy of 100 eV.

3. Boundary Activation

In a further experiment, the collisional behavior of M$^{+\cdot}$ of **2** was studied by the boundary effect. Following isolation of the M$^{+\cdot}$ ion as before, a DC voltage of −101.4 V was applied, thus moving the working point of the ion species close to the $\beta_z = 0$ boundary of the stability diagram. The daughter ion mass spectrum so obtained is reported as spectrum b in Table 3. The spectrum now consists not only of the ions of m/z 212 as described earlier, but of new fragment ions at each of m/z 198, 184, and 183, corresponding to primary losses of CHO$^\cdot$, $C_2H_3O^\cdot$, and $C_2H_4O^\cdot$, respectively. The collision efficiency, 60%, is of the same order of magnitude as that obtained by tickling.

The observation in the latter experiment (product ions of new fragmentation processes which had not been accessed by tickling) showed clearly that a higher degree of internal energy deposition is achievable by the boundary effect than by tickling.

4. Collision-Induced Dissociation in a Triple Quadrupole

In order to estimate the kinetic energies that must have been experienced by ions subject to the boundary effect and to tickling, the results described above were compared with those obtained from collision experiments carried out with a triple stage quadrupole (TSQ) instrument. The results obtained by collision with helium of M$^{+\cdot}$ of 2 having kinetic energies of 5 and 100 eV are presented as spectra c and d, respectively, in Table 3. From a comparison of these results it is apparent that the tickling data are similar to those achieved by ions having 5 eV of kinetic energy, while the boundary effect data are similar to those achieved by ions having 100 eV of kinetic energy. Further comparison of the ITMS™ and TSQ data shows an additional and important aspect of instrument performance: the daughter ion detection efficiency of 60% for the ITMS™ experiments exceeded those for the TSQ instrument, which were found to be about 20 and 10% for collisions of 5 and 100 eV ions, respectively.

C. Comparison of Resonant Excitation with Boundary Activation

In a further study, based on n-butylbenzene as a model system, and devoted to the comparison of activation by the boundary effect and by tickling,[16] it was shown that a comparable amount of internal energy is present in the ions following activation by each of the two methods. With the tickle technique, it is possible to increase the degree of internal energy enhancement, albeit at the expense of daughter ion detection efficiency, whereas for boundary excitation the conditions for optimum efficiency almost coincide with those for maximum activation.

Upon examining the effects of variation of the value for q_z at which the isolated ion species are activated, it was found that the permitted mass difference between parent and fragment ions is greater for the boundary effect method than for tickle excitation. Interesting plots of the efficiency of the collisional-induced fragmentation as a function of q_z and tickle energy are shown in Figure 8; analogous plots were produced also for the case of boundary activation, and they are shown in Figure 9. In Figures 8 and 9, the collisional-induced fragmentation efficiency is expressed as $\Sigma F/M$, where F and M are, respectively, the fragment ion and precursor ion abundances.

Two findings emerge from a comparison of the surfaces shown in Figures 8 and 9. First, in the case for boundary activation, the surface of the $\Sigma F/M$ ratio is less symmetrical than that found with tickle excitation. This observation is consistent with the experimental conditions in that the maximum was achieved under conditions of marked ion loss. Second, the dependence of the efficiency on the value of the q_z parameter was quite marked in the case of boundary activation in that the efficiency was sensibly constant for the range $0.5 \leq q_z \leq 0.6$, but diminished as the value

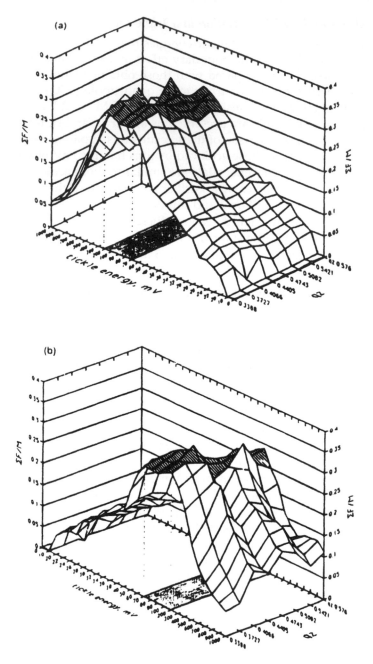

FIGURE 8

Process efficiency in tickle activation of n-butylbenzene. Plots of $\Sigma F/M$ (ratio of the sum of daughter ions' intensity over parent ion initial intensity) as a function of q_z and tickle voltage amplitude: (a) front view; (b) back view. The shaded area corresponds to conditions of maximum efficiency. (From Paradisi, C. et al., *Org. Mass Spectrom.*, 27, 1211, 1992. With permission.)

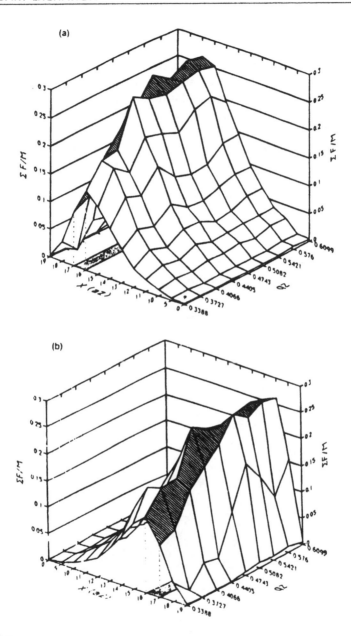

FIGURE 9

Process efficiency in boundary-effect activation of n-butylbenzene. Plots of $\Sigma F/M$ (ratio of the sum of daughter ions' intensity over parent ion initial intensity) as a function of q_z and of a_z expressed as $x = 100\,(a_z - a_{zbou})/a_{zbou}$, where, for any given value of q_z, a_{zbou} is the value of the a_z coordinate at the $\beta_z = 0$ boundary; x represents, therefore, the displacement from the $\beta_z = 0$ boundary into the instability region expressed as the percentage increment of the a_{zbou} value along the a_z axis. (a) Front view; (b) back view. The shaded area corresponds to conditions of maximum efficiency. (From Paradisi, C. et al., *Org. Mass Spectrom.*, 27, 1211, 1992. With permission.)

of q_z was decreased. The efficiency of tickle activation showed little, if any, dependence upon the value of q_z.

The ion intensity ratio, m/z 91/92, obtained from butylbenzene, may be used cautiously as a probe of internal energy deposition and graphs of this ratio, analogous to those of Figures 8 and 9, have been prepared to facilitate another comparison between tickle activation and boundary activation. These graphs are shown in Figures 10 and 11, respectively. It is seen that a higher ion intensity ratio, m/z 91/92, is obtained with tickle activation than with boundary activation, but only at tickle voltage amplitudes markedly higher (900 to 1000 mV) than those corresponding to maximum efficiency (70 to 200 mV).

The authors have resisted the temptation to assess precisely the internal energies of the dissociating parent ions on the basis of the m/z 91/92 ion intensity ratio. Such a course may be prudent in that it has been suggested that such a precise assignment of the internal energy to the n-butylbenzene molecular ion cannot be derived from the m/z 91/92 ratio

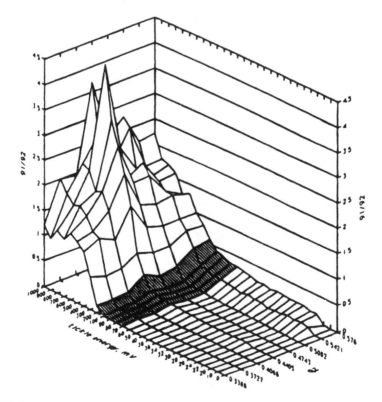

FIGURE 10

Energy deposition in tickle voltage activation of n-butylbenzene. A plot of the m/z 91/92 ion abundance ratio as a function of q_z and tickle voltage amplitude. The shaded area corresponds to conditions of maximum efficiency. (From Paradisi, C. et al., *Org. Mass Spectrom.*, 27, 1211, 1992. With permission.)

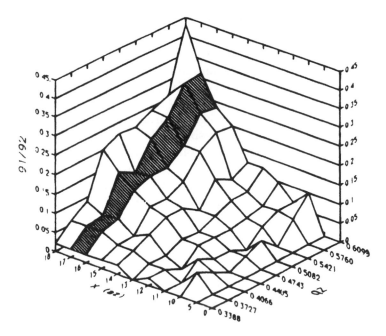

FIGURE 11

Energy deposition in boundary-effect activation of *n*-butylbenzene. A plot of the *m/z* 91/92 ion abundance ratio as a function of q_z and of a_z expressed as explained in Figure 5. The shaded area corresponds to conditions of maximum efficiency. (From Paradisi, C. et al., *Org. Mass Spectrom.*, 27, 1211, 1992. With permission.)

in a simple manner, due to the possible occurrence of the consecutive fragmentation $M^{..} \rightarrow m/z\ 92 \rightarrow m/z\ 91$. In general, however, energy deposition at a boundary exceeds that usually observed by tickle activation.[18] Analogous results have been obtained by Lammert and Cooks[12] via the application of fast DC pulses to the ion trap.

VI. BOUNDARY ACTIVATION WITH ION PRECOOLING

In a simulation study of consecutive ion isolation in the ion trap it was reported that when the working point of the mass-isolated ion is moved to the $\beta_r = 0$ boundary of the stability diagram, the ion kinetic energy is enhanced and the degree of enhancement is related to the collisonal history of the ion.[25] A detailed description of this simulation study is given in Chapter 6.

A. Simulation Study

Briefly, the simulation study showed that when ions are sufficiently cooled by collisions with a buffer gas prior to moving the working point to the vicinity of the $\beta_z = 0$ boundary of the stability diagram, their kinetic energies are moderated and their subsequent behavior can be modified. For example, for ten ions created with initial position $0.01 < R < 1$ mm and initial kinetic energy $0.05 < T < 0.3$ eV, after some 2000 µs of combined simulated ion isolation and movement of the working point to the vicinity of the $\beta_z = 0$ boundary, a kinetic energy range of 3.4 to 140 eV and an average kinetic energy of 72 eV were calculated in the absence of helium buffer gas, while with a helium pressure of 0.7 mtorr these values become 6.0 to 130 and 57 eV, respectively.

The same ten ions, when cooled collisionally (i.e., with the same randomly chosen positions and energies as previously but subjected to ca. 250 collisions with helium atoms, corresponding to a cooling time of about 1000 µs at a simulated pressure of helium at 1.5×10^{-2} torr) and subjected to the same ion-isolation process, exhibited a kinetic energy range of 0.3 to 99.0 eV and an average kinetic energy of 14 eV in the absence of helium (during the mass-isolation process), while with helium present these values were 0.8 to 118.0 eV, respectively.

From these data one can obtain valuable insight into the intimate mechanism of the boundary effect.

For randomly chosen ion coordinates and velocities as described above, the effect of helium buffer gas during and after the ion-isolation process is to moderate ion kinetic energies while the working point is close to a boundary, from an average of 72 to 55 eV. Similarly, when the same ions are subjected to prior collisional cooling, the final average kinetic energy for the subsequent helium-free system is but 14 eV. These two examples illustrate cooling behavior in the ion trap, which is now well recognized. However, the effect of helium buffer gas throughout and following the ion-isolation process for collisionally cooled ions is to enhance rather than moderate ion kinetic energies while the working point is close to a boundary, from an average of 14 to 34 eV. Thus, contrary to the normal collisional behavior when the working point is located well within the stability diagram, the presence of buffer gas leads to kinetic energy enhancement when the ions are cooled previously and the working point is moved subsequently close to a boundary.

When the working point of the collisionally cooled ions approaches a stability boundary, the time-averaged ion kinetic energy is at a low value such that the ions lie close to the bottom of the pseudopotential well. The ions gain energy from the RF field and part of this energy is converted to internal energy which, in the limit, leads to ion fragmentation; thus, the lower the kinetic energy initially, the greater the amount of energy that can be deposited as internal energy in competition with ion ejection.

In the simulation study, it was calculated[25] that as the stability boundary is approached from the q_z-axis by the application of a positive DC voltage, ion axial secular motion decreases in frequency, but increase in amplitude; in some cases, the amplitude of the axial component of the trajectory exceeds the physical dimensions of the ion trap. When the working point was positioned close to the $\beta_z = 0$ boundary, ion kinetic energy varied rapidly over the range 0 to ca. 30 eV; radial displacements extended to 1.2 mm, while axial displacements extended to 3.8 mm. The fundamental radial and axial secular frequencies were shown by calculation to be clearly different, with values of 169.4 and 13.8 kHz, respectively. Virtually all of the kinetic energy resided in axial motion. The maximum excursions of the collisionally cooled ions were smaller by at least a factor of 2 than those that did not undergo such collisional cooling. Ion kinetic energy maxima obtained in a collisional system will increase the probability of internal excitation of the ion due to conversion of ion kinetic energy in collisions with helium, and of dissociation. In conclusion, from the results of the simulation study cited above, it should be possible to select operating conditions with the working point in the vicinity of the $\beta_z = 0$ boundary so as to optimize ion kinetic energy distribution with minimal loss of ions.

B. Experimental Investigation of Boundary Activation with Ion Precooling

The availability of the above simulation data on the behavior of ionic species under different operating conditions, when the working point is close to a stability boundary, presented an opportunity to undertake an experimental study corresponding to the conditions of the simulation. Furthermore, such an experiment would expand the evaluation of various techniques for the collisionally induced decomposition of an isolated ion species.

1. Molecular Ion of a Pyrrolquinolinone Derivative

In order to study the influence of cooling time on energy deposition wrought by the boundary effect, further experiments were carried out with M[·] of compound 2.[26] The scan function employed in this investigation, reported in Figure 12, consisted of a constant ionization time of 300 μs, a two-step isolation of the ion of m/z 227, a variable cooling time, and a plateau time of 90 μs, during which a positive DC potential was applied. This sequence of events was followed by the usual analytical ramp of the RF potential. The sole scan parameter, which was changed during a given set of experiments, was the duration of the cooling time. The experiments were carried out at each of four different pressures of helium buffer gas. The cooling time was varied over the range 0 to 420 μs for each of the pressures of helium buffer gas, 1×10^{-5}, 4×10^{-5}, 8×10^{-5}, and 1.2×10^{-4} torr.

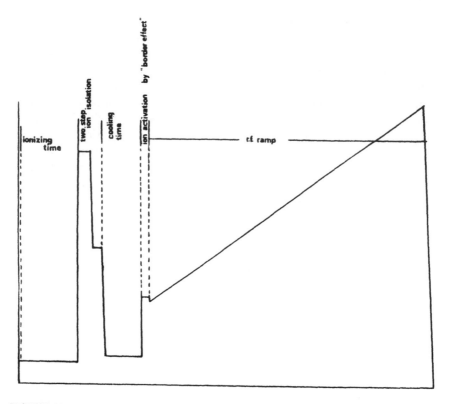

FIGURE 12
Scan function showing the sequence of operations for boundary-effect activation with an
ion precooling period. The duration of the precooling period was varied from 0 to 450 μs.
(From Paradisi, C. et al., *Rapid Comm. Mass Spectrom.*, 6, 641, 1992. With permission.)

For the two lower pressures of helium, only two fragment ions were
detected, i.e., m/z 212 and 198 originating from CH_3 and CHO· losses, respectively. The relative abundances of these two ions varied little as the
duration of the cooling time was increased, as can be seen in Figures 13
and 14. In both cases, a slow increase in the abundance of the ion of m/z
198 was observed with increasing cooling time.

The situation changed dramatically when the helium pressure was
increased from 4×10^{-5} to 8×10^{-5} torr as shown in Figure 15. Several new
fragment ion species were observed, and their relative abundances were
highly dependent on the duration of the cooling time. Ions at m/z 212 and
198, which at lower pressures of helium were the only fragmentation
products observed, still exhibited high relative abundance at zero cooling time, but as the cooling time was extended, the relative abundances
decreased rapidly to the point at which they could no longer be detected.
Concurrently, three other fragment ions, m/z 183, 184, and 154, appeared
and their relative abundances showed almost linear increases with increasing cooling time.

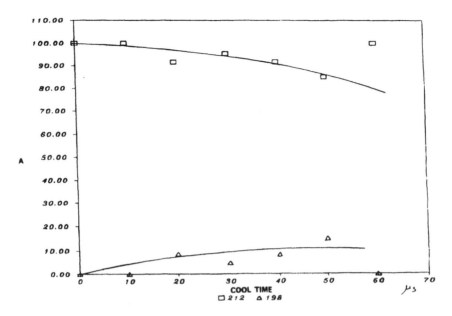

FIGURE 13
Relative abundances of fragment ions obtained from M·· of **2** as a function of the precooling period prior to moving the working point of M·· of **2** to the vicinity of the $\beta_z = 0$ boundary; the pressure of helium buffer gas was 1×10^5 torr. (□) m/z 212; (Δ) m/z 198. (From Paradisi, C. et al., *Rapid Comm. Mass Spectrom.*, 6, 641, 1992. With permission.)

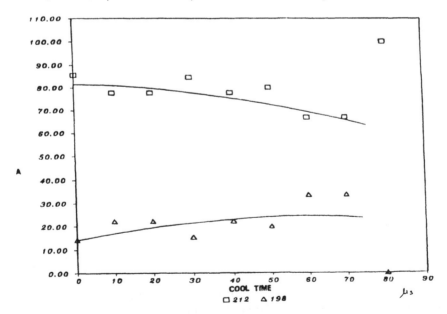

FIGURE 14
As for Figure 13, except for the pressure of helium, which was 4×10^5 torr. (From Paradisi, C. et al., *Rapid Comm. Mass Spectrom.*, 6, 641, 1992. With permission.)

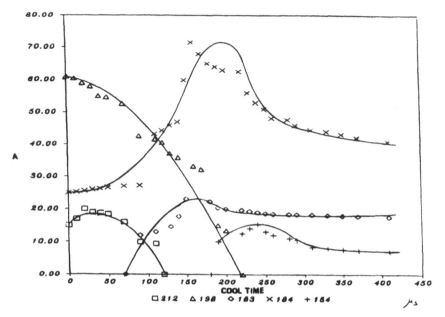

FIGURE 15
As for Figure 13, except for the pressure of helium, which was 8×10^{-5} torr. (□) m/z 212; (△) m/z 198; (×) m/z 184; (◇) m/z 183; (+) m/z 154. (From Paradisi, C. et al., *Rapid Comm. Mass Spectrom.*, 6, 641, 1992. With permission.)

At a helium pressure of 1.2×10^{-4} torr, which is the typical helium pressure employed in daughter ion spectroscopy experiments by ITMS, the situation is changed further as shown in Figure 16. Again, the intensities of $[M\text{-}CH_3]^+$ and $[M\text{-}CHO]^+$ fall with increasing cooling time, but for this helium pressure, the dependencies on cooling time of the remaining fragment ions show a maximum. This behavior is particularly clear for m/z 184 ions and indicates that a specific range of cooling time is most effective for maximizing the yield of a specific collisional product ion.

Figures 15 and 16 show a remarkable resemblance to ion breakdown curves wherein fractional ion abundances are plotted vs. energy in the center-of-mass frame. Clearly, information concerning the appearance energies of fragment ions which can be obtained from such curves should be extractable from the curves shown in Figures 15 and 16, once a suitable calibration procedure can be evolved. The dependency of the abundances of each of the four fragment ions upon helium pressure and cooling time is displayed vividly by the three-dimensional plots shown in Figures 17 to 20.

In Figure 17, the absolute abundance of the m/z 198 ion is plotted with respect to the pressure of helium and to the cooling time. It is shown clearly in this figure that the formation of m/z 198 ions is favored at high pressure of helium and low cooling time for the precursor ion. In other words, access via the boundary effect to the specific collisionally induced

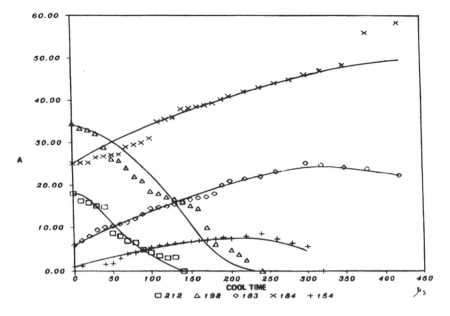

FIGURE 16

As for Figure 14, except for the pressure of helium, which was 1.2×10^4 torr. ($\square$) m/z 212; ($\triangle$) m/z 198; ($\times$) m/z 184; ($\diamond$) m/z 183; (+) m/z 154. (From Paradisi, C. et al., *Rapid Comm. Mass Spectrom.*, 6, 641, 1992. With permission.)

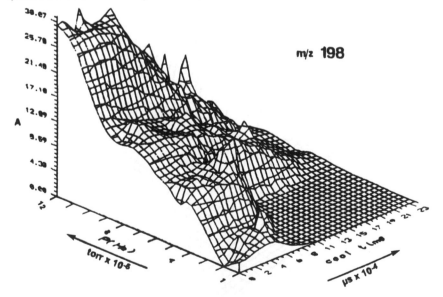

FIGURE 17

Three-dimensional representation of the signal intensity of the m/z 198 fragment ion of the $M^{+\cdot}$ of **2** as a function of helium buffer gas pressure and cooling time prior to moving the working point to the vicinity of the $\beta_z = 0$ stability boundary. (From March, R. E. et al., *Rapid Comm. Mass Spectrom.*, 7, 929, 1993. With permission.)

decomposition process which leads to the formation of m/z 198 is favored for "warm" precursor ions.

For the primary ion of m/z 183 this situation is only partially true, as is shown in Figure 18. The formation of m/z 183 is favored at higher pressures of helium, but the relative abundance of this species shows a clear increase with respect to the cooling time to which the precursor ion is subjected.

Different behavior again is observed for the relative abundances of each of the fragment ions of m/z 184 and 154, as shown in Figures 19 and 20, respectively. The appearance of m/z 154 is particularly interesting in that this ion species originates from the sequential losses of C_2H_4O and CHO from the molecular ion. The most effective conditions for the formation of these species are those of moderate cooling time (ca. 200 μs) and intermediate helium pressure (ca. 1×10^{-4} torr).

2. Molecular Ion of 4-Nitro-o-Xylene

The molecular ions of each of 4-nitro-o-xylene and 4-nitro-m-xylene were subjected to consecutive ion isolation and boundary activation, with prior cooling, in a modified commercial Varian Saturn I ion trap detec-

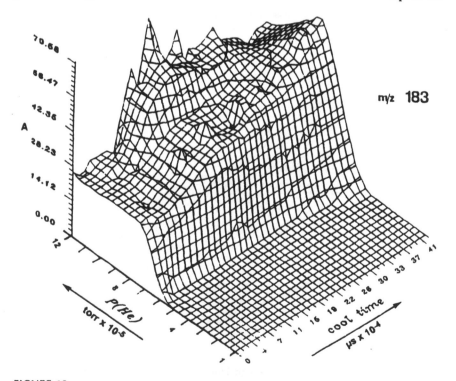

FIGURE 18

As for Figure 17, but for the m/z 183 fragment ion. (From March, R. E. et al., *Rapid Comm. Mass Spectrom.*, 7, 929, 1993. With permission.)

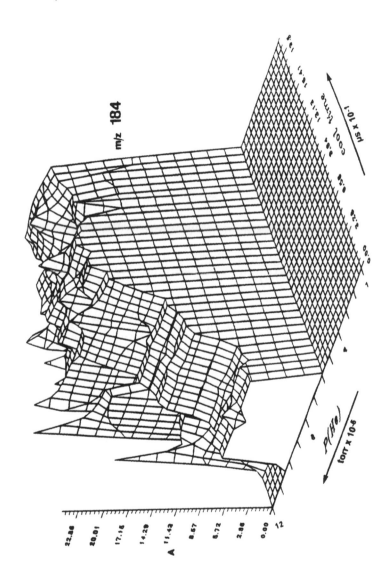

FIGURE 19

As for Figure 17, but for the *m/z* 184 fragment ion.

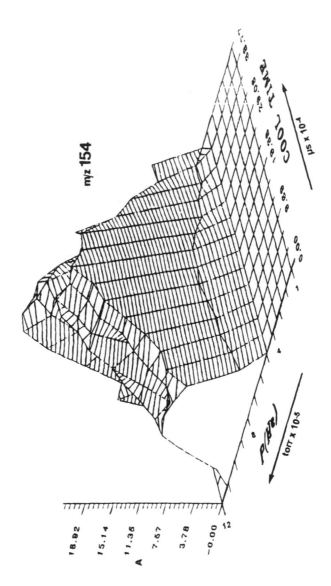

FIGURE 20
As for Figure 17, but for the *m/z* 154 fragment ion.

tor (ITD).[27] The scan function used for this work (shown in Figure 21) differed in two respects from that used in the experiments described above. First, low-mass ions were ejected at a working point on the q_z-axis with q_z = 0.86; second, the cooling period preceded the application of a DC voltage by which the working point of the mass-selected ion was moved to the vicinity of the β_z = 0 boundary. The trajectory of the working point is shown in Figure 22.

a.   Boundary-Activated Dissociation as a Function of q_z

The molecular ion, m/z 151, was isolated and subjected to boundary activation at three values of q_z with no pre-cooling. The m/z 106 fragment ion intensity was monitored as a function of DC plateau duration at each of the three values of q_z; the results are shown in Figure 23. The extent of fragmentation at q_z = 0.45 is modestly better than that at q_z = 0.5, and increases with DC plateau duration for each of these q_z values.

b.   Effects of Precooling and Direct Current Plateau Duration

Figure 24 shows the variation in fragmentation obtained by BAD of the molecular ion, m/z 151, of 4-nitro-*o*-xylene as a function of DC (95.0 V) plateau duration at q_z = 0.45 with no precooling (see Table 4). The ef-

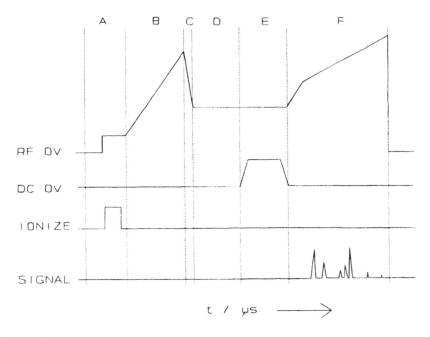

FIGURE 21
Scan function for performing consecutive ion isolation at the β_z = 1 and β_z = 0 stability boundaries with a cooling period prior to moving the working point to the β_z = 0 boundary. (From March, R. E. et al., *Can. J. Chem.*, 72, 966, 1994. With permission.)

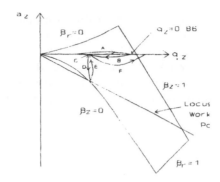

FIGURE 22
Movement of the working point corresponding to the scan function in Figure 21. (From March, R. E. et al., *Can. J. Chem.*, 72, 966, 1994. With permission.)

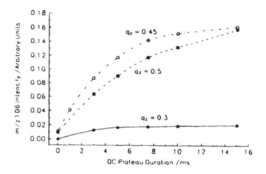

FIGURE 23
Signal intensity of *m/z* 106 from the molecular ion of 4-nitro-*o*-xylene as a function of DC plateau duration for three values of q_z. (From March, R. E. et al., *Can. J. Chem.*, 72, 966, 1994. With permission.)

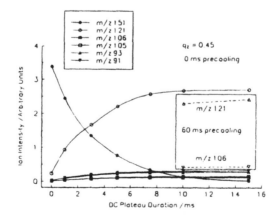

FIGURE 24
Variation in fragmentation of the molecular ion, *m/z* 151, of 4-nitro-*o*-xylene as a function of DC plateau duration at $q_z = 0.45$ with no precooling. The effects of a cooling period of 60 ms on the intensities of *m/z* 121 and 106 are shown in the inset to this figure. (From March, R. E. et al., *Can. J. Chem.*, 72, 966, 1994. With permission.)

TABLE 4

Dissociation of M· of 4-Nitro-*o*-Xylene

m/z	TSQ (10 eV)	TSQ (100 eV)	TSQ (EI)	BAD
121	78.9	61.7	20.0	64.3
106	—	—	4.0	17.5
105	18.4	31.5	56.1	2.8
93	2.7	4.4	6.4	5.4
91	—	2	13.5	10.0

Note: This is a comparison of TSQ data with those of BAD.

Source: From March, R. E. et al., *Can. J. Chem.*, 72, 966, 1994. With permission.

fects of a cooling period of 60 ms on the intensities of *m/z* 121 and 106 are shown in the inset to this figure; the intensity of *m/z* 121 (formed by loss of NO) is reduced, while that of *m/z* 106 (formed by the further loss of CH_3) is enhanced. Thus, the effect of a cooling period is to enhance the product ion intensity of the channel having the higher activation energy. This effect is seen also in Figure 25, where the intensity of the *m/z* 106 fragment ion is plotted as a function of precooling time for a variety of time periods during which the working point of the mass-selected ion was held at the DC plateau. For all finite values of the DC plateau duration, the *m/z* 106 fragment ion intensity increased with the duration of the precooling period. Similarly, the intensity of *m/z* 106 increased with DC plateau duration for each value of the precooling time.

c. *Effect of Precooling*

The fragment ions of *m/z* 93 and 91 are formed by the losses of CO and HCHO, respectively, as has been shown by collisional dissociation of *m/z* 121 in a TSQ instrument. From the variation in the fragment ion intensity ratio of *m/z* 91/93 as a function of precooling time, shown in Figure 26 and obtained at $q_z = 0.45$, it would appear that the activation energy required for the formation of the former ion species exceeds that of the latter. Variation of the precooling time on the molecular ion of 4-nitro-*m*-xylene, when subjected to consecutive ion isolation and boundary activation, showed no effect.

d. *Comparison with Triple Stage Quadrupole Data*

Again, comparison with a TSQ instrument indicated that BAD produced a fragmentation pattern that resembled both that obtained with a kinetic energy of 100 eV and upon EI.

3. *Discussion*

These results, at first sight inhomogeneous and difficult to explain, can be rationalized upon consideration of the variation of internal energy deposition in the precursor ion with respect to the two parameters,

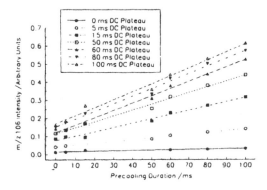

FIGURE 25

Effect of precooling time on the intensity of the fragment ion, m/z 106, produced from 4-nitro-o-xylene for various fixed periods at the DC plateau. (From March, R. E. et al., *Can. J. Chem.*, 72, 966, 1994. With permission.)

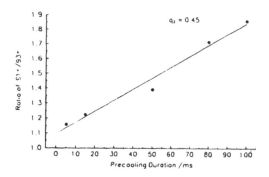

FIGURE 26

Variation of the signal intensity ratio of m/z 91/93 fragment ions obtained from 4-nitro-o-xylene as a function of precooling time; $q_z = 0.45$. (From March, R. E. et al., *Can. J. Chem.*, 72, 966, 1994. With permission.)

helium pressure and cooling time. Under the conditions of low helium pressure and cooling times in the range 0 to 420 μs, i.e., when "warm" ions are subjected to the border effect, only two fragment ions are produced representing necessarily those fragmentation processes having the lower activation energies.

The appearance of further fragment ion species at higher ambient helium pressures could be ascribed to the history of the precursor ions after the "border effect". In other words, consecutive collisions in the higher pressure domain are more efficient either with respect to energy deposition in a single ion or, from the statistical point of view, in forming greater numbers of ions exhibiting higher internal energies; each of these interpretations seeks to explain the opening of decomposition pathways which require higher activation energies.

Yet the observed increase in the production of ion species requiring higher activation energies concomitant with the increase in cooling time, as well as the observed maxima in ion abundances, demonstrate unequivocably that the history of the ions prior to movement of the working point to the vicinity of the stability boundary has a clear and pronounced effect on the kinetic energy distribution of the ions, and on the distribution of their internal energies.

The results obtained confirm the findings of the simulation study, that the ready variation of kinetic energy distribution can be effected with an appropriate choice of operating conditions prior to movement of the working point to the vicinity of a stability boundary. The present results indicate also that optimization of the collisional parameters affords a measure of control of internal energy deposition such that fragment ions can be observed from reaction pathways which are accessed by ions (in a TSQ instrument) having kinetic energies >100 eV.

The mechanism for daughter ion enhancement by collisional precooling appears to be the result of low energy collisions suffered by ions as they become focused toward the center of the ion trap. With prolonged precooling, ions are stored more effectively in that translational energy is lost and the total ion energy approaches the zero point energy level of the pseudopotential well. During the precooling period, ion internal energy may be enhanced as a direct result of the low energy collisions suffered.

When the working point of the mass-selected and precooled ions is moved to the vicinity of the $\beta_z = 0$ boundary, the ions in the cloud, tightly focused near the center of the ion trap where the field is relatively weak, experience gentle acceleration and increase in kinetic energy. During the course of suffering multiple low energy collisions with buffer gas atoms, some ion kinetic energy is converted into internal energy. Thus, the overall effect of precooling appears to be an enhancement of internal energy deposition in competition with the increase in kinetic energy. The greater the absolute deposition of internal energy (and subsequent unimolecular dissociation) prior to the ion acquiring kinetic energy in excess of the trapping potential well depth, the larger the activation energy barrier that the ion can surmount.

The collisional precooling of precursor ions has been shown to lead to the observation of a series of fragment ions having a wide range of activation energies. The results indicate that it is possible to control internal energy deposition so as to maximize the appearance of a given fragment ion species. The range of activation energies accessed in the present work for a single precursor ion species exceeded that established earlier by tickling and by the boundary effect.

VII. DISCRIMINATION BETWEEN CONSECUTIVE AND COMPETING COLLISIONALLY INDUCED DECOMPOSITIONS

A. Statement of the Problem

A general problem in MS lies in the determination of ion fragmentation patterns; this problem can be complicated by the presence of both consecutive and competing decomposition channels. A method is described by which it is possible to discriminate between consecutive and competing collisionally induced decompositions by the simultaneous application of the boundary effect and tickle activation.[28]

B. Application of Sector Instruments

A typical case, reported in Scheme 3, is that of fragment F_2^+ which can be formed directly from a precursor ion and/or through the intermediacy of a fragmention F_1^+. The solution to this problem is, in general, quite difficult when the ions are not generated by collision, and it can be attained in only some fortuitous cases by metastable ion studies.[14] The application of sector instruments using linked scans and of mass-analyzed ion kinetic energy experiments does not always yield results free from ambiguity. On the other hand, unambiguous results may be obtained with a TSQ instrument by using the reaction intermediate scan.[29] In ion cyclotron resonance (ICR) spectrometry, the intermediacy of F_1^+ can be tested easily by the double resonance technique whereby F_1^+ is resonantly ejected during the activation of M^+, thus suppressing observation of the component of F_2^+ deriving from the stepwise route.[30]

C. Application of Ion Trap Mass Spectrometry

A rather indirect indication of the stepwise process can be achieved by ITMS when the possible intermediate F_1^+ is subjected to a further activation in a MS/MS/MS experiment. By this approach it is possible to test the ability of F_1^+ to generate F_2^+. With such an experiment, a negative result will indicate that the stepwise fragmentation is excluded.

1. A Novel Method for Pathway Discrimination

A novel method, based on the simultaneous use of BAD and tickle activation in ITMS, permitted unequivocal discrimination between the two fragmentation pathways. The experiment was carried out by applying, during a single storage phase, both a DC voltage suitable to induce fragmentation of the parent ion and, simultaneously, an RF tickle voltage of sufficient magnitude to eject a selected daughter ion rapidly from the ion trap, as shown by the related scan function given in Figure 27. Such an experiment permits the establishment of a parent-to-daughter relationship unambiguously.

SCHEME 3

Consecutive and competing decomposition F_1^+ channels involving two fragment ions, F_1^+ and F_2^+, and the molecular ion.

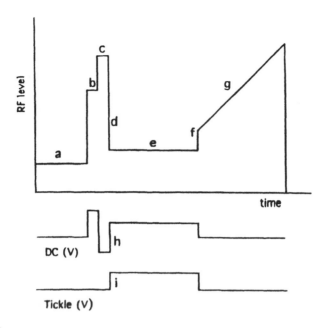

FIGURE 27

Scan function describing the variations of the DC voltage, tickle voltage, and of the RF drive potential during a cycle. a, Ionization time; b, c, separation phase; d, q_z value set; e, storage time; f, scan start set; g, acquisition phase; h, DC voltage set for parent ion boundary-effect activation; i, tickle voltage. (From Paradisi, C. et al., *Rapid Comm. Mass Spectrom.*, 6, 641, 1992. With permission.)

a. 4-Nitrophenyl-Phenyl Sulfone

For the case of competing reactions, which is the situation that pertains to the molecular ion of 4-nitrophenyl-phenyl sulfone the results shown in Figure 28 were obtained. The MS/MS mass spectrum of M$^+$ (m/z 263) obtained by BAD (and with the tickle voltage set to zero) is due to three main fragment ions: $O_2N-C_6H_4-SO^+$ (m/z 170), $C_6H_5SO^+$ (m/z 125), and $C_6H_5O^+$ (m/z 93). The same figure shows the effects of tickling the ions of m/z 170 (Figure 28a) and m/z 125 (Figure 28b). The activation phase, leading to the formation of fragment ions (m/z 170 → m/z 140 and m/z 125 → m/z 97) is followed at higher tickle voltages by the ejection of both the tickled ions and their daughter ions. The results so obtained are not substantially different than those obtained by the use of two consecutive tickle episodes, as is usually performed in (MS)^n experiments. The main difference lies in the shorter time required for performing the double activation.

b. 2-(4′-Methoxybenzoyl)benzoic Acid

For the case of consecutive reactions, such as the loss of CO_2 and the further loss of CH_3O from the molecular ion of 2-(4′-methoxybenzoyl) benzoic acid, the results reported in Figure 29 were obtained. These data

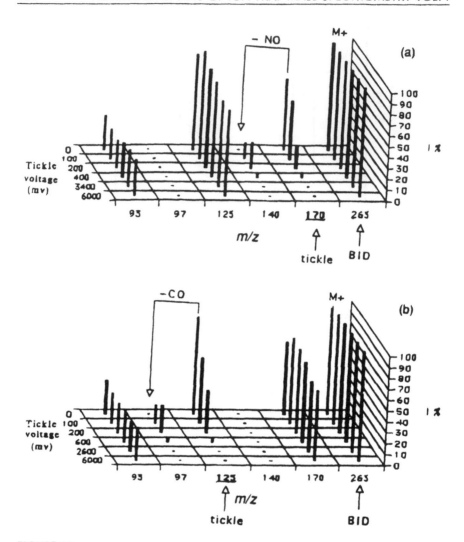

FIGURE 28

Effects of tickling selected daughter ion species during BAD of M·· of 4-nitriphenyl-phenyl sulfone. Tickling of daughter ions (a) $O_2N-C_6H_4-SO·$ (m/z 170); (b) $C_6H_5-SO·$ (m/z 125). (From Paradisi, C. et al., *Rapid Comm. Mass Spectrom.*, 6, 641, 1992. With permission.)

indicate that when the ion of m/z 212 is tickled, the first activation stage is observed at low voltages and is followed by the ejection of both ions of m/z 212 and 181. Similar data were obtained upon examination of the nitrobenzene molecular ion (compound **5**), which, by boundary dissociation shows the consecutive reactions (M·· m/z 123 → m/z 93 → m/z 65); these data are shown in Figure 30. Also in this case, ejection of the intermediate ion of m/z 93 is accompanied by the disappearance of the ion of m/z 65, thus proving the intermediacy of the former in the formation of the latter.

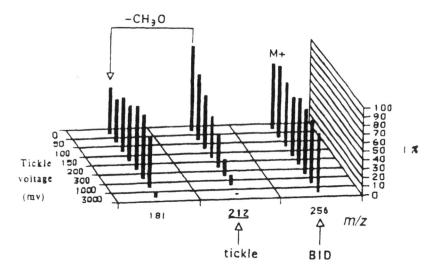

FIGURE 29
Effects of tickling selected daughter ion species during BAD of M⁺⁺ of 2-(4'-methoxyben-zoyl)benzoic acid. Tickling of daughter [M-CO₂]⁺⁺ ions (*m/z* 212). (From Paradisi, C. et al., *Rapid Comm. Mass Spectrom.*, 6, 641, 1992. With permission.)

c. Butyrophenone

In the case of a more complex situation wherein both competing and consecutive reactions occur, the approach proposed here proved to be highly effective. The boundary dissociation mass spectrum of M⁺⁺ of bu-tyrophenone, which is shown in Figure 31, shows the presence of three fragment ions, two of which are formed necessarily from the molecular ion by competitive losses of H_2O (leading to ions of *m/z* 130) and C_2H_4 (giving rise to ions of *m/z* 120), whereas the third fragment ion species of *m/z* 105 could be generated both as a primary fragment (by loss of C_3H_7) or by methyl loss from the ion of *m/z* 120. The results of tickling the fragment ion of *m/z* 120 are reported also in Figure 31; here, it was observed that the fragment ion of *m/z* 105 disappeared when the *m/z* 120 ion was ejected from the ion trap.

d. 4,4'-Dichloroazoxybenzene

The example of 4,4'-dichloroazoxybenzene is more complex and offers an opportunity to test the versatility of the method. The boundary dissociation mass spectrum of M⁺⁺ is composed of the ions of *m/z* 238 [M-CO]⁺⁺, *m/z* 237 [M-CO-H]⁺, *m/z* 203 [M-CO-Cl]⁺, *m/z* 202 [M-CO-Cl-H]⁺⁺, *m/z* 139 [ClC₆H₄N₂]⁺, and *m/z* 111 [ClC₆H₄]⁺. This last ion species could be formed, in principle, from the molecular ion as well as from the fragment ions of *m/z* 238, 203, and 139. Unfortunately, tickle activation of *m/z* 238 and 237 ions could not be performed for instrumental reasons. However, tickle activation of *m/z* 203 and 202 ions eliminated these ion species as

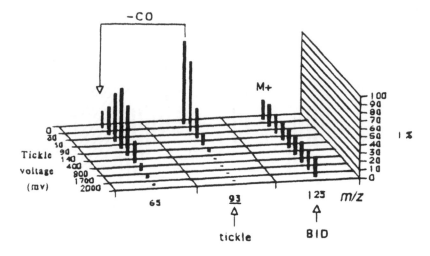

FIGURE 30
Effects of tickling selected daughter ion species during BAD of M·· of butyrophenone. Tickling of daughter [M-NO]·· ions (*m/z* 93). (From Paradisi, C. et al., *Rapid Comm. Mass Spectrom.*, 6, 641, 1992. With permission.)

precursors of *m/z* 111 ions, as the signal intensity of *m/z* 111 ions remained unchanged during this procedure. Such was not the case when *m/z* 139 ions were subjected to tickle activation, and *m/z* 139 is clearly a precursor of *m/z* 111. The mass spectra obtained at different tickle voltage amplitudes (Figure 32) still indicate the existence of an activation phase (during which the signal intensity of *m/z* 111 ions is observed to increase) followed by an ejection phase for both *m/z* 139 and 111 ions. However, a residual signal intensity due to *m/z* 111 ions, equal to about one third the original intensity, persists, thus showing that this species originates also from other competing and distinct route(s).

VIII. CONCLUSIONS

An evaluation has been made of various techniques for the collisionally induced decomposition of an ion species isolated in an ion trap. The principal techniques examined here are those of tickle voltage activation, boundary-effect activation, boundary-effect activation with ion precooling, and combined boundary-effect activation and tickle voltage activation.

While tickle voltage activation can be carried out over a relatively wide range of q_z values, the degree of internal energy deposition in the activated ion is limited. The data presented here indicate that a higher degree of internal energy deposition occurs in ions activated by the boundary effect. Furthermore, the range of product ion masses that can be captured by the ion trap is greater with boundary-effect activation than

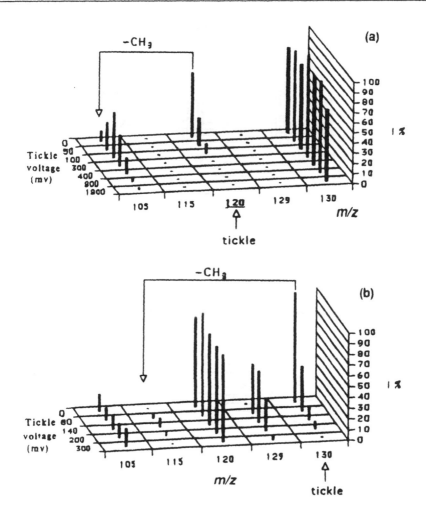

FIGURE 31

Effects of tickling selected daughter ion species during BAD of M⁺⁺ of butyrophenane. Tickling of daughter ions (a) [M-C₂H₄]⁺⁺ (*m/z* 120); (b) [M-H₂O]⁺⁺ (*m/z* 130). M⁺⁺ is not reported. (From Paradisi, C. et al., *Rapid Comm. Mass Spectrom.*, 6, 641, 1992. With permission.)

with tickle voltage activation, although the useful range of q_z values is more limited with this method compared to tickle voltage activation.

The effects of precooling ions prior to instigation of boundary-effect activation are quite considerable. The present results indicate that not only does optimization of the collisional parameters afford a significant measure of control of internal energy deposition, but that fragment ions can be observed from reaction channels which are accessed only by ions in a TSQ instrument having kinetic energies >100 eV. These results indicate an appreciable enhancement of ion trap versatility for the application of CID to the investigation of ion structures and to ion identification.

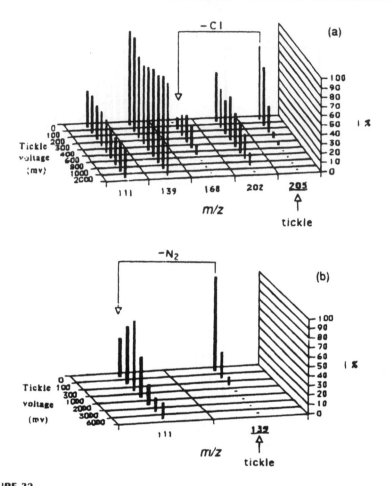

FIGURE 32

Effects of tickling selected daughter ion species during BAD of M·· of 4,4′-dichloroozoxy-benzene. Tickling of daughter ions (a) [M-CO-Cl]· (*m/z* 203); (b) [ClC$_6$H$_4$N$_2$]· (*m/z* 139); M·· (*m/z* 266). (From Paradisi, C. et al., *Rapid Comm. Mass Spectrom.*, 6, 641, 1992. With permission.)

The novel application of combined boundary-effect activation and tickle voltage activation has been demonstrated as a powerful method for discriminating between consecutive and competing CID pathways.

REFERENCES

1. March, R.E.; Hughes, R.J., *Quadrupole Storage Mass Spectrometry*, Chemical Analysis Series, Vol. 102; Wiley Interscience: New York, 1989.
2. Louris, J.N.; Cooks, R.G.; Syka, J.E.P.; Kelley, P.E.; Stafford, G.C., Jr.; Todd, J.F.J., *Anal. Chem.* 1987, *59*, 1677.

3. Fulford, J.E.; Hoa, D.-N.; Hughes, R.J.; March, R.E.; Bonner, R.F.; Wong, G.J., *J. Vac. Sci. Technol.* 1980, *17*, 829.
4. Cooks, R.G., *Collision Spectroscopy*; Plenum Press: New York, 1978.
5. March, R.E.; Londry, F.A.; Alfred, R.L.; Franklin, A.M.; Todd, J.F.J., *Int. J. Mass Spectrom. Ion Processes.* 1992, *112*, 247.
6. Gronowska, J.; Paradisi, C.; Traldi, P.; Vettori, U., *Rapid Commun. Mass Spectrom.* 1990, *4*, 306.
7. McLuckey, S.A.; Goeringer, D.E.; Glish, G.L., *Anal. Chem.* 1992, *64*, 1455.
8. Julian, R.K.; Cox, K.; Cooks, R.G., *Proc. 40th ASMS Conf. Mass Spectrom. Allied Topics.* Washington, D.C., 1992; p. 943.
9. Pannell, L.K.; Quan-Long, Pu; Mason, R.T.; Fales, H.M., *Rapid Commun. Mass Spectrom.* 1990, *4*, 103.
10. Lifshitz, C., *Int. J. Mass Spectrom. Ion Processes.* 1991, *106*, 159.
11. Louris, J.N.; Brodbelt, J.E.; Cooks, R.G., *Int. J. Mass Spectrom. Ion Processes.* 1987, *75*, 345.
12. Lammert, S.A.; Cooks, R.G., *Rapid Commun. Mass Spectrom.* 1992, *6*, 528.
13. Lammert, S.A.; Cooks, R.G., *J. Am. Soc. Mass Spectrom.* 1991, *2*, 487.
14. Creaser, C.S.; McCoustra, M.R.S.; O'Neill, K.E., *Org. Mass Spectrom.* 1991, *26*, 335.
15. Paradisi, C.; Todd, J.F.J.; Traldi, P.; Vettori, U., *Org. Mass Spectrom.* 1992, *27*, 251.
16. Paradisi, C.; Todd, J.F.J.; Vettori, U., *Org. Mass Spectrom.* 1992, *27*, 1210.
17. Creaser, C.S.; O'Neill, K.E., *2nd Eur. Meet. Tandem Mass Spectrometry.* Warwick, U.K., July 1992.
18. Curcuruto, O.; Fontana, S.; Traldi, P.; Celon, E., *Rapid Commun. Mass Spectrom.* 1992, *6*, 322.
19. Matthews, L.S., M.Sc. thesis, University of East Anglia, East Anglia, U.K., 1992.
20. Creaser, C.S.; O'Neill, K.E., *Org. Mass Spectrom.* 1993, *28*, 564.
21. Louris, J; Schwartz, J.; Stafford, G.C.; Syka, J.E.P.; Taylor, D., *Proc. 40th ASMS Conf. Mass Spectrom. Allied Topics.* Washington, D.C., 1992; p. 1003.
22. Johnson, J.V.; Pedder, R.E.; Yost, R.A., *Rapid Commun. Mass Spectrom.* 1992, *6*, 760.
23. Fontana, S.; Curcuruto, O.; Traldi, P.; Castellin, A.; Chilin, A.; Rodighiero, P.; Guiotto, A., *Org. Mass Spectrom.* 1992, *27*, 1255.
24. Cooks, R.G.; Beynon, J.H.; Caprioli, R.M.; Lester, R.G., *Metastable Ions*; Elsevier: Amsterdam, 1973.
25. March, R.E.; Tkaczyk, M.; Londry, F.A.; Alfred, R.L., *Int. J. Mass Spectrom. Ion Processes*, 1993, *125*, 9.
26. March, R.E.; Londry, F.A.; Fontana, S.; Catinella, S.; Traldi, P., *Rapid Commun. Mass Spectrom.* 1993, *7*, 966.
27. March, R.E.; Weir, M.R.; Londry, F.A.; Catinella, S.; Traldi, P.; Stone, J.A.; Jacobs, W.B., *Can. J. Chem.*, 1994, *72*, 966.
28. Paradisi, C.; Todd, J.F.J.; Traldi, P.; Vettori, U., *Rapid Commun. Mass Spectrom.* 1992, *6*, 641.
29. O'Lear, J.R.; Wright, L.G.; Louris, J.N.; Cooks, R.G., *Org. Mass Spectrom.* 1987, *22*, 348.
30. Chen, J.H.; Hays, J.D.; Dunbar, R.C., *J. Phys. Chem.* 1984, *88*, 4759.

Chapter 8

ION/MOLECULE REACTIONS

Fernande Vedel, Michel Vedel, and Jennifer S. Brodbelt

CONTENTS

0-8493-4452-2/95/$0.00+$.50
© 1995 by CRC Press, Inc.

PART A
MANIPULATION AND DETECTION OF STORED IONS
FOR THE STUDY OF ION/MOLECULE REACTIONS

Fernande Vedel and Michel Vedel

I. INTRODUCTION

In this chapter we would like to demonstrate how a sound knowledge of the ion cloud properties permits the utilization of the capabilities of the ion trap for the study of slow ion/molecule reactions.

In order to obtain signals of relatively high intensity, commercial systems generally function in the presence of rather high buffer gas pressures, whereas for specific investigations of ion/molecule reactions, ultrahigh vacuum (UHV) conditions are required. These "high" pressure conditions preclude the possibility for studying ion/molecule collisions because elastic and nonelastic collisions between the ion target and the buffer gas will introduce contributions to the ion cloud dynamics in addition to those of the process under study. Moreover, it is necessary to control the presence of the parasitic ion species formed from this neutral gas or its ions. On the other hand, some properties of ion motion, if they are well controlled, can be used in the development of very sensitive methods of ion detection; furthermore, the mass/charge selectivity of the ion trap permits the design of appropriate ion manipulations in order to suppress parasitic phenomena that can interfere with the quality of the measurement.

The study of ion/molecule collisions clearly requires a knowledge of the energy of the collisions. To this end, we have used two independent methods: (1) the chemical thermometer, which involves a rather slow reaction and requires nearly UHV conditions, and (2) a time-of-flight method. Both methods were adapted to our specific case and, as they yielded results that were in good agreement, they inspired confidence in the use of these methods.

In this chapter, following a brief review of some properties of stored ion clouds, we describe some specific investigative means which we used in order to develop very sensitive methods for ion detection. We recall here the potential of the ion trap for the manipulation and measurement of ion kinetic energy. Finally, we apply the investigative tools we have developed to specific systems for the measurement of ion/molecule reaction rate constants.

II. EXAMPLES OF THE ION MOTION EQUATION IN THE PRESENCE OF DIFFERENT INFLUENCES

Here, we show how the ideal representation of the ion motion equation is not always correct. It is necessary to modify Equation 39 of Chapter 1 by additional terms in the second member, each term representing a given perturbation. The basic Mathieu equation with respect to time for the z-component is

$$\frac{d^2z}{dt^2} + \frac{\Omega^2}{4}(a_z - 2q_z \cos \Omega t)\, z = 0 \tag{1}$$

A. Couplings Due to the Anharmonicities

As was shown above (in the description of the stretched ion trap for instance), anharmonicities in the confinement field bring about modifications of the ion motion. Despite strenuous efforts to ensure that the ion trap has the correct geometry, field defects arise, for example, even from the truncation of the electrodes. In addition to the ideal quadrupolar potential distribution expressed in cylindrical coordinates

$$V(r, z) = V_0 + A(2z^2 - r^2) \tag{2}$$

where V_0 and A are constants, with A having the same sign as the charge on the trapped particle. A general potential distribution must be considered[1]

$$V(r, z) = C_0 + C_1 H_1(r, z) + C_2 H_2(r, z) + C_3 H_3(r, z) + C_4 H_4(r, z) + \dots \quad (3)$$

with

$$H_1(r, z) = z/s$$
$$H_2(r, z) = (2z^2 - r^2)/s^2$$
$$H_3(r, z) = (2z^3 - 3zr^2)/s^3$$
$$H_4(r, z) = (8z^4 - 24z^2r^2 + 3r^4)/s^4 \qquad (4)$$

The H_j functions (spherical harmonics) are homogeneous polynomials in z/s, r/s, with the numerical coefficients chosen so that each satisfies the Laplace equation. The C_j are arbitrary constants with the units of potential, and s is a geometrical factor as, for instance, is r_0 in Equation 37 of Chapter 1. It is important to note that these higher terms must be taken into account as ion trajectories evolve further from the center of the ion trap. Therefore, it is reasonable to believe that only the first terms can induce major changes in the vicinity of the center of the ion trap.

In order to give an idea of the influence of higher terms, we integrated numerically[2] the motion equation for the case where only C_2 and C_4 are different from 0; this example corresponds to a geometry without asymmetric defects. The Fourier transform of the solution gives the spectrum of ion motion as, for instance, the fundamental frequencies. The results are shown in Figure 1, which was obtained with $a_z = 0$, $q_z = 0.4$ and $\Omega/2\pi = 1$ MHz. The numerical integration was carried out with a fast and very precise method using the Numerov algorithm[3] described in the literature;[4] the error in this method depends on the sixth power of the step. Nonlinear equations must be integrated very carefully, and numerous preliminary tests showed that this method was shorter than the equivalent Runge-Kutta methods. This is because the elementary operations number is smaller and the errors due to numerical truncature are also smaller. The Fourier transform was obtained with the IMSL routine. It is important to note that the anharmonicity induces variations of the fundamental frequencies in opposite directions with respect to the components and that this variation is not negligible; for example, the variation is close to 5% for a $C_4/C_2 = 1\%$.[2]

In addition, due to the coupling between the radial and the axial degrees of freedom, terms of higher anharmonicities will introduce new frequencies which are linear combinations of $\omega_{x,0}$ and $\omega_{z,0}$. Such couplings induce nonlinear resonances which were described initially by von Busch

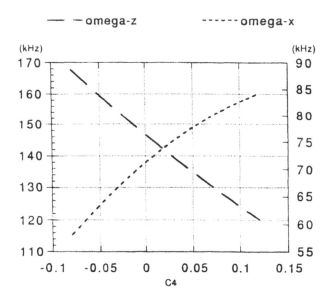

FIGURE 1
Evolution of the fundamental frequencies in the presence of anharmonicities.

and Paul[5] for the bidimensional equivalent problem of the linear electric mass filter case.

B. Terms Due to Collisions[6,7]

When ions are trapped in the presence of a neutral gas, collisions normally will occur. These collisions modify ion motion through very small variations of ion energy at each step because of the long range of the interactions. The effect of these collisions can be modeled with a stochastic force, the effect of which gives a damping term in the motion equation, as in the Brownian motion.[6] The ion motion is described by a Langevin equation[8]

$$\frac{d^2z}{dt^2} + \gamma \frac{dz}{dt} + \frac{\Omega^2}{4}(a_z - 2q_z \cos \Omega t)\, z = F(t)/m \qquad (5)$$

where γ is a damping constant determined by the rate of collisions given by $\gamma = n\sigma_{coll}v$; σ_{coll} is the cross-section for collisions with the neutral; n is the ion density; v is the ion velocity. $F(t)$ is a Gaussian random force with $\langle F(t)\, F(t') \rangle = 2\, D\, \delta(t - t')m^2$, D being the diffusion coefficient equal to $\gamma(K_BT/m)$; K_B is the Boltzmann constant; T is the thermodynamic temperature; and m is the mass of the ion.

This representation is able to describe the dynamics of the ion cloud in terms of the spatial and velocity distributions;[7] in particular, this representation shows the difference between the harmonic oscillator (where $\langle v^2 \rangle = D/\gamma$) and the ion motion, for which the statistical parameters depend on the working point because of the contribution of the micromotion.[9] This model confirms other results[4,10] and experiments[11-13] which demonstrate a Gaussian distribution in the presence of buffer gas.

In the theoretical works referred to above, the cross-section is taken as constant (hard sphere models, reasonably valid for "high" energy collisions). However, when the equation is integrated numerically, it is possible to introduce more realistic models, such as $\sigma_{coll} = \sigma_{coll}(v)$ (v is the relative velocity expressed in the center-of-mass system). This occurs with, for instance, the Langevin-Goudsmith cross-section: $\sigma_{coll}(v) = k_{coll}/v$ with $k_{coll} = 2\pi(\alpha e^2/\mu)^{1/2}$, where k_{coll} is the corresponding rate constant and α and μ are, respectively, the dipolar moment and the reduced mass.[14]

C. Space-Charge Effects

Ion/ion interactions or Coulombian interactions are no longer negligible when the ion number exceeds approximately $n \approx 10^4$ ions.[4,15] The first approach can be to add to Equation 1 a discrete sum for the ion number such as in.[16]

$$\sum_{i=0, i \neq j}^{i=n} A \frac{x_i - x_j}{r_{i,j}^3} \tag{6}$$

which represents the total electric potential formed by the sum of each individual contribution induced by the ith ion on the jth ion, of distance r from one another, and with the hypothesis that the space charge does not modify the boundary conditions. Actually, n is sufficiently high to use a continuous sum and the potential of the space-charge contribution can be written as

$$V(x, y, z) = \frac{1}{4\pi \, \epsilon_0} \rho(x, y, z) \otimes \frac{1}{|x|} \tag{7}$$

where $\rho(x, y, z)$ is the Gaussian spatial distribution density. Then, the corresponding motion equation is

$$\frac{d^2z}{dt^2} + \frac{\Omega^2}{4} (a_z - 2q_z \cos \Omega t) z = -\frac{ne}{m} \frac{\partial V(x, y, z)}{\partial x} \tag{8}$$

Because the right-hand side of this equation depends on the position-coordinate component, a sytem of coupled and nonlinear equations is obtained.

D. Concluding Remarks

We have shown some examples of ion motion equations under different influences. Other cases can be added to this list, such as laser cooling[17] which yields a dissipative term that can be responsible for crystallization of the ion,[17] and electric excitation in resonance or not with the ion motion.[18,19]

III. ION PROPERTIES AS A FUNCTION OF THE ION MOTION IN THE PRESENCE OF SPACE CHARGE

First, the usual conditions are introduced, *viz.*, when a finite pressure of neutral buffer gas occurs in the ion trap, collisions bring ions into close proximity with the result that ion/ion interactions no longer can be neglected. Then, the ion motion properties are deduced from the numerical integration of Equation 8. We have studied essentially the case of very small coupling factors, Γ,[20] where $\Gamma = e^2/(a K_B T)$ (with a the mean distance between ions); in this case, the value for Γ is of the order of 10^{-2},[21] which corresponds to cases with buffer gases at room temperature and ion numbers in the range of 10^4 to 10^6.

A. Frequency Spectrum of Ion Motion

We have computed, as discussed in Section II, ion trajectories in the presence of space charge and in the presence of a buffer gas. Very shortly after storage of ions has commenced,[22] the buffer gas induces Gaussian spatial and velocity distributions, which depend both on the mass of each species present in the ion trap and on the temperature of the neutral gas.[22,23] Then, the second member of Equation 8 is known. We used the Numerov algorithm suited to this case. In order to reduce the computation time, it is possible to reduce the order of the integral to 1, by appropriate change of variable without losing the precision in the representation of this term.[24] Because we have a nonlinear equation, the ion-motion spectrum depends on the initial conditions.[24,25]

Then, instead of the fundamental or secular frequencies designated by $\omega_{z,0}$ and $\omega_{x,0}$, as in Chapter 2, we found a set of continuous values for ω_z and ω_x. These values lie in the area corresponding to the possible initial conditions that are obtained from the distribution laws (Figure 2). Then, we define the effective fundamental frequencies, ω_z^* and ω_x^*, which play the role of averaged quantities.[4]

We computed these frequencies as a function of ion number (Figure 3) and as a function of the working point (Figure 4).[15,26] These figures show clearly that the space charge strongly modifies the frequency spectrum,

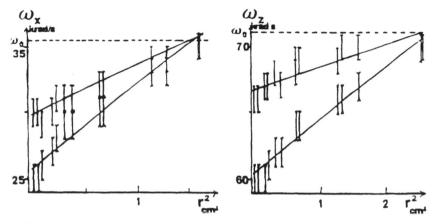

FIGURE 2

Evolution of the fundamental frequencies in the presence of space charge vs. the amplitude of the motion: Cs' ions in the presence of helium at 300 K, $a_z = 0$, $q_z = 0.2$, $\Omega/2\pi = 1$ MHz; upper line, $n = 10^5$ ions; lower line, $n = 5 \times 10^4$ ions. $r^2 = x_0^2 + y_0^2 + 2z_0^2$ (ω_x case) or $r^2 = x_0^2 + y_0^2 + 8z_0^2$ (ω_z case) where x_0, y_0, z_0 are representative of the amplitudes of ion motion.

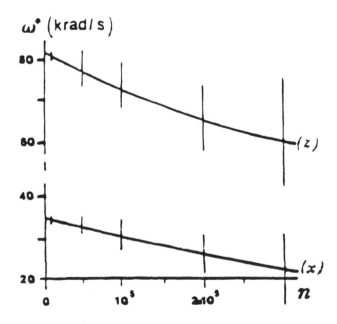

FIGURE 3

Evolution of the effective frequencies vs. the ion number, for the same conditions as in Figure 2.

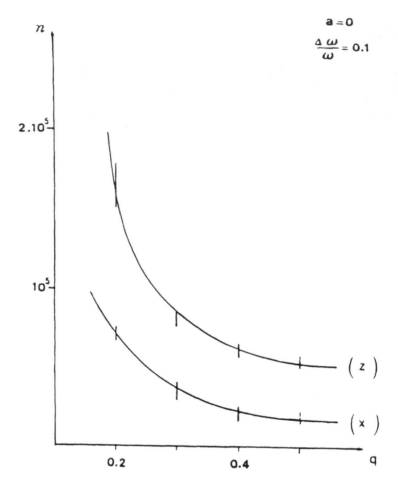

FIGURE 4
Ion number corresponding to a given relative variation of the fundamental frequencies
(x- and z-components) vs. the q_z parameter, for the same conditions as in Figure 2.

particularly when these variations are comparable to the Q-factor of the
resonant circuit needed for some nondestructive selective ion detection
methods, i.e., close to or less than 100.[27] This dependence of ω_z^* and ω_r^*
was also observed experimentally as shown in Figure 5;[28] the method
used is presented in the following section. These results are confirmed
by the observation of Doppler profiles. For wavelengths smaller than, or
of the same order as, that of the motional amplitude due to the oscillating motion, these profiles contain sidebands which are as distant from
one another as they are from the fundamental frequency of the component of the motion corresponding to the observation.[29]

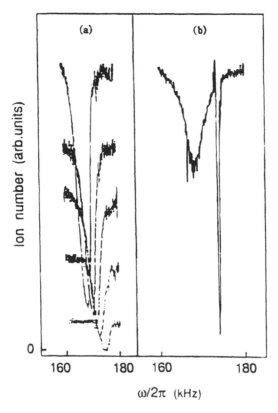

FIGURE 5
Experimental evidence of the spreading of the fundamental frequency due to space charge:
axial resonance for N^+ ions with parametric excitation at $2\omega_z$ in a pseudopotential well of
ca. 2 eV. (a) Top to bottom, the ion number is decreased by shortening the ion pulse; (b)
detailed profile which shows a very narrow response at a greater frequency due to a strong
collective effect. (From Vedel, F. and Vedel, M., *Phys. Rev.*, *A41*, 2348, 1990. With permission.)

This modification of the frequency spectrum influences the ion trajectory. The well-known ellipses in the stroboscopic phase-space representation (or Poincaré sections) are also altered (Figure 6). Nevertheless,
even for strong space charge, the stabilities of the trajectories are maintained.[15,21,26]

Space charge leads to couplings between the different degrees of
freedom, which appear as anharmonicities. Then, besides the displacements of the fundamental frequencies, new frequencies appear[5,21,24] which
are linear combinations of ω_z^* and ω_x^* (Figure 7); such linear combinations also have been found numerically.[21] Illustrations of such
couplings are found elsewhere.[30] With the bolometric technique described
below, it is possible to estimate the axial temperature. A quadrupole

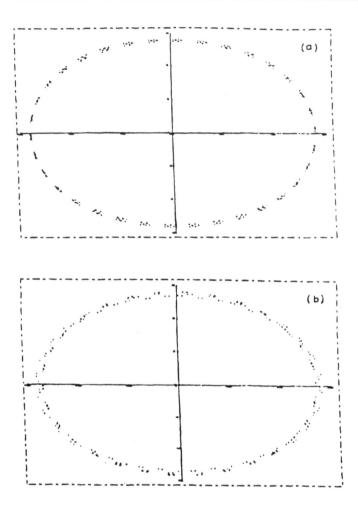

FIGURE 6

Stroboscopic phase-space representation of 10^5 Cs· ions in the presence of helium at 300 K and at the RF frequency $\Omega/2\pi = 1$ MHz, $a_z = 0$, $q_z = 0.2$; a, x-component; b, z-component; the initials conditions are standard deviations for the position components, zero for the velocity components.

alternating voltage was applied to the end-cap electrodes at $2\omega_x$ such that the electrodes were in phase. The induced excitation of the radial motion of protons stored under UHV conditions produces an increase in the axial temperature due to a transfer of the excitation energy from the two radial degrees of freedom to the axial degrees of freedom, followed by a decrease due to the reverse process until equilibrium is attained. The experimental conditions lead to time constants of roughly several seconds.

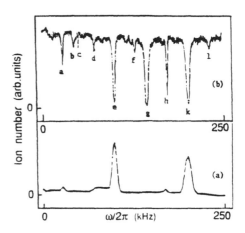

FIGURE 7

Ion excitation signals of N_2^+ ions observed with a tickle voltage for which the frequency was varied from 0 to 250 kHz and applied in the quadrupolar mode. (a) Ion ejection signals due to resonances with the z-component; (b) ion ejection signals due to resonances with the r- and z-components; a, $\omega_z - \omega_x$; b, $2\omega_z - \omega_z$; c, $\omega_z/2$; d, ω_z; e, ω_r; f, $2\omega_z - \omega_z$; g, $2\omega_z$; h, $\omega_x + \omega_z$; k, $2\omega_z$; l, $3\omega_z - \omega_x$. (From Vedel, F. and Vedel, M., *Phys. Rev.*, *A41*, 2348, 1990. With permission.)

B. Spatial Dimensions of the Ion Cloud

The repulsive ion/ion interactions give rise to distortion of the Gaussian distributions. These distortions are rather small for the usual energy conditions[15] (Figure 8). However, for stronger values of the plasma-coupling factor Γ (i.e., with higher buffer gas pressure or lower temperature), the spatial distributions tend to evolve toward uniform distributions.[29]

C. Energy Distribution

The Gaussian shape is due to relaxation mechanisms that are always present. Although the self-coherent field is not negligible, computations[15,21,26] and experiments[28,29] show that the energy distributions are not modified (nor is the RF breathing or "ripple"), even when the space charge is great enough to modify the frequency spectrum and the spatial distribution. Actually, the presence of coupling terms in the equations of ion motion decorrelates the ion position and the ion velocity distribution.

IV. ION DETECTION BY RESONANT EJECTION

The capabilities of the ion trap can be fully exploited only when it is possible to detect the stored ions. To this end, many methods of detection exist, each of them having its own particular specificity. The methods are either destructive or not, mass-selective or not, and, in the case of mass-selectivity, the mass range attainable must be ascertained. Finally, the detection method can be absolute or relative, and may or may not show the temporal dependence of the parameter under observation.

The simplest method for detection is optical detection[13] because it can be used in such a way that it is, practically, uncoupled from the ion cloud.

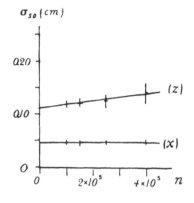

FIGURE 8

Spatial distribution parameter, σ_{s0}, vs. ion number. Cs$^+$ ions in the presence of helium at 300 K, $\beta_x = 0.4$, $\beta_z = 0.2$, $\Omega/2\pi = 1$ MHz.

However, the ions under investigation must be optically observable; hence, optical methods require suitable lasers with which to interrogate the ions. Other methods use the dynamics of the stored ions. Rettinghaus[27] had the idea of coupling the ion motion with a resonant circuit at a frequency that exists in the ion spectrum,[9] such as ω_z or $\Omega \pm \omega_z$. This method is sensitive, it can be used to give absolute measurements,[31] and it is nondestructive. The detection circuit induces a coherent ion motion which is superimposed on the incoherent one;[9] therefore, the spatial and energy distributions are modified, and the conditions of the experiment are no more constant.

Some other methods entail destabilizing the ions by stopping the confinement, or storage, voltage. One such method is to eject all the ions and to analyze them with a mass spectrometer.[32] This method is interesting because it indicates the different ion species that are present. However, it does not permit precise determination of the ion number of a given ion species because the detection efficiency is both mass dependent and ion energy dependent, and because of geometrical factors.[33] A well-known method in the ion trap mass spectrometry (ITMS) is the destabilization of ion trajectories by increasing the RF voltage.[34] The various ion species are ejected mass-selectively out of the ion trap either in order of increasing or decreasing m/z ratio. For optimum sensitivity, strong buffer–gas pressures are needed to concentrate the cloud in the center of the cage; such collisional focusing grossly perturbs the energy conditions of the experiment between the time of introduction of buffer gas and the time of detection.

In Marseilles, the authors were interested in developing a sensitive detection method which uses the spectral properties of ion motion to measure very small decreases in ion number of a given ion species with the least possible modification of the selected ion species. We present here a description of this method together with some of the aspects of ion dynamics which were observed.

A. Description of the Method[28,35]

The chosen method employs resonant ejection; that is to say, ions are ejected from the ion trap with the application of an alternating potential in resonance with a frequency of the ion motion spectrum. The pulse of resonant excitation is applied while the confinement voltage is on and after a given delay following the instant of the ion creation; this period corresponds to the reaction time. Once the ion species has been resonantly ejected, the other ion species in the ion trap remain confined until such time as the confinement voltage is interrupted and the ion species contribute to the signal amplitude that is observed subsequently. The process is represented in Figure 9. The ions that are received on the electron multiplier and form the signal are but a certain fraction of the number because of the geometry of the system. The signal is representative of the resonant ejection, both in the axial direction and in the radial direction; it is formed from ions that are not excited and are consequently all ejected in the same way. The ion number is estimated by the difference of the signal, with and without additional excitation. It is not altered by the errors introduced by the determination of a "zero level".

Any peak of the spectrum can be used to measure the stored ion number. It is possible to fix the frequency and to measure the variation of the maximum, or to integrate the signal around the broadness of this maximum. The second procedure, while more precise, requires more time.

B. Presentation of the Experiment

We used an hyperbolic steel trap ($r_0^2 = 2z_0^2 = 2$ cm^2). Its end-cap electrodes are formed from steel mesh. N$^+$ ions are created inside the trap by electron impact on N$_2$ gas and stored under UHV conditions (10^{-7} to 10^{-9} mbar). The working point is close to $a_z = 0$, $q_z = 0.5$; the RF radial frequency Ω is equal to $2\pi \times 500$ krad s^{-1}. The electron multiplier placed above the upper end-cap electrodes collects ions which are ejected out of the trap along the positive z-axis.

FIGURE 9

Timing sequence: (a) time scale; (b) ionization pulse starting at T_c; (c) RF drive voltage interrupted between T_{off} to T_{on}; (d) tickle voltage starting at T_t, which analyses the ion cloud; (e) electron multiplier signal at T_{ob}. (From Vedel, F. and Vedel, M., *Phys. Rev.*, *A41*, 2348, 1990. With permission.)

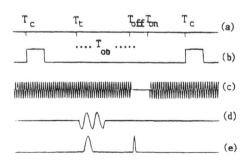

The additional alternating tickle voltage, V_t, (0.01 to 0.5 $V_{(0-p)}$) is applied in either the dipolar or quadrupolar mode for 10 ms, with a frequency ω which is varied continuously from 0 to $\Omega/2$ (Figure 10). For each step of ω, a run is performed as shown diagrammatically in Figure 9. Ions are created at T_c by an ionization pulse lasting from 0 to 100 ms, are stored during a further period of 10 ms, and are subjected to the resonant excitation pulse V_t at time T_t; the confinement voltage is collapsed between T_{off} and T_{on}. The observation time T_{ob} (the gated time when the signal on the electron multiplier is observed) is adjustable. When the observation is made at T_t, the signal is due to the ions that are ejected along the z-axis while, when T_{ob} is set at T_{off}, the signal represents ions that have not responded to the tickle and have stayed in the trap until T_{off}. At this time ions leave the trap by destabilization due to the cessation of the confinement voltage. Hence, when $T_{ob} = T_t$, peaks occur at those tickle frequencies which induce axial excitation; when $T_{ob} = T_{off} > T_t$, the ion signal forms a spectrum with inverted peak, absorption signal, indicating those tickle frequencies that earlier led to either radial or axial excitations.

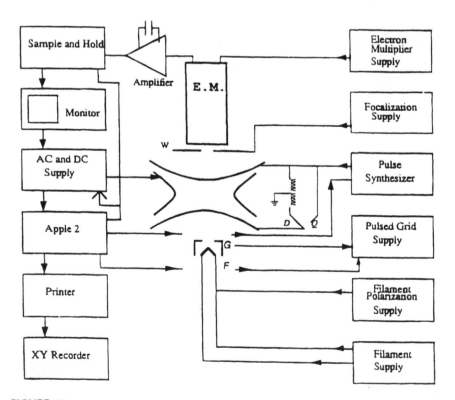

FIGURE 10
Schematic representation of the electronic apparatus used for the confinement of ions and detection by resonant ejection.

By varying the tickle frequency, it is possible to inspect the spectrum of the ion motion as presented in Volume II, Chapter 7, Figure 6.

C. Evidence for Black Holes

Despite the prevailing understanding that this expression generally designates the inefficiency of storage for daughter ions,[16] we have observed directly this phenomenon for N^+ ions. In both cases, the observed deficiencies in the confinement properties are due to nonlinear resonances; therefore, the causes are different. Similar black holes (as in Reference 36 and in related works for working points that correspond to a fundamental frequency which is a submultiple of the RF frequency Ω) have been observed previously in a quadrupole mass filter by von Busch and Paul.[5] These submultiples of the RF drive frequency were interpreted as resonances due to the presence of a multipolar contribution of the 12th order, which couples macromotion with micromotion. In our case, we observed (Figure 11) a strong diminution in the ion storage capability for particular working points, those for which β_x and β_z are in a simple ratio, β_x/β_z = 3/4 or 2/3.

Here, ions are created inside the trap under conditions that can lead to relatively long confinement times (<10 s). The detected ions are not excited and do not suffer high-order anharmonicities due to the resonance

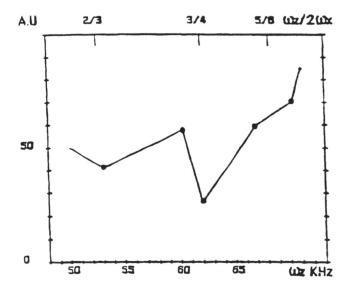

FIGURE 11
Relative number of ions stored as a function of ω_z; minima were obtained at the rational values of $\omega_z/2\omega_z$.

term. The couplings between the axial and the radial degrees of freedom due to higher anharmonicities lead to some instabilities because "magic" frequencies are present in the spectrum. Actually, additional resonances arise from linear combinations due to the arithmetic computations of the trigonometric functions in the second member of the equation of motion, as the fundamental frequencies are then commensurate. Kinetic energy transfer from one direction to another occurs and can overtake the value corresponding to the acceptance ellipse.

D. Sensitivity Improvement Using Broad-Band Excitation

In this new method, the excitation frequency is changed both slightly and rapidly instead of applying a reference frequency during a fixed period of time.[37] Each point of the measurement was obtained from the application of an excitation frequency for a period of 100 ms. For each point, the applied excitation frequency, ω_{exc}, was composed of ω_m plus a variable quantity, $\delta\omega$, which was changed 50 times during the time interval of 100 ms. In order to avoid memory effects which could lead to phenomena such as hysteresis, the magnitude of $\delta\omega$ was created randomly. While in the method described immediately above, V_{exc} was equal to 1 V at ω_m, in this broad-band excitation method, $V_{exc} = 0.1$ V and for the same duration of 0.1 s. V_{exc} can be represented by a sum of $V_{exc}(\omega_j)$; 256 different frequency values were possible, so that the set of these values covered 12.5 kHz, corresponding approximately to the width of the peaks obtained in the first method.

We obtained an increase of the efficiency of the ion detection. The point of optimum detection corresponds to the minimum of the absorption peak signal. This depth gives the ion number exactly. Once locked at this value, the experiment was done in both cases using either only one frequency or a variable frequency. Figure 12 shows peaks of the same depth, in spite of the voltage for the broad-band excitation 10 times smaller than that in the case of the single frequency. The advantage of this method is evident; because the voltage of the excitation is smaller, ion trajectories will evolve in an area in which the field is closer to the ideal case, and the signal will be of higher purity. The explanation for this observation is described in detail elsewhere.[37] The reader will recall that the principle of this method is based on optimum filtering.

E. Application to the Measurement of Ion Lifetimes

Once the frequency for the measurement has been chosen, in order to obtain the ion number with good precision it is necessary to accumulate the results of some 40 scans for a given experiment, i.e., for a given

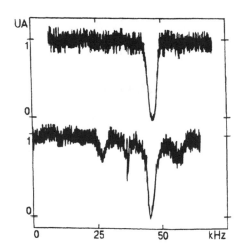

FIGURE 12

Absorption spectra of N_2^+ in response to a dipolar tickle. Lower spectrum, $V_{exc} = 1$ V, monochromatic excitation; upper spectrum, $V_{exc} = 0.1$ V, wide-band excitation. Note that the anharmonic components are strongly reduced due to reduction of the excitation amplitude.

confinement duration t_c. Then, the ion signal intensity is plotted vs. t_c. The ion lifetime, under a given set of conditions such as the presence of a given neutral, can be extracted from the decreased ion signal intensity when the confinement lifetime, which is limited by the imperfections of the system, is sufficiently great in comparison to the lifetime of the ion species under study.

V. SELECTIVE EJECTION BY RESONANT EXCITATION

This section shows the capability of a given tickle voltage to eliminate unwanted ion species from the ion trap.

A. Principle of Double-Resonance Excitation

In order to study the extent to which ion motion is modified by the application of a tickle voltage, we performed another experiment for which the time sequence is shown by Figure 13. Here, two pulses are applied between T_c and T_{off}. Ions are subjected to an initial pulse of a chosen but adjustable frequency, while the second pulse, for which the frequency is swept, analyzes the ion motion spectrum as described above.

B. Results

The ion motion spectrum was analyzed when the ions were subjected to a quadrupolar tickle of a frequency, T. As is readily predicted by a numerical computation,[2] the spectrum contains, in addition to the coupling

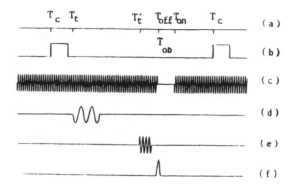

FIGURE 13

Timing sequence: (a) time scale, (b) ionization pulse starting at T_c, (c) RF drive voltage interrupted between T_{off} and T_{on}, (d) first tickle voltage starting at T_t permits selective excitation, (e) second tickle voltage of variable frequency, starting at T_t, which analyzes the ion cloud, (f) electron multiplier signal at T_{ob}.

already observed, new frequencies at $\omega_x \pm T$ and $\omega_z \pm T$, as shown in Figure 14. These frequencies exist in the ion motion only during the application of the tickle, so the time of the observation must coincide with the application of the tickle.

During simultaneous storage of N^+ and N_2^+, a quadrupolar excitation is carried out at the frequency ω_z relative to N^+. The efficiency of the selective ejection is shown in Figure 15, which corresponds to the excitation at ω_z for 10 ms.

VI. TIME-OF-FLIGHT (TOF) METHOD

Time-of-flight methods are often used in ion trap experiments. Once confined, ions are ejected following different protocols, and the profile is formed by the response of the electron multiplier vs. the time of arrival of the ions at the first dynode. The profile permits the measurement of either the ion number or, with the help of models of the ion cloud (based on spatial and velocity distributions), its kinetic energy. In the laboratory the authors developed a TOF method in order to study the ion position and velocity distributions.[38] In this method, the progression of the RF voltage was suddenly arrested, thus allowing the ions to escape. The measured profile of the ion TOF is related to the ion distribution in position and velocity at the moment when the confinement voltage ceases to vary and then is replaced by a direct current (DC) voltage with the same amplitude as the RF voltage at the time of interruption. The phase of interruption at this moment was adjustable from $-\pi/2$ to $+\pi/2$, in increments of $\pi/64$.

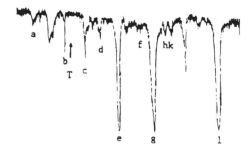

FIGURE 14

Absorption spectrum N· in the case of a double resonance quadupolar tickle. The frequency of the first tickle is T, the frequency of the second tickle is varied continuously. a, $\omega_x - T$; b, $\omega_z - T$; c, ω_x; d, $2\omega_x - T$; e, ω_z; f, $\omega_x + T$; g, $2\omega_x$; h, $2\omega_z$; j, $2\omega_z - T$; k, $\omega_z + T$; l, $2\omega_z$.

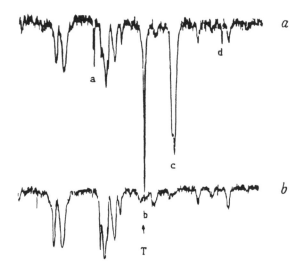

FIGURE 15

(a) Absorption spectrum of N· and N$_2^+$ for the case of a tickle of variable frequency in the range 0 to 250 kHz; (b) double resonance experiment as in Figure 14. The tickle frequency is ω_x(N·). The peaks labeled a, b, c, and d correspond to frequencies ω_x, ω_z, $2\omega_x$, and (ω_x +

A. Description of the Chosen Protocol

In the experiment described here, the authors chose to stop the confinement voltage at a value of the RF phase very close to $3\pi/2$. As consideration of the x, ẋ distribution shows,[21] it is for such an RF phase that the absolute value of the degree of correlation between the position and the velocity is at a maximum, and leads to a more clear TOF profile. Indeed, in such cases, for an ion with a positive, and respectively negative, velocity a good probability exists that it will have a positive, and respectively negative, position. This is the best condition for the prediction of the value of one quantity from the value of the other.

Due to the difficulty encountered in trying to suppress the confinement voltage rigorously at $3\pi/2$, we decided to superimpose on the confinement potential an additional DC voltage V_x. We worked with a residual potential that lies between 0.1 and 0.5 V. An example of the TOF profile obtained can be seen in Figure 16.

B. Model for the Expected Profile[39]

We have developed both analytical and simulation computation in order to calibrate the energy of the ion of interest. The simulation was three dimensional so that the computed ion trajectory realistically took account of the geometry of the system. An example of such profiles is shown in Figure 17.

C. Application to the Measurements of the Stored Ion Number

By integration of the TOF profile or by consideration of the profile maximum it is possible to measure the relative total ion number as a func-

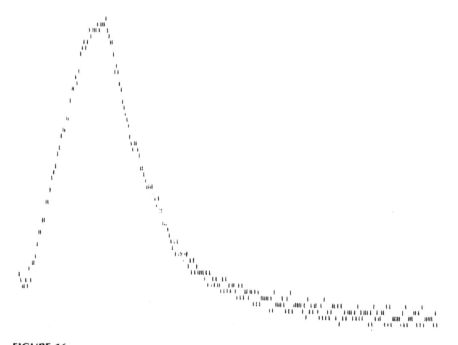

FIGURE 16

Experimental TOF profile of N· ions stored for 2 s. Data were obtained at intervals of 200 ns; 200 runs were accumulated.

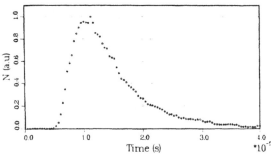

FIGURE 17
Simulated three-dimensional
TOF profile obtained for
3000 samples; the mean ion
kinetic energy is equal to
0.75 eV.

tion of a parameter, usually time. Of course, one must be assured of the mass purity of the ion cloud.

Parenthetically here it is of interest to note that when we tested our TOF method by measuring the ion number simultaneously with a classic resonance detection, we observed that the working point was displaced as if a negative voltage was applied to the end-cap electrodes. This effect is due to the presence of a strong voltage for accelerating the ions to the electron multiplier. This additional weak dipolar voltage adds a small anharmonicity.

D. Other Time-of-Flight Techniques

A TOF technique was developed at McGill University (Montreal, Canada).[40] An extraction pulse of 100 to 300 V, at any controllable RF phase, is used. Ions are driven into an extraction cavity at 600 V, with a ramp cavity DC bias at 200 V; the voltage ramp is -100 Vμs^{-1}. The net effect is to dilate the velocity scale. Typical TOF profiles are of some microseconds in width, i.e., somewhat shorter than ours.

VII. ION ENERGY MEASUREMENTS

Many methods can be used to investigate ion kinetics. In the following, we describe those methods for which the application to ion/molecule collisions is relatively easy. We do not intend to refer to Doppler measurements, which are specially devoted to spectroscopy.[41]

A. Ion Energy Measurements by Means of a Specific Ion/Molecule Reaction[42]

The basic idea of using a chemical thermometer was applied first in 1976.[43] Ion/molecule reactions in methane, water, and ammonia were studied, and it was found that ion energies lay in the range 1 to 3 eV in a Paul trap with an ambient pressure in the range of 10^4 to 10^5 mbar. Recently, the Ar^+/N_2 charge exchange reaction[44] and the reaction of O_2^+ with CH_4

were proposed[45] as having sufficiently well-established energy dependence. These measurements were made under high buffer–gas pressure conditions (10^{-3} mbar), with manipulation of the ions by moving the working point inside the stability diagram (i.e. by increasing the RF voltage until destabilization of the trapped ions occurs).

In order to determine the temperature of the N^+ ion species of interest, we used the N^+/CO charge exchange reaction. The cross-section for this reaction was measured using the techniques of guided beams;[46] the cross-section was found to vary from 100 to 1 Å^2 between 0.01 and 10 eV. A three-dimensional model of the ion cloud (Gaussian spatial and velocity distributions), which depends on the RF frequency, permits computation of this rate constant in the case in which the reaction occurs in an RF trap.

The total rate constant k of the reaction under study, the ion energy distribution $f(\mathbf{v})$, and the reaction cross-section are related as follows:

$$k = \int \sigma(\mathbf{v}) \, f(\mathbf{v}) \, |\mathbf{v}| \, d\mathbf{v} \tag{9}$$

where the vector $\mathbf{v}$ is the velocity in the center-of-mass system of the two collision partners.

When the ion velocity distribution is very simple (for instance, in the case where a beam is used) the calculation of k by means of Equation 9 is facile. In the RF trap, the velocity distribution is more complex due to the nature of the confining field.

The ion velocity distribution is represented by Gaussian laws. Therefore, these distributions carry the signature of the RF confining potential. They are not isotropic and depend on the RF phase of the alternating voltage V_{RF}; they can be expressed at zero RF phase ($t = kT_m$), and at any time t by using the transformations between $(x(t), \dot{x}(t))$ and $(x_0, \dot{x}_0)$.[21]

The three-component velocity distribution is given by

$$f_{\dot{X}(t), \dot{Y}(t), \dot{Z}(t)} \left(\dot{x}(t), \dot{y}(t), \dot{z}(t) \right)$$
$$= f_{\dot{X}(t)}(\dot{x}(t)) f_{\dot{Y}(t)} \left(\dot{y}(t) \right) \cdot f_{\dot{Z}(t)} \left(\dot{z}(t) \right)$$
$$= f_{V_i(t)} \left(\mathbf{v}_i(t) \right) \tag{10}$$

where

$$f_{\dot{X}(t)} \left(\dot{x}(t) \right) = \frac{1}{\sqrt{2\pi}\sigma_{vx}(t)} \exp\left(-\left(\dot{x}(t)^2 / 2\sigma_{vx}^2(t) \right) \right) \tag{11}$$

Because these instantaneous distributions are Gaussian, instantaneous temperature and instantaneous mean ion kinetic energy $E(t)$ can be expressed from the standard deviation of the velocity. For example, the part of $E(t)$ relative to the x component, $E_x(t)$, is equal to

$$E_x(t) = \frac{1}{2} m_{\text{ion}} \sigma_{vx}^2(t) \tag{12}$$

If we assume a Maxwell-Boltzmann distribution for the neutral reactant

$$f_{M_{VX \cdot VY \cdot VZ}}(V_x, V_y, V_z) = f_{V_M}(\mathbf{V_m}) \tag{13}$$

then the instantaneous value of k is obtained from

$$k(t) = \int \sigma(g)\, g f_{\mathbf{v_r}}(t)\, (\mathbf{v_i}(t))\, f_{V_M}(\mathbf{v_m})\, d\mathbf{v_m} d\mathbf{v_i} \tag{14}$$

where g is the modulus of the relative velocity; $g = |\mathbf{v_m} - \mathbf{v_i}|$.

Equations 10 and 11 show that the ion velocity distribution is characterized by two time-dependent parameters, $\sigma_{vx}(t)$ and $\sigma_{vz}(t)$; however, experiments give only values averaged over an RF period, thus $k(t)$ must be also averaged over an RF period, as

$$k = (\Omega/2\pi) \int k(t)\, dt \tag{15}$$

For physical significance, k must be understood as depending on the total ion kinetic energy. Let Σ_{vx} be the average over the RF phase of the root-mean-square velocity, for example, for the x-component

$$\Sigma_{vx} = (\Omega/2\pi) \int \sigma_{vx}(t)\, dt \tag{16}$$

The total time-averaged mean ion kinetic energy (in the laboratory system) is defined as

$$E = \frac{1}{2} m_{\text{ion}}\left(2\Sigma_{vx}^2 + \Sigma_{vz}^2\right) \tag{17}$$

Then the dependence of $k(t)$ on $\sigma_{vx}(t)$ and $\sigma_{vz}(t)$, as expressed by Equation 14, can be regarded as depending on E only, by virtue of Equation 17.

The precision of the calibration of E will depend on the precision of the experimental measurement of k on the one hand, and on the sample number used for the Monte Carlo simulation and the precision of the test reaction cross-section values, on the other hand. In order to evaluate the influence of the amplitude of the RF voltage, Equation 14 was calculated for different working points corresponding to the same global velocity parameters Σ_{vx} and Σ_{vz}. Although the instantaneous velocity of the ion and then the energy of the collision can be rather different, the simulated averaged rate constant k (which was of comparable value to the experimentally determined quantity) was found to be nearly constant for different q_z values on the q_z-axis, showing that the amplitude of the micromotion was neutralized.

B. Ion Energy Measurements by Means of a Time-of-Flight Method

The TOF techniques presented above are used with the help of simulation profiles[39,40] to determine the kinetic energy of the ion species of interest. In experiments described in Reference 39, N⁺ ions were stored with V_{DC} = 0.5 V, V_{RF} = 25 V, and $\Omega/2\pi$ = 500 kHz. Figure 18 shows that under our conditions, the N⁺ kinetic energy is equal to 0.7, +0.4, and –0.3 eV. In previously described work,[40] wherein the ion trap and the working conditions were different, as was the ion species, ^{39}K, and the conditions of ion creation required helium buffer gas cooling, the measured energy was of the same order, 0.3 V. The uncertainty was not reported. While this recent work[40] confirms the ion kinetic energy results obtained in Reference 29, a temperature variation vs. the ion number N, found in Reference 29 as $N^{2/3}$, was not confirmed by recent work.[39]

C. Ion Energy Measurements By Means of a Bolometric Method

A bolometric method was first discovered by Dehmelt and co-workers.[47,48] The method uses the resonant absorption of a passive circuit tuned to the axial oscillation frequency, consisting basically of an inductor connected across the capacitance between the ring electrode of the ion trap and the end-cap electrodes. The randomly varying potential differences due to the axial motion across this tank circuit give rise to a noise proportional to the square of the product n (ion number) and T (the ion temperature, with the assumption that the ions are thermalized). For such measurements, the noise voltage is amplified, square-law detected, and filtered to yield a DC voltage proportional to the noise temperature of the tank circuit.

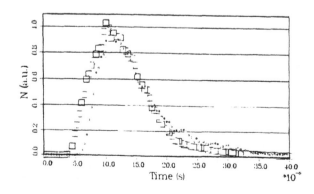

FIGURE 18
Adjustment of the experimental TOF profile with respect to the simulated profile.

VIII. ION ENERGY MANIPULATION

The versatility of the device is fully exploited when it is possible to vary ion kinetic energy. Actually, it can be interesting to cool ions in order to concentrate them near the center of the trap, to enhance the time of confinement, or to achieve enhanced sensitivity of ion detection. Otherwise, in the case of ion/molecule collisions studies, it is enough to fit the ion kinetic energy with the relevant values of the collisions.

In this section we discuss only those methods that are well suited for ion/molecule collision studies. For this reason, we do not take into account buffer gas cooling, which introduces unsuitable particles that lead to parasitic interactions and is largely presented elsewhere.[41] Neither do we treat the case of laser cooling, which is overly restrictive with respect to the attainable ion species and which leads only to cases of limited application (long distance interaction) and very low temperatures in the Kelvin or the sub-Kelvin range.

A. Ion Energy Variation by Means of Different Potential Well Depths

There are different modes by which one may vary potential well depth. One mode is to keep the same RF drive frequency and to move the working point in the stability diagram, for instance, along the q_z-axis by increasing the amplitude of the confinement voltage. The second mode is to change both the RF drive frequency and the confinement voltage.

An example of the first mode is illustrated by a second-order Doppler profile, which corresponds to the hyperfine transition of the fundamental level of the $^{173}Yb^+$,[12] vs. the q_z-axis (Figure 19), which leads to an averaged kinetic energy of 0.126 eV/eV of potential well depth.

FIGURE 19
Second-order Doppler profile of one hyperfine transition of the ground state of ^{173}Yb vs. the potential well depth (of spherical shape). From this result it is seen that ion kinetic energy, KE, is given by the relation KE = 0.126 D. (from Reference 12).

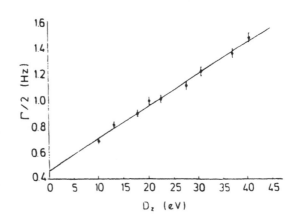

B. Ion Energy Variation By Means of Resonant Absorption of the Axial Kinetic Energy[49,50]

The bolometric method presented earlier can be used to monitor the kinetic (translational) energy of the ion cloud. In the absence of external heating, the Joule heating of the loss resistance of the circuit causes the temperature of the ions to approach exponentially that of the tank circuit with a time constant proportional to the mass and to the size of the trap. The time constants are rather long and the method can be inefficient when collision rates are of the same order.

C. Ion Energy Variation By Means of Resonant Excitation

Resonant excitation of the secular motion of an ion along the axis of a quadrupole ion trap can be affected by the absorption of power from an auxiliary generator. This process was employed initially by Paul et al.[51] and Fischer[52] as a method of detection of trapped ions. The onset of energy absorption by charged particles from an auxiliary oscillator can be observed visually.[53] While resonant excitation can be affected by monopolar, quadrupolar, and dipolar fields, dipolar excitation is used more commonly with commercial ion traps.[54] The supplementary or auxiliary RF potential oscillating at a selected frequency has been referred to as a tickle as opposed to the main RF potential, which is known as the drive potential. Dipolar excitation is discussed in some detail in Chapters 3 and 6.

Resonance excitation can be used to enhance stored ion kinetic energy and in the limit to eject ions from an ion trap. Mass-selective ion ejection may be brought about by prolonged resonant excitation. When the rate of enhancement of ion kinetic energy by resonance excitation is moderated, either endothermic ion/molecule reaction pathways can be accessed or ion kinetic energy can be transferred to internal energy through collisions with helium buffer gas and accumulated in the resonantly excited species to the point of ion fragmentation. Such fragmentation processes induced collisionally are commonly used in tandem mass spectrometry.

Thus far, we have discussed situations in which resonant excitation is carried out at constant RF drive voltage amplitude. However, a common mode of operation in commercial ion traps is to hold constant the resonant excitation frequency, corresponding to a q_z value slightly less than 0.908, and to ramp the RF drive amplitude. In this manner, several or all ion species can be ejected mass-selectively (at a working point close to the $\beta_z = 1$ stability boundary) as each species comes into resonance with the chosen tickle frequency and is ejected. This procedure is known as axial modulation. When the resonant excitation frequency is decreased, corresponding to a much lower value for q_z, axial modulation permits extension of the mass range of the ion trap.[55]

IX. A STUDY OF THE N⁺/N₂ REACTION[42]

A. Motivation and Description for the Chosen Ion/Molecule Reaction

Ion/molecule reactions are used for modeling the chemistry of gases, which is useful in many studies in aeronomy and astrophysics. The Paul trap, which is able to detect very small variations in ion number in a known and controlled environment, is certainly very convenient for the measurement of rate constants of such ion/molecule recations.

Our present interest lies in the measurement of the N^+/N_2 charge exchange reaction, because this reaction is one of the 24 reactions involved in physical-chemical models for high speed air flow-field computations. Therefore, it is necessary to produce an N^+ ion ensemble having the same energy properties as under atmospheric conditions and then to be as close as possible to the temperature conditions of the neutral molecule. The first step consists in measuring the kinetic energy of the cloud of stored N^+ ions.

If we assume that the N^+ ions created by electron impact are formed usually in the ground state, $N^+(^3P)$, in the presence of relaxed N_2, the four lowest-lying channels, together with the relative energy defects, are

$$N^+ + N_2 \rightarrow N_2^+\left(X^2\Sigma_g^+\right) + N\left(^4S\right) \quad -1.046 \; eV \tag{18}$$

$$N^+ + N_2 \rightarrow N_2^+\left(A^2\Pi_u\right) + N\left(^4S\right) \quad -2.17 \; eV \tag{19}$$

$$N^+ + N_2 \rightarrow N_2^+\left(B^2\Sigma_u^+\right) + N\left(^4S\right) \quad -4.22 \; eV \tag{20}$$

$$N^+ + N_2 \rightarrow N_2^+\left(B^2\Sigma_u^+\right) + N\left(^2P\right) \quad -4.62 \; eV \tag{21}$$

These different paths illustrate that various product channels may be accessed depending on the magnitude of reactant ion kinetic energy. Thus, a knowledge of ion kinetic energy is essential for the determination of a specific reaction rate constant under study.

B. Analysis of the Different Ion Loss Processes and Principle of a Rate Constant Determination[42]

As a rate constant is deduced from the rate of ion signal intensity diminution, it is necessary to determine all of the ion loss channels that contribute to this overall decrease. The other causes of N^+ consumption must be measured or known, or they must be eliminated. First, the stored ions do not experience ideal stable trajectories, thus they escape from the ion trap after a given time. Due to anharmonicities and space charge, the ion trap and the stored ion cloud cannot be described by an ideal representation; thus, one may define an "intrinsic lifetime", T_0, for ions in the trap. Second, residual pressure and neutral reactants are responsible for elastic collisions which can destabilize N^+ ions with the lifetime, T_c.

Finally, let τ_i be the lifetime corresponding to each contribution to the N^+ consumption at each instant. The remaining N^+ ion number, $N_r(t)$, is then given by

$$N_r(t) = A \exp\left[-t\Sigma\left(1/\tau_i\right)\right] \qquad (22)$$

Because the observed ion number, $N_r(t)$, can be approximated by an exponential law, it is possible to measure a decay time T_m given by

$$1/T_m = \Sigma k_{N^+ - A_i} p_{A_i} + k_{N^+ - N_2} p_{N_2} + 1/T_0 + 1/T_c \qquad (23)$$

where p_{A_i} are the partial pressures of the different species A_i present in the trap besides N_2 (pressure p_{N_2}). The eventual presence of CO (needed for the energy calibration) can be underlined by considering an additional term in Equation 23 as $k_{N^+-CO}p_{CO}$.

The slope k_m of the decrease of $1/T_m$ vs. p_{N_2}, when the partial pressures of all the components are maintained constant (eventually at the residual pressure), is

$$k_m = k_{N^+ - N_2} + k_c \qquad (24)$$

where k_c corresponds to the elastic collisions between N^+ and N_2 only.

As seen from Equation 24, a value for T_0, which is related to the confinement imperfections, is not required for the determination of the rate constants under study. Nevertheless, it must be significantly larger than the lifetimes corresponding to the phenomena investigated.

In order to determine the reactant N^+ ion kinetic energy in the N^+/CO charge exchange reaction, N^+ ions were stored in the presence of N_2 and CO.

The formation of N_2^+ and CO^+ ions during the electron pulse increases the charge density, which, in turn, can perturb the observed N^+ ion number. In order to improve the signal-to-noise ratio, it was decided to avoid the simultaneous confinement of N_2^+ and CO^+ ions. Selective ejection of mass 28 (N_2^+ and CO^+) was preferable to a displacement of the working point in the stability diagram, as the latter choice does not permit using the best zone within the stability diagram ($a_z \approx 0$, $q_z \approx 0.5^{13,23}$) for confinement and is not simple experimentally. Thus, a suitable tickle was applied at T_i as described above. The tickle, for which the frequency corresponds to one component of the motion spectrum of the N_2^+ and CO^+ ion species is of some volts in amplitude, depending on the well depth. The tickle voltage is applied for 1 s prior to the beginning of the reaction time.

C⁺ and O⁺ ions exist in small quantities, but the mass resolution of our detection method is sufficiently large to separate these species from N⁺. It was verified experimentally that in this mass range we are able to have a mass resolution for singly charged ions of unity. At the prevailing low ion density, the different species are not coupled by space charge. The N_2^{2+} ions (not discernible from N⁺) react very rapidly with N_2 and are eliminated during this period as has already been explained.

The observed value for T_0 is >10 s, while the N^+/N_2 charge exchange is very slow (ca. 10^{-12} cm³/s). The value of k_c due to N_2 and CO is determined by investigating the contribution of pure elastic ion/molecule collisions.[42] The N⁺ ion lifetime was measured in the presence of noble gases (He, Ne, and Ar), in which collisions are essentially elastic and k_c was evaluated by interpolation.

Once the relaxation is achieved, the components of the ion position and velocity components are Gaussian as mentioned above, and the ion population decrease is exponential due to "natural evaporation" because of the imperfections of the ion trap. In the experiment presented here, this exponential decrease is observable after a storage time close to 3 s (Figure 20).

For a pseudopotential well depth of 4 eV, the value of the measured rate constant was found to be $k_m = 0.43 \times 10^{-9}$ cm³/s $\pm 0.04 \; 10^{-9}$ cm³/s. The quoted errors are the statistical uncertainties from the ion number measurement. This result leads to a value for $k_{N^+-CO^+}$ once allowance has been made for the principal reaction under study and corrected for the 10% contribution corresponding to the NO⁺/C channel; $k_{N^+-CO^+} = 0.29 \times 10^{-9}$ cm³/s ($\pm 0.07 \times 10^{-9}$ cm³/s). Here, the quoted errors include the pressure uncertainties.

FIGURE 20
Decrease in N⁺ ion number vs. time in the presence of N_2 for different pressure conditions (O, 3.8×10^{-9}; Δ, 7.6×10^{-9}; and □, 2.3×10^{-8} mbar), in a semilogarithmic representation. The changes in the slopes of the curves indicate when equilibrium has been attained. (From Münch, A. et al., *Phys. Rev.*, A35, 4148. 1987. With permission.)

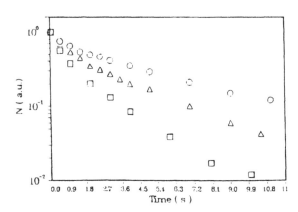

C. Results

The TOF measurement led to an ion kinetic energy of 0.65 ± 0.07 eV. This value is very close to the value found with the chemical thermometer. Figure 21 represents the computations of Equation 15 via Equation 14. Then the ion kinetic energy which corresponds to the measured k_{N-CO} rate constant is equal to 0.7 eV (+ 0.4, −0.3 eV).

X. CONCLUSION

We have shown how a homemade ion trap is extremely well suited to the study of ion/molecule collisions under UHV conditions. To this end, it is necessary to have a good knowledge of the ion properties and to employ them in the development of specific tools for experimental investigations. We have presented, in particular, an original ion detection method that is very sensitive; this method does not require displacement of the working point and yet gives sufficient mass resolution. Moreover, this method is able to furnish (for a given experiment, i.e., for a given collision study) important information on the ion cloud and on the quality of the storage device used. We have described two methods for the investigation of the ion kinetic energy; these methods were used on the same ion species under the same conditions of confinement, and led to identical ion kinetic energies.

Our experience has been applied on the N^+/N_2 charge transfer reaction. The N^+/N_2 collision is endoergic by 1.04 eV when the reactants are in the ground state. Our results show that reactant N^+ ion energy is insufficient to overcome the endothermicity of the collision; therefore, heating of the reactant ion species by one of the methods described above is imperative.

Study of such collisions from the point of view of investigating hypersonic physics also requires measurement in nonequilibrium conditions.

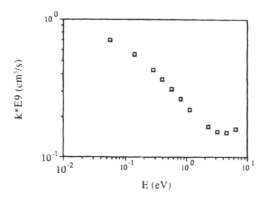

FIGURE 21

Computed rate constants for the N^+-CO charge exchange reaction in a RF quadrupole ion trap vs. the N^+ incident energy in the laboratory frame.

Thus, it will be of interest to investigate methods for the introduction of known amounts of excited N_2 into the ion trap.

PART B
SELECTIVE ION/MOLECULE REACTION
CHEMISTRY IN A QUADRUPOLE ION TRAP

Jennifer S. Brodbelt

XI. INTRODUCTION

Over the past two decades, increasing attention has focused on both the fundamental study of ion/molecule reactions and the use of selective ion/molecule reactions for analytical applications.[56] The understanding of the mechanisms, kinetics, and thermochemistry, as well as the details of substituent effects, functional group interactions, and sites of reactions, provides a foundation for the rational design of selective ion/molecule reactions for a variety of analytical applications. The advent of ion trapping methods[57] has allowed many features of ion/molecule reactions to be investigated in detail and has offered further evidence that the use of ion/molecule reactions offers new approaches to the solution of complex problems in mass spectrometry. One feature of the quadrupole ion trap that makes it an especially versatile device for the study of ion/molecule reactions is the ability to operate at a wide range of pressures, making pressure-resolved studies facile. Several supplementary techniques have proven particularly valuable for undertaking such studies of ion/molecule reaction chemistry. The mass-selective isolation mode[58] allows reactive ions to be evaluated individually for mapping of specific reaction pathways. MS" techniques,[59] including collisionally activated dissociation (CAD), permit elucidation of dissociation routes of product ions formed from selective ion/molecule reactions. Moreover, selective deuterium labeling and ligand exchange techniques are important for the determination of reaction sites, mechanistic details, and thermodynamic parameters. In this part of the chapter, the evaluation of both fundamental and analytical aspects of gas-phase bimolecular reactions are illustrated in conjunction with quadrupole ITMS. The topics covered in this part include the characterization of novel selective chemical ionization reagents, the determination of specific sites of reaction, the examination of structural factors that mediate reactivity, the evaluation of host–guest chemistry in the gas phase, and the study of functional group interactions and kinetic energies of ions.

XII. CHARACTERIZATION OF SELECTIVE CHEMICAL IONIZATION REAGENTS: DIMETHYL ETHER

The popularity of chemical ionization[60] for analytical applications has promoted interest in the characterization of novel reagent gases for more selective and sensitive analysis.[61] One reagent, dimethyl ether, recently was proven to show special selectivities toward substituted aromatic compounds in a quadrupole ion trap.[62] The population of reactive ions generated by electron ionization of dimethyl ether in a quadrupole ion trap at nominally 5×10^{-6} torr dimethyl ether consists of ions at m/z 45 $(CH_2=OCH_3)^+$ and m/z 47 (CH_3OCH_3) H^+. When these reagent ions were permitted to react with neutral aromatic compounds possessing various substituents introduced into the trap, three types of products were formed: $(M + H)^+$, $(M + 13)^+$, and $(M + 15)^+$ (see Table 1). By selective isolation of each reactive ion from dimethyl ether, it was shown that the (CH_3OCH_3) H^+ ions led to the formation of protonated analyte molecules, whereas the $(CH_2OCH_3)^+$ ions led to the formation of $(M + 13)^+$ and $(M + 15)^+$ analyte ions.

All the aromatic compounds protonated readily, but those with an ether or alcohol functional group preferentially formed $(M + 13)^+$ via [M + (CH_2OCH_3) – $CH_3OH]^+$, while those with a carbonyl functionality instead formed $(M + 15)^+$ via [M + (CH_2OCH_3) – $CH_2O]^+$. No monofunctional aromatic compound produced both $(M + 13)^+$ and $(M + 15)^+$ adducts. The ortho-, meta-, and para-isomers of the bifunctional aromatic compounds reacted via either the formation of $(M + 13)^+$ or $(M + 15)^+$, while only one of the meta-isomers, m-methoxyacetophenone, generated both types of adducts. The abundance of these adducts relative to the protonated products was typically 20 to 30%. For those compounds with multiple functional groups, $(M + 15)^+$ was produced for the meta- and para-isomers if at least one of the functionalities was a carbonyl group, while $(M + 13)^+$ was formed if the two functional groups were in ortho-positions or if neither functionality was a carbonyl group.

TABLE 1

Reactions of Organic Substrates with Dimethyl Ether Ions

Compound	$(M + 1)^+$	$(M + 13)^+$	$(M + 15)^+$
Anisole	+	+	NO
Phenol	+	+	NO
Benzaldehyde	+	NO	+
Acetophenone	+	NO	+
ortho-Hydroxyacetophenone	+	+	NO
meta-Hydroxyacetophenone	+	NO	+
para-Hydroxyacetophenone	+	NO	+

ᵃ + indicates products are formed.

Source: Adapted from Brodbelt, J. et al., Anal. Chem., 63, 1205, 1991.

It was shown that this positional selectivity was influenced by the deactivating and activating directional properties of the substituents toward electrophilic aromatic substitution. Carbonyl substituents, which are *meta*-directing and deactivating substituents, promoted the formation of (M + 15)⁺ ions, whereas the hydroxy and methoxy substituents, which are *ortho-/para*-activators, induced the formation of (M + 13)⁺ ions. The *ortho*-isomers (o-hydroxybenzaldehyde and o-hydroxyacetophenone) have two directors that both enhance the same aromatic position (adjacent to the hydroxy group) for methylene addition. For the *para*-substituted isomers, the functional groups are spatially separated, and the favored product was directed by the absence or presence of a carbonyl group.

The unusual selectivities observed for dimethyl ether chemical ionization in a quadrupole ion trap were not preserved in a conventional chemical ionization source. For example, from studies undertaken in a standard quadrupole mass spectrometer with dimethyl ether admitted to the ion source at nominally 1 to 2 torr, a mixture of (M + H)⁺, (M + 13)⁺, and (M + 15)⁺ ions were observed for many of the aromatic substrates. The absence of selectivity was attributed to the higher pressure of the conventional ion source, which promoted efficient stabilization of all the adducts formed, and thus diminished the distinction of small kinetic or energetic differences in reaction pathways.

XIII. RATIONALE FOR SELECTIVITY OF CHEMICAL IONIZATION REACTIONS

From a fundamental perspective, it is important to understand the mechanisms of chemical ionization and the details of the selectivity of ion/molecule reactions involved in adduct formation. Such an understanding can provide a basis for a rational approach to developing alternative chemical ionization agents. To elucidate the basis for the functional group-selective methylene substitution reactions of dimethyl ether observed in the quadrupole ion trap, an evaluation of the selective behavior of other methylene-donating reactive species, including ions from ethylene and ethylene oxide, was recently reported.[63] Ion/molecule reactions between these three reagent gases and simple aromatics, including both mono- and disubstituted oxyaromatics, were investigated in detail to determine functional group selectivity.

Based on CAD studies of adduct ions formed from ion/molecule reactions of dimethyl ether ions, ethylene oxide ions, or ethylene ions, the pathways leading to the formation of structurally selective product ions for the aromatic substrates (M) were elucidated. For example, the (M + 45)⁺ adducts from dimethyl ether were determined to be covalently bound precursors to both (M + 13)⁺ and (M + 15)⁺ ions, and the (M + 41)⁺ adducts of ethylene were precursors to (M + 13)⁺ and (M + H)⁺ ions. The formation of the various aromatic adduct ions by using ethylene or ethylene oxide as alternative reagent gases did not follow an obvious pattern based on substituent selectivity, nor did the formation of $(M + 13)^+$ duplicate the selectivity trends observed for the formation of $(M + 13)^+$ using dimethyl ether.

The methylene substitution selectivity was observed only for dimethyl ether ions and not with ethylene oxide or ethylene ions. When using $CH_2 = CHCH_2^+$ from ethylene or protonated ethylene oxide to promote methylene substitution reactions in the quadrupole ion trap, all of the monosubstituted aromatic compounds produced $(M + 13)^+$ ions, regardless of the nature of the substituents. Because of the unusual methylene substitution selectivities, the exothermicities of the different methylene substitution reactions were calculated from estimated heats of formation of the reactants and heats of formation of the products.[64] In general, the overall exothermicities for the methylene substitution reactions involving the $CH_2 = CHCH_2^+$ and $(ETOX + H)^+$ ions were about 8 to 10 kcal/mol more exothermic than the corresponding reactions involving $CH_3OCH_2^+$. Additionally, methyl addition was a competitive exothermic process accessible to the $CH_3OCH_2^+$ ions, which was not an available reaction channel for the reactive $CH_2 = CHCH_2^+$ and $(ETOX + H)^+$ ions. In general, the methoxy and hydroxy substituents are electron-donating groups that enhance the electrophilic substitution of the aromatic ring. However, the aromatic compounds with electron withdrawing substituents (those containing carbonyl) deactivated the electrophilic addition process. This latter effect made the methylene substitution reaction less exothermic than the competing methyl addition reaction for acetophenone and benzaldehyde.

Thus, the selectivity of methylene substitution of the three reactive species ($CH_3OCH_2^+$, $CH_2 = CHCH_2^+$, $(ETOX + H)^+$) was attributed to the overall substituent-directed favorabilities of the electrophilic aromatic additions coupled with the exothermicities of the reactions. The methylene-donating ions generated from ethylene or ethylene oxide underwent reactions with the aromatic substrates that were more exothermic, and thus were better able to drive the methylene substitution process for all the aromatic compounds, regardless of the electron-donating or releasing capabilities of the substituents. For the reactive ions from dimethyl ether, the methylene substitution reactions were less exothermic, and simple methyl transfer was an effective competing process.

XIV. METHYLATION REACTIONS OF LACTONES

Minor structural differences of related organic molecules may promote large variations in their reaction and dissociation characteristics. Thus, the use of selective ion/molecule reactions to derivatize molecules in the gas phase may lead to analytically useful methods of differentiating classes of compounds or even isomers. For example, the addition of a methyl cation (in contrast to a proton) to an organic substrate may affect specifically those reactions that involve sterically hindered transition states or formation of entropically unfavorable products. With this principle in mind, gas-phase methylation reactions of an array of lactones were examined using the ITMS™ with the objective of correlating the differences in product ion structures based on characteristic dissociation pathways.[65] The characterization of lactones is significant for several reasons. First, they are biologically relevant molecules, having particular importance in the metabolism of carbohydrates, adenosine triphosphate (ATP), and NADPH, and they have antitumor activities and cytotoxic properties.[66] Mass spectrometric studies of lactones generally have been limited to characterization of the fragmentation behavior of lactones upon electron ionization (EI),[67] spectra with few differentiating features. In order to develop an improved means of distinguishing related lactones, the lactone were ionized in a quadrupole ion trap by gas-phase methyl cation addition, an alternative to protonation, and characterized by CAD techniques.[65]

The structures and dissociation reactions of gas phase protonated and methylated four-, five-, six-, and seven-membered ring lactones, some with methyl substituents in various positions, were studied in order to correlate ring size and substituent effects. Protonated lactones showed much more complex energy-resolved breakdown curves than did protonated cyclic ketones or ethers of similar ring size. Losses that appeared consistently included water at 28, 42, 44, and 46 u, attributed to sequential losses of H_2O and CO. The structures of the resulting ions were determined through a combination of tandem mass spectrometry (MS/MS) experiments, deuterium labeling studies, and comparison of energy-resolved breakdown curves. For example, sequential activation (MS^n) of protonated γ-valerolactone (m/z 101) demonstrated that dehydration of the protonated analyte was followed by decarbonylation (see Figure 22 and Scheme 1). The resulting fragment ion at m/z 55 was shown to be an unsaturated hydrocarbon ion based on its CAD pattern (including loss of H_2, CH_4, and C_2H_2). In general, the protonated lactones could be grouped into two categories based on one aspect of their dissociation behavior. The γ-valerolactone and β-methyl–γ–butyrolactone (six- and seven-membered rings) did not dissociate via loss of 44 u, C_2H_4O; whereas, the four- and five-membered rings and α-methyl–γ–butyrolactone dissociated via loss of 44 u. This distinction proved to be an important feature in the comparison of the gas-phase methylated lactones.

Reactions of the lactones with $CH_3OCH_2^+$, a methyl cation donor, led to formation of $(M + CH_3)^+$ ions, and structural differences of lactone isomers were characterized by CAD. Representative CAD spectra of the $(M + H)^+$ and $(M + 15)^+$ ions are shown in Figure 23 for two of the methyl-substituted five-membered ring lactones, α-methyl–γ–butyrolactone and γ-

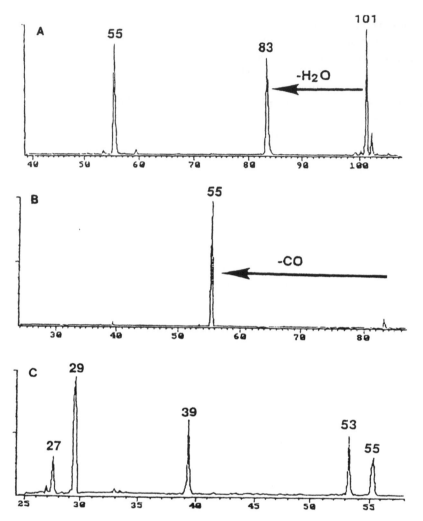

FIGURE 22
Sequential activation of m/z 101, then m/z 83, then m/z 55 from protonated γ-valerolactone. (From Donovan, T. and Brodbelt, J., *J. Amer. Soc. Mass Spectrom.*, 3, 52, 1992. With permission.)

SCHEME 1
Proposed mechanism for dehydration/decarbonylation route of protonated γ-valerolactone. (From Donovan, T. and Brodbelt, J., *J. Amer. Soc. Mass Spectrom.*, 3, 52, 1992. With permission.)

valerolactone (Table 2). Based on their dissociation behavior, the gas-phase methylated lactones were grouped into two distinct categories. For the six- and seven-membered rings, and β- and γ-methyl-substituted five-membered ring lactones, the dissociation behavior of the methylated species was directly analogous to that of the protonated species. For example, for the gas-phase methylated γ-valero, β-methyl–γ-butyro, δ-valero, and ε-capro lactones (represented by Figure 23b), the predominant loss was via 60 u (net loss of methanol and CO), and the resulting "fragment" ion had the same m/z ratio as the predominant ion from dissociation of the protonated molecule (net loss of 46 u, water and CO). This behavior suggested that proton and methyl attachment sites were the same, and both protonated and methylated species dissociated by analogous pathways. The methylation reaction apparently did not dramatically alter the energetically or entropically accessible dissociation routes of this group of lactones.

However, for the remaining lactones (most of the smaller ring sizes including β-propio, β-butyro, γ-butyro, and α-methyl–γ-butyro lactones) the effect of gas-phase methylation on the dissociative behavior was dramatic. The major product ion from collisional activation of the $(M + 15)^+$ adducts was always m/z 59, and alternative dissociation routes were not competitive. This behavior was not analogous to that of the corresponding $(M + H)^+$ ions, as evident from the example shown in Figure 23a. The structure of the intriguing ion of m/z 59 was elucidated by MS/MS/MS experiments. For example, the $(M + CH_3)^+$ ion of α-methyl-γ –butyrolactone was activated, producing the fragment of m/z 59, as shown in Figure 24. Activation of the ion of m/z 59 led to a CAD spectrum that matched the CAD spectrum obtained from a model compound, 3-methoxy-1-butanol. A mechanism for the formation of m/z 59 is illustrated in Scheme 2.

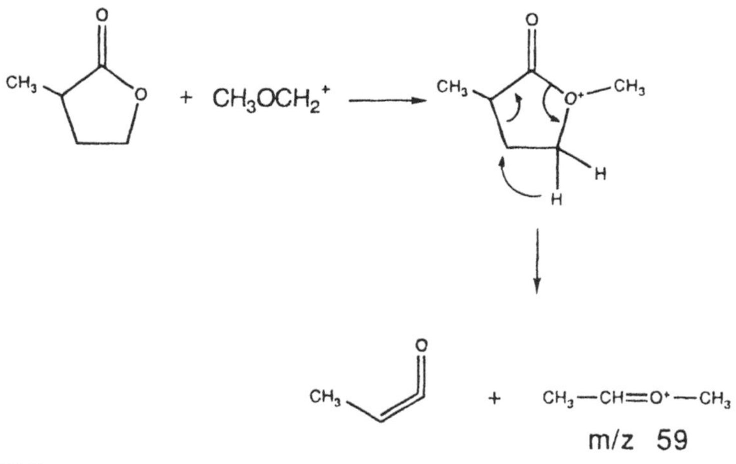

SCHEME 2
Formation of m/z 59 from methylated α-methyl-butyrolactone. (From Donovan, T. and Brodbelt, J., *J. Amer. Soc. Mass Spectrom.*, 3, 58, 1992. With permission.)

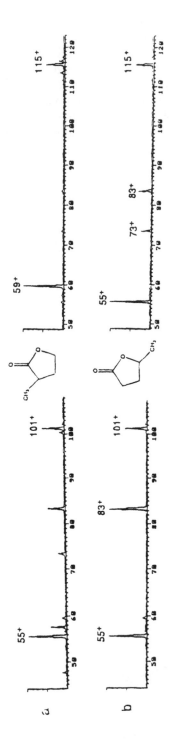

FIGURE 23

CAD spectra of protonated (*m/z* 101) and methylated (*m/z* 115) (a) α-methyl–γ–butyrolactone and (b) γ-valerolactone. (From Donovan, T. and Brodbelt, Jr., J. *Amer. Soc. Mass Spectrom.*, 3, 57, 1992. With permission.)

TABLE 2

CAD Spectra of (M + H)⁺ and (M + 15)⁺ Ions of Lactones[a]

Lactone (m/z of (M + H)⁺, m/z of (M + 15)⁺)	m/z of fragment ions							
	(M + H)⁺			(M + 15)⁺				
β-Propiolactone (73, 97)	43	45	*55*	41	43	45	55	*59*
β-Butyrolactone (87, 101)	*43*	45	69	*59*				
γ-Butyrolactone (87, 101)	*43*	45	69	*59*	83			
α-Methyl–γ–butyro (101, 115)	*55*	57	59	73	83	*59*		
β-Methyl–γ–butyro (101, 115)	*55*	59	83	*55*	73	83		
γ-Valerolactone (101, 115)	*55*	59	83	*55*	73	83		
δ-Valerolactone (101, 115)	*55*	59	83	*55*	73	83		
ε-Caprolactone (115, 129)	*69*	73	97	*69*	87	97		

[a] Acquired with an ITMS™; most abundant ions are italicized.

Source: From Donovan, T. and Brodbelt, J., J. Amer. Soc. Mass Spectrom., 3, 58, 1992. With permission.

The groupings of the gas-phase methylated lactones based on dissociation behavior duplicated the groupings used to distinguish the protonated lactones based on dissociation behavior. Thus, there was a striking dissociative distinction related to ring size and location of the methyl substituent among the series of lactones. Several explanations for the striking differences in the reaction and dissociation behavior of the two groups of lactones were proposed. One group of the lactones either had a smaller ring size or a methyl substituent occurred adjacent to the carbonyl position. It was suggested that these structural differences result in very different energetic and entropic barriers which affected the preferred site of methyl addition relative to the favored site of protonation. For example, protonation could have occurred at the carbonyl oxygen, whereas methylation occurred at the ether oxygen for the group of smaller ring-size lactones. However, it was suggested that the group of lactones of larger ring size protonated and methylated at the same site, whereas the results did not support this for the lactones of smaller ring size. Second, it was proposed that the addition of a bulky, highly polarizable methyl cation to the lactones of smaller ring size could have been more favorable for inducing ring opening, resulting in acyclic products. In general, the methyl cation addition process provided more information about distinguishing

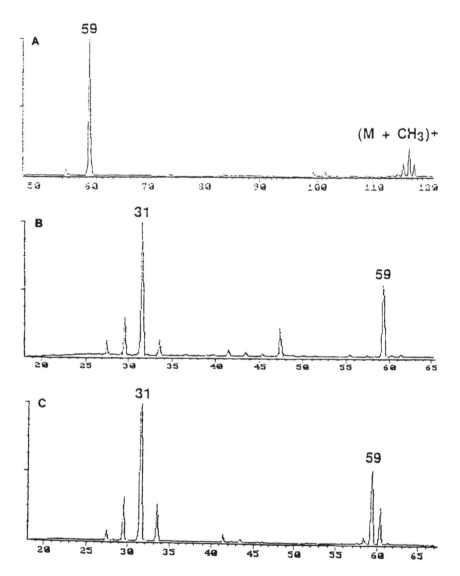

FIGURE 24

Sequential CAD spectrum for isolation and dissociation of m/z 59 from dissociation of m/z 115 of methylated α-methyl–γ–butyrolactone and CAD spectrum of the m/z 59 ion from α-cleavage of 3-methoxy-1-butanol. The ion at m/z 47 in spectrom B is due to protonated dimethyl ether. (From Donovan, T. and Brodbelt, J., *J. Amer. Soc. Mass Spectrom.*, 3, 58, 1992. With permission.)

groups of related lactones than about distinguishing individual structurally similar lactones.

XV. SITES OF REACTIONS: BICYCLIC SUBSTRATES

Gas-phase ion/molecule reactions are analytically useful not only because of the molecular structural derivatization that occurs, but also because small differences in the structures of complex molecules may result in dramatic differences in reactivities. In spite of the increasing number of studies that have addressed the use of selective ion/molecule reactions for analytical applications,[68] the sites of gas-phase protonation and cation attachment in the formation of products of ion/molecule reactions still remain largely undetermined, and only a few reports have addressed this problem.[69] To assist in the design of ion/molecule reactions that are structurally diagnostic for complex molecules, it is useful to understand the specific sites of reaction. Typically, the reactions of smaller building-block units of larger molecules must be examined to pinpoint the reaction site. In fact, from examination of subunits of complex molecules, sites of reactions, the influence of neighboring functional groups, and other structural factors that mediate reaction pathways can be evaluated.

A recent study[70] examined whether 2,5-dimethylpyrrole–γ–butyrolactone underwent the bimolecular and dissociation reactions that were characteristic of both simple lactones and pyrroles, or whether one of the heteroatomic rings preferentially mediated the bimolecular and dissociation behavior, and the site of ion/molecule reaction initiation was assigned based on comparative studies of the selective bimolecular and dissociation reactions of model compounds. 2,5-Dimethylpyrrole–γ–butyrolactone, an analog to a potent plant growth stimulator, consists of two component rings that can be represented easily by model compounds: γ-butyrolactone and 2,5-dimethylpyrrole.

2,5-Dimethylpyrrole–γ–butyrolactone formed $(M + 13)^+$ ions upon reaction with dimethyl ether ions, ($CH_3OCH_2^+$), in the ion trap mass spectrometer. A mechanism was proposed to illustrate the process (see Scheme 2). In order to determine the site of methylene substitution in the formation of $(M + 13)^+$ ions, experiments in which ion/molecule reactions with model compounds representative of each half of the pyrrole lactone (2,5-dimethylpyrrole and butyrolactone) were compared. Upon dimethyl ether chemical ionization, both model compounds protonated, but 2,5-dimethylpyrrole formed an $(M + 13)^+$ adduct, whereas γ-butyrolactone formed an $(M + 15)^+$ adduct (see Figure 25). Because the 2,5-dimethylpyrrole–γ–butyrolactone and 2,5-dimethylpyrrole exhibited similar reaction chemistry, the pyrrole ring was determined to be the director and substrate for formation of $(M + 13)^+$ ions.

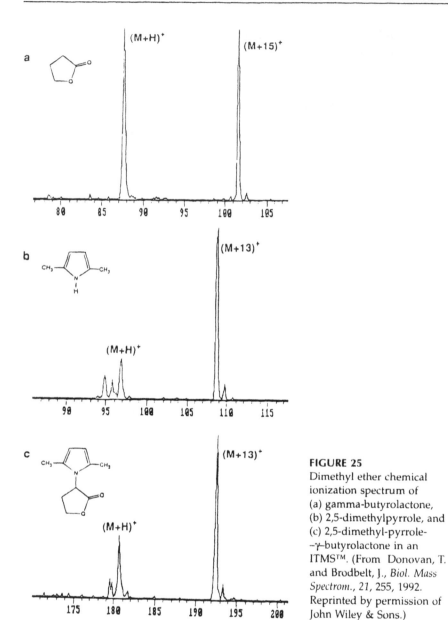

FIGURE 25
Dimethyl ether chemical
ionization spectrum of
(a) gamma-butyrolactone,
(b) 2,5-dimethylpyrrole, and
(c) 2,5-dimethyl-pyrrole-
–γ–butyrolactone in an
ITMS™. (From Donovan, T.
and Brodbelt, J., *Biol. Mass
Spectrom.*, 21, 255, 1992.
Reprinted by permission of
John Wiley & Sons.)

To elucidate the specific site on the pyrrole ring of 2,5-dimethylpyr-role–γ–butyrolactone which attacked the methoxymethylene cation, various saturated and substituted analogs of 2,5-dimethylpyrrole were allowed to react with dimethyl ether ions in the ion trap. Potential reactive sites included the nitrogen atom and the double bonds. Methylene substitution products at (M + 13)· were formed only for compounds con-

TABLE 3

Product Distribution for Reactions of Dimethyl Ether Ions
with Model Compounds

Compound	(M + 13)⁺	(M + 15)⁺	(M + 45)⁺
2,5-Dimethylpyrrole-γ-butyrolactone	Yes	No	No
2,5-Dimethylpyrrole	Yes	No	No
γ-Butyrolactone	No	Yes	Yes
1-Methylpyrrole	Yes	No	No
2,5-Dimethylpyrrolidine	Yes	No	No
1-Methylpyrrolidine	No	Yes	Yes
Cyclohexene	Yes	No	No
Pyrrole	Yes	No	No

Source: From Donovan, T. and Brodbelt, J., *Biol. Mass Spectrom., 21,*
256, 1992. Reprinted by permission of John Wiley & Sons.

taining a double bond or a secondary amine (see Table 3). For those compounds with neither a double bond nor a secondary amine (1-methylpyrrolidine), (M + 15)⁺ and (M + 45)⁺ products were formed. Interestingly, the (M + 45)⁺ adduct, which had been identified from a previous study[62] as the appropriate precursor to (M + 13)⁺ product ions, did not dissociate to produce (M + 13)⁺ ions after collisional activation of the adduct. Presumably this failure was because of the absence of an appropriately positioned proton needed for the elimination of methanol. For methylene substitution at a nitrogen site, a proton must be attached directly to the heteroatomic site to facilitate the specific elimination of methanol. Such site-selective behavior precluded the formation of (M + 13)⁺ ions from any simple tertiary amine that did not possess an alternate reactive site such as a double bond. On the basis of this evaluation of model compounds, it was apparent that the only structural feature of , 5-dimethylpyrrole-γ-butyrolactone that was capable of promoting the methylene substitution reaction was one of the double bonds, indicating that it was the reactive site. As shown in Scheme 3, a double bond of the pyrrole lactone attacks the positively polarized methylene group of the dimethyl ether ion. This step is fol-

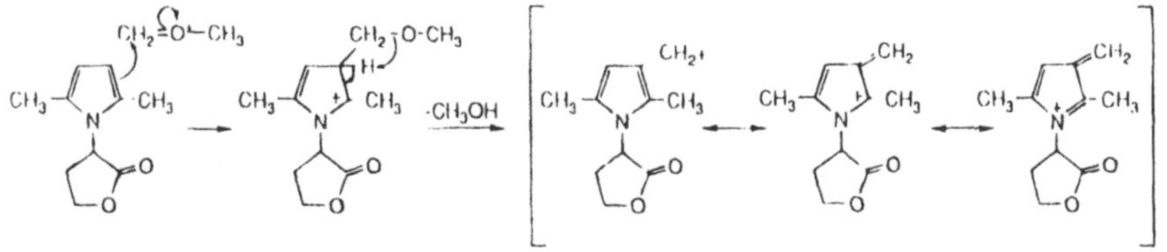

SCHEME 3
Mechanism of formation of (M + 13)⁺ ions for 2,5-dimethylpyrrole–γ-butryolactone. (From Eichmann, E. S. et al., *Amer. Soc. Mass Spectrom., 3,* 537, 1992. With permission.)

lowed by spontaneous elimination of CH_3OH, resulting in a resonance-stabilized $(M + 13)^+$ ion.

XVI. FUNCTIONAL GROUP INTERACTIONS: DIOLS AND AMINO ALCOHOLS

Historically, the influence of neighboring group interactions has been an active area of research in organic chemistry in solution.[71] Likewise, there have been numerous accounts of the importance of functional group interactions of many types of difunctional molecules,[72] including diols, diacids, diesters, and amino alcohols, in gas phase reactions with positive ions. For example, the hydroxyl and amino groups create nucleophilic sites in molecules and may participate in hydrogen-bonding interactions. Simple diols and amino alcohols can serve as first-order models of more complex biologically relevant molecules which contain multiple hydroxyl groups such as glycosides.

The effects of various functional group interactions on ion/molecule reactions of simple diols[73] and amino alcohols[74] in a quadrupole ion trap were recently reported. Product ions were characterized via CAD techniques and by comparison with the fragmentation behavior of model compounds. The primary objectives were to correlate functional group interactions with the product distributions of selected ion/molecule reactions involving dimethyl ether ions and to determine if the difunctional molecules and their structurally modified product ions dissociated in ways that reflected differences in functional group interactions.

The reactive ions generated from dimethyl ether, $CH_3OCH_2^+$, at m/z 45 promoted methylene-substitution reactions of the diol and amino alcohol substrates. A representative mechanism for the reaction of the methoxymethylene cation with 1,2-propanediol was proposed (see Scheme 4). It was suggested that the methylene substitution likely proceeded via the initial formation of an ion/molecule complex of $CH_3OCH_2^+$ and the diol (see Scheme 4), with the primary electrostatic interactions occurring between the lone pair electrons on the oxygen atoms of the diol and the positively polarized methylene group of the methoxymethylene cation. The reaction proceeded by nucleophilic attack on the $CH_3OCH_2^+$ by eiher oxygen atom of the diol, resulting in a complex, $(M + CH_3OCH_2)^+$, which was not observed as a stable species in the quadrupole ion trap. Elimination of methanol occurred spontaneously by abstraction of a proton from an oxygen atom by the methoxy group, resulting in the product ion at $(M + 13)^+$. It was suggested that stabilization of the acyclic product ion may be provided by intramolecular interaction of the unmodified hydroxy group with the positively polarized methylene end group, resulting in a cyclic structure.

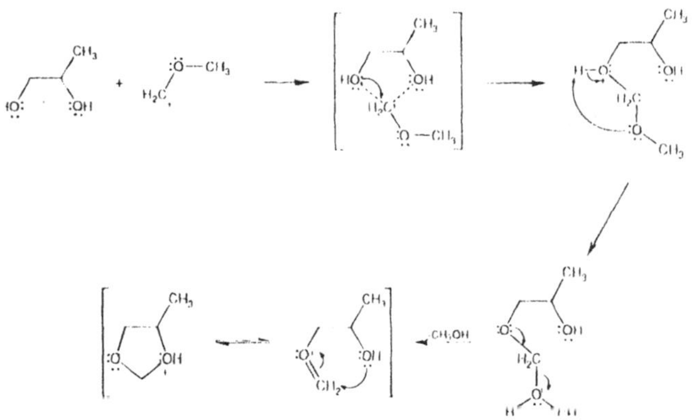

SCHEME 4

Proposed mechanism for the formation of (M + 13)· ions of 1,2-propanediol.(From Donovan, T. and Brodbelt, J., *Biol. Mass Spectrom., 21,* 255, 1992. Reprinted by permission of John Wiley & Sons.)

The relative product distributions of ions generated from ion/molecule reactions with dimethyl ether ions are listed in Table 4 for eight diols having variable hydroxyl group separation and carbon chain length. As the number of carbon atoms formally separating the hydroxyl groups increased, the extent of methylene substitution as compared to proton addition decreased. For example, methylene substitution and proton addition were equally favorable for 1,2-ethanediol and 1,2-propanediol, but the methylene substitution was not observed at all for 1,5-pentanediol or 1,6-hexanediol. When the total carbon chain length was held constant but

TABLE 4

Product Distributions for Ion/Molecule Reactions of Diols with Dimethyl Ether Ions

| Diol | % Total ion current | |
	(M + H)·	(M + 13)·
1,2-Ethanediol	50	50
1,2-Propanediol	50	50
1,3-Propanediol	70	30
1,2-Butanediol	60	40
2,3-Butanediol	60	40
1,3-Butanediol	70	30
1,4-Butanediol	85	15
1,5-Pentanediol	100	0
1,6-Hexanediol	100	0

Source: From Eichmann, E. S. et al., *J. Amer. Soc. Mass Spectrom.,* 3, 537, 1992. With permission.

the interfunctional distance was varied, as represented by the butanediol series, the formation of $(M + 13)^+$ decreased in the order 1,2 butanediol $\approx$ 2,3-butanediol > 1,3-butanediol > 1,4 butanediol. These results indicated that the proximity of the functional groups had a dramatic effect on the favorability of competing reactions. The ion/molecule product distributions for reactions of monofunctional alcohols with dimethyl ether ions were also examined, but no formation of $(M + 13)^+$ ions was observed. This result served to highlight the necessity for the presence of two proximate nucleophilic hydroxyl groups in a molecule in order for the methylene substitution reaction to be possible.

These observations were rationalized in several ways. First, it was proposed that the favorability of the methylene substitution reactions for the diols with hydroxyl groups on adjacent carbon atoms was influenced by secondary functional group participation by the proximate hydroxyl groups, as shown in Scheme 4. Such interactions would also stabilize the ion/molecule complex initially formed during the bimolecular reaction between $CH_3OCH_2^+$ and the diol and then assist in stabilizing the resulting product ion by intramolecular cyclization. In fact, as evidenced from CAD spectra and by comparison to model compounds, the final product ions were shown to be cyclic in nature.

For each diol that reacted with dimethyl ether ions to form products at $(M + 13)^+$ due to $(M + CH_3OCH_2 - CH_3OH)^+$, CAD spectra were acquired to assist in the structural characterization of the $(M + 13)^+$ product ions, the CAD spectra of cyclic and acyclic model ions were compared to the CAD spectra of the diol ions, and MS^n (sequential activation) experiments were done for more detailed characterization of dissociation pathways. For example, ions were generated from 3-ethoxy-1-propanol (EI-induced α cleavage of the terminal methyl radical) and from protonation of 1,3-dioxane to model the $(M + 13)^+$ ion of 1,3-propanediol. The CAD spectra of these ions indicated that the spectrum for the acyclic model was significantly different, but the spectra for the 1,3-dioxane and the 1,3-propanediol models were satisfactory matches, suggesting that the cyclic form of the $(M + 13)^+$ ion indeed predominated over the acyclic form.

From these studies, functional group interactions were shown to be particularly important in mediating the bimolecular and dissociation reactions of diols and amino alcohols. The proximity of the functional groups was related to the relative favorability of the methylene substitution reactions due to stabilization of the transition states and/or products via intramolecular cyclization.

XVII. SELECTIVE REACTIONS OF AROMATIC SUBSTRATES: *ORTHO* EFFECTS

Functional group interactions classified as "*ortho* effects" have been characterized for many disubstituted aromatic ions.[75] These effects are manifested as specific dissociation pathways that are observed for *ortho*-substituted aromatic compounds that are not observed for the other isomers. A recent study compared the dissociation routes of various closed-shell ions generated via ion/molecule reactions of the *ortho-*, *meta-*, and *para*-isomers in order to evaluate *ortho* effects.[76] Five disubstituted aromatics were selected because each had a combination of electron withdrawing and releasing substituents, functional groups that can exert significant electronic effects in chemical reactions, especially electrophilic aromatic substitution processes, and presumably would be especially effective for enhancing functional group interactions such as *ortho* effects. The ions of interest were generated by proton transfer from dimethyl ether, or by methylene addition or methyl addition from ion/molecule reactions with dimethyl ether or ethylene oxide. The closed-shell ions formed by these different ionization processes were compared in order to evaluate the magnitude of *ortho* effects based on the size and polarizibility of the cationic species attached to the aromatic. The structures and dissociation pathways of the ions were characterized by low-energy CAD in a quadrupole ITMS™ and by comparison to the behavior of model ions. In some cases, the open-shell ions generated by EI were also characterized to allow comparison of the extent of *ortho* effects in closed-shell vs. open-shell systems.

The CAD spectra of the M^+, $(M + H)^+$, and $(M + 13)^+$ or $(M + 15)^+$ ions for five compounds containing various combinations of the electron-releasing and electron-withdrawing substituents were examined: methoxyacetophenone, hydroxyacetophenone, anisaldehyde, methoxyphenol, and hydroxybenzaldehyde. From comparisons of the CAD spectra of the closed-shell aromatic ions generated by ion/molecule reactions with dimethyl ether ions and the open-shell molecular ions generated by EI, it was apparent that *ortho* effects were more pronounced for the closed-shell species. For example, the molecular ions, M^+, of the methoxyacetophenone isomers produce qualitatively similar CAD spectra, but the CAD spectra of the protonated and methylated species were distinct. With respect to only closed-shell ions, the $(M + 13)^+$ ions often showed *ortho* effects even when the simple protonated species did not show qualitative differences (hydroxybenzaldehyde and methoxyphenol). By contrast, the $(M + 15)^+$ adducts only demonstrated *ortho* effects when the simple protonated species also showed significant *ortho* effects. These differences in the spectra of $(M + H)^+$, $(M + 13)^+$, and $(M + 15)^+$ ions were attributed to the type of ionization processes used to generate the ions. The $(M + H)^+$ and $(M + 15)^+$ ions were formed by simple direct attachment of a proton or methyl cation, but the $(M + 13)^+$ ions were formed via an electrophilic aromatic substitution reaction which involved substantial rearrangement of the aromatic substrate.

As an example, the methoxyacetophenone isomers contain an electron-withdrawing acetyl substituent and a moderately electron-releasing methoxy substituent. The CAD spectra of the open-shell radical cations, $M^{+\cdot}$, of *ortho-*, *meta-*, and *para*-methoxyacetophenone generated by EI were qualitatively similar, all showing extensive loss of methyl radical and formation of a variety of aromatic-type ions ($C_6 H_5^+$, $C_7 H_7^+$); however, the CAD spectra of isomeric closed-shell $(M + H)^+$ methoxyacetophenone ions showed several striking differences (Figure 26). The protonated *meta-* and

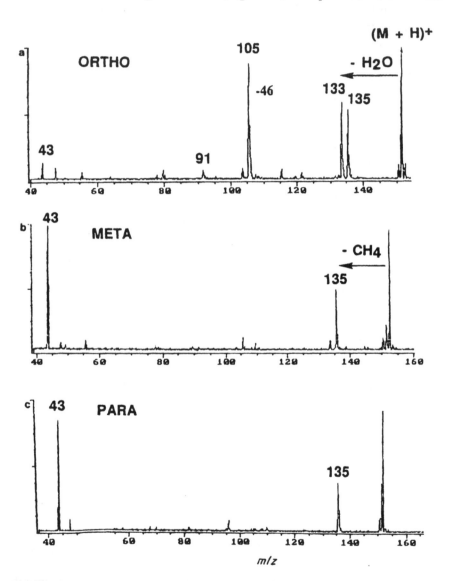

FIGURE 26

CAD spectra of protonated *o-*, *m-*, and *p*-methoxyacetophenone (m/z 151) acquired with a quadrupole ITMS™. (From Donovan, T. and Brodbelt, J., *Org. Mass Spectrom.*, 27, 12, 1992. Reprinted by permission of John Wiley & Sons.)

para-isomers both dissociated via loss of methane and via formation of acetyl radical (formation of m/z 135 and 43, respectively), fragmentation routes that were similar to those noted for protonated unsubstituted acetophenone. This implied that the electron-withdrawing acetyl substituent played a bigger role in mediating the dissociation reactions than the electron-releasing methoxy substituent. These dissociation processes were not the most significant observed for the *ortho*-isomer. Instead, the *ortho*-isomer produced m/z 133 via dehydration and m/z 105, the most abundant fragment ion. These dissociation pathways (loss of water or 46 u) were not analogous to any of the dissociation pathways reported for either protonated anisole or protonated acetophenone. This striking difference was attributed to the *ortho*-positioning of the substituents which permitted functional group interactions not possible for the *meta*- and *para*-isomers. These interactions allowed access to different types of dissociation reactions and/or promoted ionization at a different site than for the *para*- and *meta*-isomers (ring vs. substituent).

XVIII. EFFECTS OF REACTANT ION KINETIC ENERGY ON ION/MOLECULE REACTIONS

The outcomes of selective ion/molecule reactions are not only determined by the structures of the reactants but also the kinetic and internal energies of the reactants. For example, the reactant ion kinetic energy may exert significant effects on the promotion of endothermic, and the abridgement of exothermic, ion/molecule reactions.[77] The kinetic energy of an ion in a quadrupole ion trap may be increased by the application of a supplementary RF voltage at the resonance frequency of the ion. Sufficient conversion of kinetic energy to internal energy may occur upon collision of a translationally energized ion with a neutral reactant, resulting in the promotion of an endothermic reaction. Thus, the ability to drive endothermic reactions opens up new avenues for applications of selective ion/molecule reactions.

A recent survey of various charge-exchange and proton-transfer reactions was performed in order to obtain an initial evaluation of the feasibility of driving endothermic ion/molecule reactions in a quadrupole ion trap.[78] From the array of experiments performed for that study, the charge exchange reaction of maximum endothermicity that could be driven was 1.11 eV endothermic, and the proton transfer reaction of maximum endothermicity that could be driven was 25.8 kcal/mol (1.11 eV) endothermic. These upper limits represented experimentally determined boundaries based on a limited number of experiments and did not indicate an absolute upper limit. As more energetic activation conditions were used (higher activation voltage, higher q_z value, longer activation time) CAD, neutralization, and ion ejection processes became increasingly competitive compared to endothermic charge exchange or proton-transfer reactions. The maximum conversion values, ranging from 5 to 15% were reported, and no apparent systematic rules governed the optimal conditions for promoting endothermic ion/molecule reactions.

For example, the reaction between naphthalene ions and phenol (optimized at $q_z = 0.35$ with a 20-ms activation period and a 600 mV$_{(p-p)}$ activation voltage (141.9 kHz)) resulted in 12% efficiency, whereas the reaction between naphthalene ions and benzene (optimized at $q_z = 0.29$ with a 35-ms activation period and a 500 mV$_{(p-p)}$ activation voltage) resulted in 6% efficiency. Some reactions that had endothermicities within the established energy limits were not successfully driven under any activation conditions (the benzene/hexafluorobenzene and 2-naphthol/benzene systems). Several factors contributed to the rationale for these results. First, it was stated that an absolute measure of reaction endothermicity did not define other entropic factors involved in the reaction, and some reactions that were minimally endothermic had very unfavorable steric or configurational requirements. Second, some reactant ions had dissociation pathways with relatively low activation barriers, so that CAD processes became competitive with endothermic proton-transfer or charge exchange reactions. Thus, the cross-section for endothermic dissociation was much greater than that for endothermic ion/molecule reaction. Each of these factors tended to cause depressed or zero-measured conversion efficiencies. Thus, competitive dissociation channels and ion ejection processes were two factors that limited the observed favorability of endothermic ion/molecule reactions.

XIX. HOST–GUEST CHEMISTRY IN THE GAS PHASE

Molecular recognition is a central theme of many important biological and chemical phenomena, and host–guest complexation has been studied extensively in solution.[79] The ability to examine host–guest chemistry without the influence of solvent effects by application of various mass spectrometric techniques has brought a new perspective[80-83] to understanding some of the fundamental details of molecular recognition. Host–guest complexes were formed in a quadrupole ion trap by ion/molecule association reactions between model hosts (such as macrocyclic polyethers) and guests (such as metal ions or ammonium ions). The structures of the host–guest complexes were characterized by CAD, and orders of binding affinities of different hosts for selected guest ions were determined by using ligand exchange (bracketing) techniques. For example, the orders of relative gas-phase basicities and ammonium ion affinities were established by the ligand exchange method in which a polyether molecule bound to a cation is allowed to interact with a different neutral polyether molecule. The observation of a transfer of the cation to the second molecule indicates that the latter has a higher affinity for the cation. A complete order of relative cation binding affinities can be determined by repetition of the exchange experiment with many different pairs of polyethers.

In the quadrupole ion trap, typically each polyether was admitted to 2×10^{-6} torr, and the transfer reaction was monitored for 10 to 100 ms. An example of the method is illustrated in Figure 27. As shown, the ammonium ion complex of tetraethylene glycol dimethyl ether (tetraglyme) was isolated and allowed to interact with tetraethylene glycol (tetraglycol) for 100 ms. The ammonium ion does not transfer to tetraglycol. In contrast, the reaction shown in Figure 27B indicates that the ammonium ion efficiently transfers from tetraglycol to tetraglyme. The relative orders of gas phase basicities and ammonium ion affinities are shown in Table 5 for a series of polyethers, including both cyclic and acyclic model hosts. The polyethers that were larger demonstrated greater cation affini-

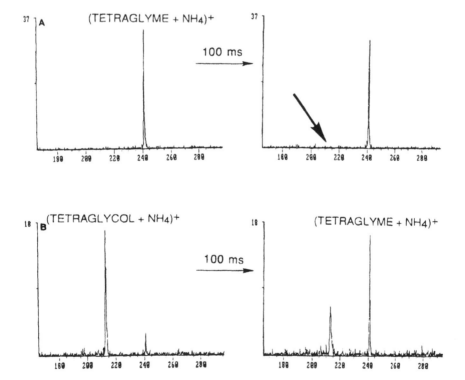

FIGURE 27
Ammonium ion transfer between tetraethylene glycol dimethylether and tetraethylene glycol. (A) Isolation of the ammonium ion complex of tetraglyme, followed by reaction with tetraglycol. (B) Isolation of the ammonium ion complex of tetraglycol, followed by reaction with tetraglyme. (From Wu, H.-F. and Brodbelt, J., *J. Amer. Soc. Mass Spectrom.*, 4, 721, 1993. With permission.)

TABLE 5

Orders of Relative Affinities of Polyethers[a]

H[·]	NH$_4^+$
12-C-4	12-C-4
3-GLYME	3-GLYME
15-C-5	4-GLYCOL
4-GLYME	15-C-5
18-C-6	5-GLYCOL
4-GLYCOL	4-GLYME
21-C-7	18-C-6
5-GLYCOL	21-C-7

[a] The term a-C-b represents the a-crown-b ether. The term n-GLYCOL represents an *n*-ethylene glycol molecule, and the term n-GLYME represents an *n*-ethylene glycol dimethylether molecule. Order of increasing affinity down the column.

Source: From Wu, H.-F. and Brodbelt, J., *J. Amer. Soc. Mass Spectrom.*, 3., 720, 1993. With permission.

ties, and the largest crown ethers showed particularly high ammonium ion affinities relative to their acyclic counterparts. The trends were rationalized based on the size-selective favorability of hydrogen-bonding interactions between the polyethers and the cations. In general, the quadrupole ion trap has been shown to be a versatile device for the investigation of host-guest chemistry in the gas phase.

XX. CONCLUSIONS

The quadrupole ITMS™ has proven to be particularly useful for investigations of mechanistic aspects of gas phase ion/ molecule reactions. The ability to trap selectivity and then monitor the reactions of specific ions allows reaction pathways to be mapped as a function of time, pressure, or other experimental conditions. Sequential isolation and activation techniques (MS^n) are valuable for structural characterization of product ions and for the elucidation of reaction and dissociation mechanisms. Because of its selectivity and sensitivity, ion/molecule reaction chemistry in the quadrupole ion trap affords many future possibilities for analytical applications in mass spectrometry.

REFERENCES

1. Beaty, E.C., *Phys Rev.* 1986, A33, 3645.
2. Vedel, F.; Vedel, M.; Beneux, A., unpublished work, 1989.
3. Neviere, M.; Cadilhac, M.; Petit, R., *IEEE Trans. Antennas Propag.* 1973, 27, 37.
4. Vedel, F.; André, J.; Vedel, M., *J. Phys.* 1981, 42, 1.

5. von Busch, F.; Paul W., Z. Phys. 1961, 164, 588.
6. Blatt, R.; Zoller, P.; Holzmüller, G,; Siemers, I., Z. Phys. 1986, D4, 121.
7. Siemers, I.; Blatt, R.; Sauter, Th.; Neuhauser, W., Phys. Rev. 1988, A38, 5121.
8. van Kampen, N.G., Stochastic Processes in Physics and Chemistry. North Holland: Amsterdam, 1981.
9. Dehmelt, H.G., Adv. Atom. Mol. Phys. 1967, 3, 53; ibid, 1969, 5, 109.
10. Vedel, F.; André, J.; Vedel, M.; Brincourt, G., Phys. Rev. 1983, A27, 2321.
11. Schaaf, U.; Schmeling, U.; Werth, G., Appl. Phys. 1981, 25, 249.
12. Münch, A.; Berkler, M.; Gerz, C.H.; Wilsdorf, D.; Werth, G., Phys. Rev. 1987, A35, 4147; Werth, G., Phys. Scr. 1988, T22, 792.
13. Ifflander, R.; Werth G., Metrologia. 1977, 13, 1167.
14. Gioumousis, G.; Stevenson, D.P., J. Chem. Phys. 1958, 29, 294.
15. Vedel, F.; André, J., Phys. Rev. 1984, A29, 2098.
16. Chattopadhyay, A.P.; Ghosh, P.K., Int. J. Mass Spectrom. Ion Phys. 1983, 49, 253.
17. Quint, W.; Schliech, W.; Walther, H., Recherche. 1989, 20, 1194; Blümel, R,; Chen, J.M.; Peik, E.; Quint, W.; Schliech, W.; Shen, Y.R.; Walther, H., Nature. 1988, 334, 309.
18. March, R.E.; McMahon, A.W.; Londry, F.A.; Alfred, R.L.; Todd, J.F.J.; Vedel, F., Int. J. Mass Spectrom. Ion Processes. 1989, 95, 119.
19. March, R.E.; McMahon, A.W.; Allinson, E.T.; Londry, F.A.; Alfred, R.L.; Todd, J.F.J.; Vedel, F., Int. J. Mass Spectrom. Ion Processes. 1990, 99, 109.
20. Bollinger, J.J.; Wineland, D.J., Phys. Rev. Lett. 1984, 53, 348.
21. Vedel, F. Thèse d'Etat, Université de Provence: Marseille, 1985.
22. André, J.; Vedel, F., J. Phys. 1977, 38, 1381.
23. Vedel, M.; André, J.; Chaillat-Negrel, S.; Vedel, F., J. Phys. 1981, 42, 541.
24. Landau, L.; Lifchitz, E., Mécanique. MIR: Moscow, 1966; pp. 111–129.
25. Hayashi, C., Nonlinear Oscillations in Physical Systems. McGraw-Hill: New York, 1964.
26. Vedel, F; André, J., Int. J. Mass Spectrom. Ion Processes. 1985, 65, 1.
27. Rettinghaus, G., Z. Angew. Phys. 1966, 13, 321.
28. Vedel, F.; Vedel, M., Phys. Rev. 1990, A41, 2348.
29. Cutler, L.S.; Giffard, R.P.; McGuire, M.D., Appl. Phys. B. 1985, B36, 137.
30. Church, D.A., Phys. Rev. 1987, A37, 277.
31. Gaboriaud, M.N.; Desaintfuscien, M.; Major, F.G., Int. J. Mass Spectrom. Phys. 1981, 41, 109.
32. Lawson, G.; Bonner, R.F.; Todd, J.F.J., J. Phys. E: Sci. Instrum. 1973, 6, 357.
33. Vedel, M., Thèse d'Etat, Université de Provence: Marseille, 1987.
34. March, R.E.; Hughes, R.J., Quadrupole Storage Mass Spectrometry. Chemical Analysis Series, Vol. 102. Wiley Interscience: New York, 1989.
35. Vedel, F.; Vedel, M.; March, R.E., Int. J. Mass Spectrom. Ion Processes. 1990, 99, 125.
36. Guidugli, F.; Traldi, P., Rapid Commun. Mass Spectrom. 1991, 5, 343.
37. Vedel, F.; Vedel, M.; March, R.E., Int. J. Mass Spectrom. Ion Processes. 1991, 108, R11.
38. Mosburg, E., Jr.; Vedel, M.; Zerega, Y.; Vedel, F.; André, J., Int. J. Mass Spectrom. Ion Processes. 1987, 77, 1.
39. Vedel, M.; Vedel, F.; Rebatel, I., Phys. Rev. 1995, A51, 2294.
40. Lunney, M.D.N.; Buchinger, F.; Moore, R.B., J. Mod. Optics. 1992, 39, 349.
41. Vedel, F., Int. J. Mass Spectrom. Ion Processes. 1991, 106, 33.
42. Vedel, F.; Vedel, M., J. Mod. Optics. 1992, 39, 431.
43. Lawson, G.; Bonner, R.F.; Mather, R.E.; Todd, J.F.J.; March, R.E., J. Chem. Soc., Faraday Trans. 1976, 72, 545.
44. Basic, C.; Yost, R.A.; Eyler, J.R., Proc. 38th Conf. Mass Spectrom. Allied Topics. Tucson, AZ, 1990; p. 858.
45. Nourse, B.D.; Kenttämaa, H.I., J. Phys. Chem. 1990, 94, 5810.
46. Gerlich, D., Electron. Atom. Collisions. Elsevier: Amsterdam, 1986; p. 541.

47. Dehmelt, H.G.; Walls, F.L., *Phys. Rev. Lett.* 1968, *21*, 127.
48. Wineland, D.J.; Dehmelt, H.G., *J. Appl. Phys.* 1975, *46*, 919.
49. Walls, F.L.; Stein, T.S., *Phys. Rev. Lett.* 1973, *31*, 975.
50. Church, D.A.; Dehmelt, H.G., *J. Appl. Phys.* 1969, *40*, 3421.
51. Paul, W.; Osberghaus, O.; Fischer, E., *Forschungsberichte des Wirtschaft und Verkehrministeriums Nordrhein Westfalen*. No. 415. Westdeutscher Verlag: Cologne, 1958.
52. Fischer, E., *Z. Phys.* 1959, *156*, 1.
53. Jungmann, K.; Hoffnagle, J.; DeVoe, R.G.; Brewer, R.G., *Phys. Rev. A.* 1987, *36*, 3451.
54. Paul, W.; Reinhard, H.P.; Von Zahn, U., *Z. Phys.* 1958, *152*, 143; Fulford, J.E.; Hoa, D.-N.; Hughes, R.J.; March, R.E.; Bonner, R.F.; Wong, G.J., *J. Vac. Sci. Technol.* 1980, *17*, 829; Armitage, M.A.; Fulford, J.E.; Hoa, D.-N.; Hughes, R.J.; March, R.E., *Can. J. Chem.* 1979, *57*, 2108.
55. Kaiser, R.E., Jr.; Louris, J.N.; Amy, J.W.; Cooks, R.G., *Rapid Commun. Mass Spectrom.* 1989, *3*, 225.
56. Farrar, J.; Saunders, W., Eds., *Techniques for the Study of Ion/Molecule Reactions*. John Wiley & Sons: New York, 1988.
57. Buchanan, M., Ed., *Fourier Transform Mass Spectrometry: Evolution, Innovation, and Application*. ACS Symp. Ser., Vol. 359, American Chemical Society: Washington, D.C., 1987; Marshall, A.G., *Acc. Chem. Res.* 1986, *5*, 167; Gross, M.L.; Rempel, D.L., *Science.* 1984, *226*, 261; March, R.E.; Hughes, R.J., *Quadrupole Storage Mass Spectrometry*. Chemical Analysis Series, Vol. 102. Wiley-Interscience: New York, 1989.
58. Weber-Grabau, M.; Kelley, P.J.; Syka, J.E.P.; Bradshaw, S.C.; Brodbelt, J.S., *Proc. 35th Ann. Conf. ASMS Allied Topics*. Denver, CO, 1987; p. 114.
59. Louris, J.; Brodbelt, J.; Cooks, R.G.; Glish, G.L.; Van Berkel, G.; McLuckey, S.A., *Int. J. Mass Spectrom. Ion Processes*. 1990, *96*, 117.
60. Harrison, A.G., *Chemical Ionization Mass Spectrometry*. CRC Press, Boca Raton, FL, 1983.
61. Vairamani, M.; Ali Mirza, U.; Srinivas, R., *Mass Spectrom. Rev.* 1990, *9*, 235.
62. Brodbelt, J.; Liou, C.-C.; Donovan, T., *Anal. Chem.* 1991, *63*, 1205.
63. Donovan, T.; Liou, C.-C.; Brodbelt, J. *J. Am. Soc. Mass Spectrom.* 1992, *3*, 39.
64. Lias, S.G.; Liebman, J.F.; Levin, R.D.J., *Phys. Chem. Ref. Data*. 1984, 13.
65. Donovan, T.; Brodbelt, J., *J. Am. Soc. Mass Spectrom.* 1992, *3*, 47.
66. Rawn, J.D., *Biochemistry*. Harper & Row: New York, 1983.
67. McFadden, W.H.; Day, E.A.; Diamond, M.J., *Anal. Chem.* 1965, *37*, 89; Friedman, L.; Long, F.A., *J. Am. Chem. Soc.* 1953, *75*, 2832; Millard, B.J., *Org. Mass Spectrom.* 1968, *1*, 279; Chen, P.H.; Kuhn, W.F.; Will, F.; Ikeda, R.M., *Org. Mass Spectrom.* 1970, *3*, 149.
68. Heath, T.G.; Allison, J.; Watson, J.T., *J. Am. Soc. Mass Spectrom.* 1991, *2*, 270; Pachuta, R.R.; Kenttämaa, H.I.; Cooks, R.G.; Zennie, T.M.; Ping, C.; Chang, C.J.; Cassady, J.M., *Org. Mass Spectrom.* 1988, *23*, 10; Cole, M.J.; Enke, C.J., *J. Am. Soc. Mass Spectrom.* 1991, *2*, 470.
69. Lau, Y.K.; Nishizawa, K.; Tse, A.; Brown, R.S.; Kebarle, P., *J. Am. Chem. Soc.* 1981, *103*, 6291; Wood, K.V.; Burinsky, D.J.; Cameron, D.; Cooks, R.G., *J. Org. Chem.* 1983, *48*, 5236; Nakata, H.; Suzuki, Y.; Shibata, M.; Takahashi, K.; Konishi, H.; Takeda, N.; Tatematsu, A., *Org. Mass Spectrom.* 1990, *25*, 649.
70. Donovan, T.; Brodbelt, J., *Biol. Mass Spectrom.* 1992, *21*, 254.
71. Capon, B.; McManus, S.P., *Neighboring Group Participation*. Vol. 1. Plenum Press: New York, 1976.
72. Fenselau, C.C.; Robinson, C.H., *J. Am. Chem. Soc.* 1971, *93*, 3070; Reich, M.; Schwarz, H., *Org. Mass Spectrom.* 1977, *12*, 566; Claeys, M.; Van Haver, D., *Org. Mass Spectrom.* 1977, *12*, 531; Respondek, J.; Schwarz, H.; Van Gaever, F.; Van de Sande, C.C., *Org. Mass Spectrom.* 1978, *13*, 618.
73. Eichmann, E.S.; Alvarez, E.; Brodbelt, J.S., *J. Am. Soc. Mass Spectrom.* 1992, *3*, 535.

74. Eichmann, E.S.; Brodbelt, J.S., *J. Am. Soc. Mass Spectrom.* 1993, *4*, 230.
75. Riley, J.; Baer, T.; Marbury, G., *J. Am. Soc. Mass Spectrom.* 1990, *2*, 69; Ramana, D.; Ramakrishna, N., *Org. Mass Spectrom.* 1989, 24, 317; Ramana, D.; Sundaram, N.; George, M., *Org. Mass Spectrom.* 1988, *23*, 63; Bursey, J.T.; Bursey, M.M.; Kingston, D.G.I., *Chem. Rev.* 1973, *73*, 191; Schwarz, H., in *Topics in Current Chemistry.* Vol. 73, Dewar, M.J.S.; Hafner, K.; Heilbronner, E.; Ito, S.; Lehn, J.-M.; Niedenzu, K.; Schafer, K.; Wittig, G., Eds. Springer-Verlag: New York, 1978; Filges, U.; Grutzmacher, H.F., *Org. Mass Spectrom.* 1986, *21*, 673; Filges, U.; Grutzmacher, H.F., *Org. Mass Spectrom.* 1987, *22*, 444.
76. Donovan, T.; Brodbelt, J., *Org. Mass Spectrom.* 1992, *27*, 9.
77. Kinter, M.T.; Bursey, M.M., *Org. Mass Spectrom.* 1987, *22*, 775; Bensimon, M.; Houriet, R., *Int. J. Mass Spectrom. Ion Processes.* 1986, *72*, 93; Forbes, R.A.; Lech, L.M.; Freiser, B.S., *Int. J. Mass Spectrom. Ion Processes.* 1987, *77*, 107.
78. Wu, H.-F.; Brodbelt, J.S., *Int. J. Mass Spectrom. Ion Processes.* 1992, *115*, 67.
79. Vogtle, F.; Wever, E., Eds., *Host Guest Complex Chemistry. Macrocyles.* Springer-Verlag: New York, 1985.
80. Brodbelt, J.; Maleknia, S.; Liou, C.; Lagow, R., *J. Am. Chem. Soc.* 1991, *113*, 5913.
81. Zhang, H.; Chu, I.; Leming, S.; Dearden, D.A., *J. Am. Chem. Soc.* 1991, *113*, 7415.
82. Maleknia, S.; Brodbelt, J., *J. Am. Chem. Soc.* 1992, *114*, 4295.
83. Liou, C.-C.; Brodbelt, J.S., *J. Am. Chem. Soc.* 1992, *114*, 6761.

INDEXES

Author Index

403

CHEMICAL INDEX

SUBJECT INDEX

A

A, coefficient, 32
A_n, 70
A_u, 36
$A_{2,ideal}$, 70
a_r, 38,
a_r, definition, 33
a_u, 8, 34–40, 45, 61
a_u, definition, 33
a_x, definition, 8
a_y, definition, 8
a_z, 8, 13–14, 38–39, 42, 73–74, 78–79, 102, 114
a_z, definition, 8, 33
a_z, q_z, 102
a_z, q_z plane, 73
Absorption peak, 16
Absorption spectrum, 363
AC dipole field, 122
Acceptance mountains, 81
  canyon depth, 81
  height, 81
  profile, 81–82
Accumulation efficiency, 183
Accurate mass assignment, 48
Accurate mass measurement, 12,
Activation energy, 332
Activation process, 217
Additional AC voltage, 63
Adducts, 376
Adduct formation, 377
Adiabatic approximation, 59
AGC, *see* Automatic gain control
Ages of ion trap, 28–29
Air, 142–143

Alternative modes of operation, 12
Alternative scan functions, 12
American Society for Mass
  Spectrometry, 171
Amino alcohols, 388
Ammonium ion affinities, 395
Ammonium ion transfer, 395
Amplitude dependence, 62–63
Amplitude of ion oscillation, 303
Analyte ions, 12, 19, 209, 220
Analyte molecules, 20
Analyte peaks, 10,
Analytical,
  applications, 10, 21
  mass spectrometer, 12,
  methods, 60
  protocol, 182
  scan, 211–212
Angle parameter, 153
Angular frequency, 33–34
Angular momentum, 192–193
Anharmonicities, 346
Animated displays, 238
Anions, *see* Negative ions, 47
Apex isolation, 42, 215–216, 258
Apple computer, 180
Applied potential, note, 36
Approximation method, 224
Arrested motion, 19
ASGDI/ion trap system, 19
ASGDI source, 18–19
Assembly tolerance, 53
Asymptote, 68, 149
  angle, 65, 137, 147, 235
  cone, 152